Eating Smoke

Eating Smoke

Fire in Urban America, 1800–1950

Mark Tebeau

The Johns Hopkins University Press
Baltimore and London

Printed in the United States of America on acid-free paper
9 8 7 6 5 4 3 2 1

The Johns Hopkins University Press
2715 North Charles Street
Baltimore, Maryland 21218-4363
www.press.jhu.edu

Library of Congress Cataloging-in-Publication Data

Tebeau, Mark.
Eating smoke : fire in urban America, 1800–1950 / Mark Tebeau.
p. cm.
Includes bibliographical references and index.
ISBN 0-8018-6791-6 (hardcover : alk. paper)
1. Fire extinction—United States—History. I. Title.
TH9503 .T43 2002
363.37'0973'091732—dc21
2002151370

A catalog record for this book is available from the British Library.

For Kristi

Contents

Preface and Acknowledgments

My father was a firefighter. I remember feeling pride when he visited my second grade class on fire prevention day. I remember telephoning and speaking to him at the engine house every night before I went to bed. I remember the "firestorm" of 1976 and the burn he received. I remember his coming home in the morning to fix me breakfast, sometimes smelling of smoke.

Although he never spoke of his work, I knew what he did. In a world where job titles mean little, my father's said a lot. After twenty years on the job, he retired, abruptly and without much fanfare. He took another job with one of those meaningless titles—control operating engineer, or something like that, at an electrical utility. We moved to a larger house, in the suburbs. We achieved the American dream, but at what cost? I wondered, at the time, what did he give up? What did the change mean to him? Can you stop being a fireman?

I returned to these questions as I foraged for a dissertation topic. In that I was interested in gender, technology, and cities, it occurred to me that writing a history of firefighting and fire protection might be a good idea. Amazed when I discovered that no historian had examined the topic in great depth, I found an engrossed audience when I sheepishly suggested the vague outlines of this project. My father even inspired the title when we had one of our few conversations about firefighting. He mentioned off-handedly that many firefighters really disliked wearing their breathing apparatus, and that "old-timers" in particular believed that real firemen would not wear them at all.

The personal meaning of this work increased when, as I made final revisions on the manuscript, terrorists attacked the World Trade Center. My first thought as I watched the buildings collapse was of the hundreds of firefighters who were dying, and of their fatherless children. Although I was a child of a former firefighter, my research had taught me something that I never knew as a kid. After years of studying fire protection, I had learned about the job's many dangers: I had read countless stories in which falling walls had killed firemen who struggled to penetrate to the interior of buildings to perform rescues and throw water on the fire.

This project began as a dissertation, and was nurtured also by public historians working in history museums. Joel Tarr, John Modell, and Rick Maddox advised me and read this work. They encouraged me to freely explore issues that other professional historians thought uninteresting and unimportant. Each brought something special to the project and made it better. Joel has guided my professional development, and helped to ground this story in the history of American cities; John's eclectic but rigorous approach to scholarship promoted creative thought and allowed me to experiment with ideas, methods, and sources; Rick's ability to move between theory and history proved inspiring; his close reading helped me to find this book's argument. As the project developed, I also discovered sources and colleagues in historical societies and libraries. Public historians shared with me their craft, and they enhanced this work by making me more sensitive to sources other than the printed documents on which historians almost exclusively rely. By introducing me to the richness of American material culture, these historians—such as Cory Amsler at the Bucks County Historical Society—transformed my understanding of this subject and improved this work.

The project received financial support from Carnegie Mellon University, the Missouri Historical Society, the National Museum of American History (at the Smithsonian Institution), the Pennsylvania Historical and Museums Commission, and the Pennsylvania Humanities Council, both of which supported my work as curator on a firefighting exhibition at the Bucks County Historical Society, the Graduate School of Business at Harvard University, and Cleveland State University.

Over the past decade many historians, researchers, friends, and colleagues influenced my training and thinking and/or read parts of this manuscript or reviewed pieces of the larger project at different times. I thank: Keith Allen, Cory Amsler, Mary Blewett, Eileen Boris, Tom Buchanan, Liz Cohen, Herrick Chapman, Jim Grossman, David Hounshell, Tom Humphrey, Andrew Hurley, Meg Jacobs, Elizabeth Johns, Miriam Kaprow, Linda Kerber, Bill Kenney, Carol Lasser, Steve Lubar, Robert McCarl, Clay McShane, Larry Peskin, Carroll Pursell, Steven Pyne, Don Ramos, Peter Rutkoff, Eric Sandweiss, Mark Santow, Will Scott, Barbara Clark Smith, Peter Stearns, Shirley Wadja, Sam Watson, Mark Wilkens, Helena Wright, Marilyn Zoidis, Michael Zuckerman. At the Johns Hopkins University Press, Bob Brugger, Melody Herr, Marie Blanchard, and an anonymous reader exhibited patience and shepherded this project to press.

Numerous archives, museums, and historical societies provided access and graciously helped me in my research: Baker Library at Harvard University, the National Museum of American History at the Smithsonian Institution, the Historical Society of Pennsylvania, the Library Company of Philadelphia, the New York Historical Society, the New York Public Library, Bancroft Library at the University of California, Berkeley, the Free Library of Philadelphia. The Philadelphia Center for Early American Studies provided me access to the libraries at the University of Pennsylvania, the Bucks County Historical Society, Carnegie Mellon University libraries, the Missouri Historical Society, the Mercantile Library of St. Louis, the St. Louis Public Library, the Carnegie Library of Pittsburgh, the University of Pittsburgh libraries, the Library of Congress, the Hall of Flame in Phoenix, Fireman's Hall Museum in Philadelphia, and the CIGNA Archives and the Museum and Art Collection.

I received special assistance from a number of individuals in the library and firefighting communities: Sue Collins at CMU, Marianne Nolan at Cleveland State, Betsy Smith at the Spruance Library of the Bucks County Historical Society, Emily Miller and Martha Clevenger at the Missouri Historical Society, Leslie Simon, archivist at CIGNA, Sue Levy and Melissa Hough at the CIGNA Museum and Art Collection, Ellen Morfei at ACE America, Allan Haddox at the American Insurance Association, Ward Childs and Jefferson Moak of the Philadelphia City Archives, Harry Magee and Dan Kenney at Fireman's Hall, Pete Molloy at the Hall of Flame, Peter St. Clair of the Arlington Fire Department, Neil Svetanics, former fire chief of the St. Louis Fire Department, Harold Hairston, commissioner of the Philadelphia Fire Department.

Friends and family have been supportive and understanding; they have offered a place to stay, entertained me, and shared their lives: Mark Schleinitz and Melissa Fischer, Neil Chriss, Jeff and Val Cohen, Scott Feldman and Stephanie Lubin, Mark Kochvar and Vicki Helgeson, Cory Amsler and Eileen Shapiro, Cheryl Wheeler, Karl Kilgore, and Charly Gibbons, Mark and Shana Santow, Mark Wilkens, Madi Goodman and Andrew Mossin, and Kian Wright. Dan and Sandy Mickelson provided emotional and economic support as I followed this project across the country. They believed in it and me, and they knew when not to ask if it was finished.

In many respects, this book is about my family; it is about the world my parents—Ralph and Lois Tebeau—created. They respected knowledge and encouraged me to follow my heart. They made many sacrifices so that I might have an easier path. Few of us have the privilege of knowing our parents; working on this project allowed me to learn more about their lives, especially part of my father's.

Kristin Mickelson has lived with this project longer than anyone. She saw it arrive, take its first steps, and then meander in the academic wilderness. She supported me in more ways than I can count, offered her expertise as a statistician, and encouraged me to find a voice with her own wit and style. She makes the world more imaginative, beautiful, and moral; best of all, she gave me Amelia and Eli.

Eating Smoke

Introduction

The Problem of Fire

Shortly after 1 A.M. on August 10, 1887, clanging alarm bells roused the men of St. Louis's Hook and Ladder Company No. 6, led by foreman Christian Hoell. Minutes earlier, a night watchman had pulled the handle on a fire alarm box, signaling that a building in the city's commercial district had caught fire. Beginning in the cellar of a warehouse at the intersection of Second and Main Streets, the blaze spread up open stairwells and through the hollows in the walls. Second and third alarms rang in quick succession. Firemen rushed to the scene, searched for endangered citizens, and fought the blaze. Teams of men dragged hoses into the mammoth structure, seeking out the base of the inferno. Hook and Ladder Company No. 6 positioned its aerial ladder apparatus precariously close to the brick-and-iron-fronted structure, pouring water onto the holocaust from above. When after eight hours firemen had regained control, suddenly and without warning the warehouse's exterior walls tumbled. The building entombed the company and its truck. Several men suffered severe injuries. Three perished instantly, among them foreman Hoell.

The tragedy struck a deep nerve among city residents, especially as they learned the cruel irony underlying Christian Hoell's death. Described as the father of "modern life-saving services," Hoell could not save himself, ultimately crushed under the weight of the unstable and increasingly dangerous urban landscape. Nearly a decade earlier—following St. Louis's devastating Southern Hotel fire in which dozens of guests perished despite extraordinary acrobatics by firefighters—Hoell had formulated a special scaling ladder and climbing corps to facilitate rescuing people trapped by fire. His innovation transformed firefighting in the United States, helping to make the nation's fire departments more effective at saving lives in increasingly taller and more densely populated urban centers. Hoell trained new recruits in locales as distant as New York City in the skills of using the ladders, in the craft of saving lives. He played a critical role in the process through which firefighters and the National Association of Fire Engineers defined firefighting as a heroic life-saving occupation. As St. Louis residents mourned the loss by lining the route of a funeral procession and attending public wakes, they recognized not only the hazards faced by firemen but also their own vulnerability to fire.[1]

As the *Evening Chronicle* mourned the "martyrs to duty," a St. Louis business journal offered a different description of the blaze and its aftermath. The *Whipple Daily Fire Reporter* peered into the ledgers of those affected, revealing how another group of Americans confronted the problem. Written for an audience of businessmen, the *Fire Reporter* article recorded the details of the fire and provided a comprehensive accounting of the economic repercussions. The heaviest losses afflicted Bishop & Spears, a company that made and distributed the popular "Boss" brand peanuts. The firm occupied most of the building, and, moreover, most of its loss was uninsured. Falling walls also had damaged adjoining structures, and with exacting precision the *Fire Reporter* inventoried the financial consequences for those businesses as well. The economic aftermath reached into the coffers of insurance companies around the globe. Firms headquartered in the financial centers of the United States and Europe—New York, Hartford, Philadelphia, London, and Hamburg—suffered about 75 percent of the loss, which exceeded $350,000.

If fire underwriters winced at the substantial loss, they nonetheless understood the risks of their business, just as Hoell knew the hazards of his job. Soberly, insurers tallied the losses in ledgers and searched for patterns in the statistical rubble. They expected that their actuarial tables, with multiple classifications of risk, would help them to acquire insights about the problem and to improve their ability to profit from indemnifying the capital that flowed along the canals, railroads, and streets of industrial America. When firms underwrote insurance, they made calculated decisions to place financial capital in jeopardy. Insurers balanced considerable amounts of information concerning their portfolios of risk, the nature of the problem of fire, the insurance market, and their financial situation. By the last decades of the nineteenth century, underwriters managed data using a variety of tools and organizational strategies, including a broad network of industry associations, company representatives, and business service firms, such as the Whipple Insurance Protective Agency, which published the *Fire Reporter*. Whether they inspected cities, mapped hazards, or compiled loss data, specialists provided information that helped firms to know fire dangers in distant places. In their home offices, insurers collected, organized, and classified the information using the latest management technologies. They established categories of safety and danger, which guided risk taking and shaped their ability to offer economic security to clients. The *Fire Reporter* reveled in the details of the industry—the means through which international capitalism defended itself and the accumulated wealth of urban America.[2]

Efforts to contain the problem of fire in urban America developed through the conjunction of firefighters' and underwriters' labor—through the heroic and the banal, through concerns about people and property, through the conflict between labor and business. Firefighters and underwriters struggled against the same environmental hazard, but they conceived the threat in starkly different terms. Firefighters battled fire up close, understanding the problem as a real, physical danger. Insurers worked to control the hazard in their ledgers, providing security by taking abstract, economic risks. Whereas firefighters preserved human life and the built landscape against environmental devastation, fire underwriters primarily protected financial value. Just as the type of security that each offered stood at opposing ends of the spectrum—the physical and the abstract—firefighters' and underwriters' methods and tools also varied. These contrasting approaches to hazard and risk give structure to this story about the elemental struggle to shield American cities.[3]

Eating Smoke examines patterns of historical change, especially the dynamic process of urbanization, by posing straightforward questions about the social and cultural significance of the battle against fire. How did firefighters and underwriters perceive the hazard of fire? How did firefighters and underwriters seek to control the problem? How did their efforts change over time and give shape to American cities between 1800 and 1950? Emanating from this fundamental line of inquiry, a number of related questions provide structure for each chapter. How did these actors remove or mitigate the danger? How did the work of underwriters and firefighters relate to one another and contribute to the development of fire protection more broadly? How did their quotidian routines produce safety and refashion societal order? And, how did their labor make urban America into a physical and cultural reality? To what degree did the interactions between firemen and underwriters structure urban society—its institutions, landscapes, and cultural practices?[4]

If war, famine, disease, and death were the four horseman of the apocalypse, fire surely could have been a fifth. Controlling fire long posed a fundamental historical challenge to human society, but the growth of cities, especially after industrialization, caused fire safety to become more critical. Cities held an increasing and changing danger, and the greater concentration of people and property raised the stakes in the battle to hold fire within human control. In 1800 most Americans dwelled in small communities, but by 1950 nearly two-thirds lived in cities. By that time the population of twenty-five cities was greater than one million inhabitants, and more than one hundred cities had at least twenty-five thousand residents. During this period of extraordinary growth and change, fire posed a profound and sustained threat. No other environmental danger jeopardized the entirety of the city-building process—encompassing human life, property, and the dreams of city boosters—in such a sweeping or intense fashion. Even before a conflagration reduced Chicago to rubble in 1871, over four hundred large fires destroyed nearly $200 million worth of property in thirty of the nation's principal cities. After the Chicago blaze, into the first decade of the twentieth century, sweeping fires struck countless cities. When nineteenth-century firemen noted in their ledgers that "if it had not been for the activity of our firemen the whole block would have burned," they did not idly boast. Likewise, the smallest of blazes posed problems for the fire insurers, who struggled to meet the terms of fire insurance contracts, and hundreds of companies went bankrupt because they could not pay claims. Curiously, as fire threatened American society, it served as mass entertainment. Thousands of onlookers gathered to watch the action at fire scenes; stories of blazes filled newspapers, provided the subject of theatrical performances, and appeared in the early films of Thomas Edison. Yet into the twentieth century, fire remained a scourge. Losses mounted yearly, even though the danger of citywide conflagration gradually diminished. Fire loss began to level off about 1920. When adjusted for commodities prices and variation in the amount of property exposed, fire loss diminished considerably after 1930, although loss per capita actually began to increase after World War II.[5]

Ironically, fire is not only a "bad master" but also a "good servant." This incongruity exposes a fundamental aspect of human interaction with fire—one that has cultural as well as physical dimensions. Metaphorically, fire reflects this complexity, at once connoting multiple paradoxes between giving life and taking it, between renewing and destroying, between good and evil. With an extraordinarily rich historical symbolism, fire's figurative quality is fraught with ambiguity that only enhances its inscrutability and makes controlling its expression all the more urgent. In nineteenth-century America, this allegorical dimension materialized with great force as fire became ever more critical to driving industrial and urban development. In these processes, millions of people moved from country to city, converting raw materials into commodities and pushing cities to new heights. The migrants also produced dynamic neighborhoods and volatile economic markets. Shifting populations and increased capital investment created stunning opportunities for fire loss, with huge concentrations of fuel, often in novel forms and combinations. The changing built infrastructure and its human elements produced entirely new "fire environments" and altered existing hazards. Not only did unstable urban landscapes and markets increase their susceptibility to fire, especially to conflagration, but they made controlling the devouring element more difficult. Mastering fire became a test with dramatic stakes—with its cultural dimensions underscoring its material imperatives and vice versa.[6] Thus, as fire threatened the physical landscape, it jeopardized the social and cultural infrastructure of American cities.

Seemingly an undeniable social good, extinguishing blazes, paying insurance claims, and physical rebuilding signaled not just the restoration of order but also the reassertion of particular regimens of political, economic, and social authority. For example, although preexisting arrangements of political and economic power typically reasserted themselves after fires—especially after conflagrations—societal fears and prejudices were reflected in the rebuilding process. Further complicating matters, the problem of fire did not remain static over time. Each time the material infrastructure or cultural fabric of urban America changed, so too did the danger. For example, late in the nineteenth century the threat of fire intensified and changed as cities grew taller, contained new combustible materials, and were more densely populated. Establishing order in environments so routinely in flux represented an enormous challenge. Architects, engineers, capitalists, legislators, and ordinary Americans held an interest in controlling the problem of fire, but firefighters and fire insurers took primary leadership in these efforts. Although other Americans helped to develop technical expertise about the fire hazard, ultimately solutions to the problem emerged only with the support of firefighters and the insurance industry. Battling a dynamic material and cultural danger, firefighters and underwriters constructed and reinvented strategies that drew upon the most basic social, cultural, political, and economic currents of American life. Among other things, attitudes about gender, labor, business, technology, the environment, efficiency, and public responsibility shaped firefighters' and insurers' labor and the safety they created. In developing the urban fire protective infrastructure, firefighters and underwriters reordered the lived spaces and collective consciousnesses of Americans; they also made urban industrial society both physically and culturally practicable.[7]

Just as moments of crisis expose the inner workings of a culture, so too the everyday activities through which people protected themselves from fire reveal how Americans ordered their lives. For example, when individuals began to purchase fire insurance from joint-stock corporations at the turn of the nineteenth century, this marked a trend toward looking to economic markets for security against environmental calamity. Although this shift is not surprising given that the United States was in the early stages of a market revolution, the dramatic growth of the insurance industry during the nineteenth century recommends the importance of studying this rather mundane solution to a dramatic problem. Indeed, fire insurance played a primary and increasingly important role in the economy by protecting property and supporting the credit system. Additionally, it helped to fuel the growth of cities with fresh infusions of capital, and eventually the fire insurance industry became central to diminishing the impact of fire on everyday urban life. The history of fire insurance clearly belongs in any account of fire protection, urbanization, or industrial development, but it remains largely unwritten. Fortunately, research into life insurance and business history provides guidance, emphasizes the connections between the fire insurance industry and broader societal issues, and suggests several questions: What role did the management of information play in the industry's development? How did underwriters organize, classify, and understand fire danger? How did that hazard become a quantified "risk?" How did making fire protection into a commodity shape public safety? How did understandings of class and gender structure the business of fire insurance?[8]

Of course the merits of studying firefighting seem obvious, but relatively little work has been completed in this area either, with most of it focusing on volunteer fire departments prior to the Civil War. In the pages of numerous dissertations, journals, popular histories, and a recent monograph, researchers primarily have debated its ethnic, class, and gender dimensions. As a result, researchers have not studied closely changes in firefighting work and organization. This implies that the methods of firefighting are transparent, as obvious as the danger itself, and it fosters the impression that firefighting methods changed little after the advent of steam engines. Nothing could be further from the truth, as suggested by the work of anthropologists and sociologists studying firefighting, as well as the broader scholarship in the history of work and cities. Indeed, rigorously attending to minute changes in firefighting work, technology, and organizational structure reveals firefighting to be complicated labor, constantly changing in direct relation to the urban environment and the broader program of fire protection. This is not to say that the social composition of fire departments is not critical; rather it recommends asking basic questions about the work of fighting fires, making the tactics and techniques of firefighting of central importance. What methods did firefighters use to fight fires; why and how did those tactics and techniques change? How did the experiences of firefighters—in facing fire, in engine houses, and in their communities—change over time? How did firefighters organize their labor differently over time? Who were firefighters and how did they understand their service? Perhaps most importantly, how did their work affect public safety in the face of the problem of fire?[9]

The narrative structure employed in *Eating Smoke* facilitates comparison of firefighting and fire insurance: how did firefighters *and* underwriters understand and seek to control the problem of fire? Juxtaposing the work of firefighters and underwriters in alternating chapters renders each group clearly, highlights the interplay between them, and exposes the tension inherent in their different visions of hazard and risk. These paired chapters are divided into four chronological groupings. Each part explores a distinctive moment in the struggle to define and to confront the problem of fire, capturing the dynamics of historical change as well as emphasizing continuities. The four parts—"Smoke," "Fire," "Water," and "Paper"—suggest the connections between fire protection and the process of urbanization by metaphorically linking them to broader social changes in the nature of community, societal organization, and the economy. "Smoke" explores the period from 1800 through 1850, "Fire" from 1850 to 1875, "Water" from 1875 through 1900, and "Paper" from 1900 to 1950.

As cities began to experience dramatic growth early in the nineteenth century, *smoke* emanating from manufactories signaled the possibilities and hazards of urban and industrial growth. Both firefighters and underwriters identified and sought to understand the emerging problem; they remade old institutions, invented new approaches to controlling danger, and tested them in the urban laboratory. By midcentury, *fire* emerged as a central threat to urbanization, and efforts to reorganize fire protection intensified. Insurers gathered and analyzed data about the problem, seeking to know it better. They remade the hazard into a quantifiable risk—into a commodity—as they looked toward the nascent industrial order for guidance in organizing fire protection. Volunteers remade their institutions, emphasizing their expertise as firefighting specialists, and began to transform their avocation into an occupation.

During the last decades of the nineteenth century, firefighters used *water* and ladders to attack fires more aggressively, and took the lead in providing urban safety. For the first time, firefighters routinely ventured inside structures, "eating smoke" as they battled fires from the inside. Firefighters began to rescue citizens trapped in a built landscape that had grown taller and more densely populated. At the same time, insurers more intensively collected and manipulated information, created standard methods for classifying fire risk, and erected an extraordinary program of surveillance to apprehend the risks of the ever-changing urban environment. However, they struggled to remain solvent and disavowed any responsibility for public safety. As result, firemen became icons of safety, providing the first systematic standard of fire safety in the increasingly urban nation. During the twentieth century, *paper* provided the means of triumphing over fire and trumped the use of water in making cities materially safer. The fire insurance industry shifted direction; underwriters translated their knowledge about fire risk into expansive business, legal, and behavioral practices. Drawing upon the latest techniques of business management and marketing, as well as five decades of engineering and architectural research, these codes transformed the American landscape. Interestingly, firefighters remained the most prominent symbols of safety. Like other Americans, they chose to discipline themselves in order to combat fire more effectively. Fire departments developed extensive bureaucracies and standard work rules. As firefighters established a more regimented approach to fighting fires, they also secured better wages and benefits and developed highly patterned careers. Even so, eating smoke and saving lives remained critical markers of firefighting prowess, as firemen became, somewhat paradoxically, disciplined heroes.[10]

Broadly speaking, the shift from smoke to paper metaphorically suggests the triumph of organizational technologies and intellectual labor over firefighting tools and physical work in containing the problem of fire. Additionally, the transitions from smoke to fire to water to paper reflect the degree to which the problem of fire grew more standard, systematic, and rational from the nineteenth century to the twentieth century, mirroring broader trends in the work, community, and everyday lives of ordinary Americans. Not imposed from the outside, this drive for improved efficiency and systematic protection developed within the work cultures of firefighters and fire insurers. From the early nineteenth century, firemen and underwriters repeatedly introduced new technologies, reorganized their labor, created written standards of behavior and procedure, and introduced measures that quantified the efficiency of their work. They made the organization of fire protection more complex and bureaucratic and their work activities more standard and regimented, usually according to written procedural guidelines. Likewise, the provision of fire protection shifted from local, often geographically circumscribed communities united by face-to-face social relations, to an array of impersonal networks mediated by professional and economic organizations that crossed neighborhood, city, regional, and even national boundaries. Of course neither attempts to organize fire protection more rationally nor the gradual shift of authority away from local communities occurred completely. Indeed, fire protection continued to vary by locale well into the twentieth century, and its provision continued to depend upon a mix of expert knowledge and local understandings. Throughout the twentieth century, firefighters remained intimately tied to neighborhoods, and the fire insurance industry depended upon a steady stream of information from well-connected local organizations and representatives. The tension between local experience and broader efforts at standardization remained a salient issue in fire protection well after the 1950s.

Capturing the dynamic interaction between day-to-day experience—especially of firefighters and insurers—and the increasing rationalization of fire protection posed a difficult methodological challenge. As a social historian, I am particularly interested in understanding the particular historical experiences of firefighters and underwriters, without obscuring the broader implications of their efforts. Therefore, *Eating Smoke* examines both *local* and *national* stories simultaneously—a balancing act made more complicated by my decision to examine the history of fire protection over such a considerable period of time. As a result, I made several significant choices about source materials and narrative style, including especially the decision to tell this story through the lens of firefighters and underwriters, within the context of their local work environments (which was outlined above). After much deliberation, I also decided to develop detailed accounts of firefighting in Philadelphia and St. Louis and to examine fire insurance by studying the Aetna Fire Insurance Company and the Insurance Company of North America. However, this study is more than a comparative analysis of two cities and two firms, although these cases are exceptionally well suited to telling this story. Critically, I have placed these particular stories into a much broader context by closely examining the major firefighting and fire insurance organizations—the International Association of Fire Engineers and the National Board of Fire Underwriters. In addition, I have studied firefighting and fire insurance by reading countless professional journals, work manuals, conference proceedings, and accounts of fire protection in dozens of other cities and towns—all of which helped to shape my analysis, even though they are not always detailed explicitly. Nonetheless, it would be inaccurate to claim that *Eating Smoke* reveals, definitively, all the facets of the nation's changing means of protecting itself from fire. There are numerous exceptions, in part because local structural conditions varied and produced subtly yet significantly different fire environments. In fact, the challenge of creating a systematic program to deal with such a dynamic danger—variable over place and time—bedeviled underwriters and firefighters, and it is a central theme of this book. Ultimately, however, this study should be viewed as the sum of its parts; it is as much about firefighting as it is about fire insurance, and it is as much about the process of urbanization in the United States as it is about fire protection.[11]

The near invisibility of fire protection, except for moments of extreme crisis, attests to how dramatically firefighters and underwriters transformed the American landscape during the nineteenth and twentieth centuries. Until the unsettling events of September 11, 2001, most Americans believed in the permanence of the cities around them, despite evidence to the contrary. That confidence in the integrity of the physical environment, and in the ability of the capitalist economy to maintain itself, has rested on faith in institutions and technologies like those deployed by firefighters and underwriters. When trapped by flames, we expect to be rescued. When our property is damaged, we expect to be compensated. After all, property insurance is an unquestioned part of every home mortgage. We are inured to the pervasiveness of the fire protective infrastructure by which firefighters and insurers have helped to remove the danger from our consciousness. Although the fashion in which this was brought about at first blush seems normative and uncontroversial, this was not always the case; the methods with which Americans have controlled fire did not appear suddenly or without conflict. Fire safety extended gradually, and though it was the project of many Americans, it developed primarily at the intersections of the work performed by firefighters and underwriters. As these groups physically and metaphorically battled fire, they not only brought the problem largely under control, but they also helped to make urban America a reality.

Part I / Smoke

CHAPTER ONE

Workshops of Democracy

The Invention of Volunteer Firefighting

Shortly before midnight on a tranquil May evening in 1849 the bells of docked steamboats and ringing of fire bells awoke St. Louis residents to the danger of fire. Hardly stirred by the commotion and grumbling about the late hour, residents slowly turned out to watch firefighters battle yet another steamboat fire along the city's main artery, the Mississippi River. Indifference quickly gave way to interest and excitement, and then to dread. A burning steamboat, the *White Cloud*, had broken from its moorings and, propelled by the current, floated down the wharf. It crashed into other boats, setting them ablaze. As boat after boat flared, the riverfront became a raging furnace for nearly a mile. Then, suddenly, twenty-two kegs of gunpowder loaded on one vessel exploded in a pyrotechnic display. The repeated blasts transformed the city into "a scene of horrifying confusion." Embers set structures along the riverfront ablaze. Northeasterly winds sent flames along the levee, toward the warehouses located adjacent to the river. Sparks alighted "on the merchandise lying on the wharf and the housetops." Fearing the worst, bystanders joined firefighters working to check the blaze. The battle surged back and forth for hours.

Unfortunately, by the time that night had turned to day the blaze had gained the upper hand. When the city's water supply failed, firefighters could do little to stop the conflagration as it raced from square to square. Given the dire situation, a forty-one-year-old merchant and the leader of the Missouri Fire Company, Thomas Targee, took drastic measures to save the city. Targee ordered that gunpowder be brought forward as he prepared to destroy buildings in the path of the fire. As the chaos mounted, men carrying kegs of powder to the foot of the blaze along Market Street dodged flaming embers, which rained down and spread the fire. Racing against time, volunteers used axes to open buildings. Targee rushed into fiery structures, threw in the powder, and quickly darted away. He blew up five buildings before he died. As one witness recalled, "I watched Captain Targee, smoke begrimed and haggard, stagger and run past me with another keg of gunpowder on his shoulder. . . . Targee entered. Almost immediately there was a terrific explosion." Targee became the first firefighter in St. Louis to die in the line of duty, and his efforts failed to check the fire. The blaze expired only after the wind changed direction, and not before it claimed over 450 buildings.[1]

Such conflagrations were not extraordinary events in nineteenth-century America; their relative frequency speaks volumes about one by-product of urban and industrial growth in the United States. Fire threatened the social order of the urbanizing nation literally and metaphorically, and, as a consequence, this problem greatly interested urban residents. Efforts to bring the fire hazard under control demanded sustained attention. They sometimes required extraordinary measures, like Thomas Targee's last-ditch and fatal effort. Celebrated in the newspapers and later on canvas by local St. Louis artist Mat Hastings, Targee's heroics represent the remarkable public service that volunteer firefighters provided to their neighbors. Although relatively few volunteers died in the line of duty in the years before the Civil War, and although their deaths were primarily related to saving property, firefighters protected society by endangering their bodies. The juxtaposition of virile manhood and horrible death, captured in Hastings's painting of Targee, can be found in countless images and stories from the period, including the celebrated prints of Currier & Ives and the poetry of Walt Whitman. Whitman's elegiac tribute to firemen in *Song of Myself* expressed well the particular danger of fire and the nobility of the service performed by the firemen attempting to master nature run amok. With great indifference and surprising swiftness, fire could threaten to engulf the built landscape, sap the vitality of the social order, and destroy the vibrancy of the men fighting against it. In the early years of the nineteenth century American urban dwellers constructed a system of fire protection in which the skilled bodies of volunteer firemen—as individuals and companies—assumed the risks that fire posed to industrial and urban growth.[2]

Volunteer firefighters celebrated their service quite publicly, creating mythic

The Great Fire at St. Louis, Missouri, Thursday Night, May 17th, 1849, 1849, Nathaniel Currier, lithograph. A sweeping fire nearly destroyed St. Louis in 1849. Conflagrations routinely plagued North American cities as they grew dramatically in the nineteenth and early twentieth centuries. Courtesy, Missouri Historical Society, St. Louis

narratives that valorized their efforts on behalf of their neighbors, communities, and the state. Such stories have been burned indelibly into the popular imagination as they have been retold and reinterpreted for nearly two centuries. In St. Louis, accounts of Thomas Targee's death have been repeated often since the city's great fire. In the decades following the Civil War, his fellow volunteers remembered their youthful service by forming a veterans association, saving memorabilia, and commissioning the painting of Targee. As the former volunteers died in the early twentieth century, their collections were donated to the local historical society, where Targee's portrait periodically graced the walls. By this time, professional firemen in St. Louis had appropriated Targee's image and his story as their own and commemorated him as a forebear in a line of men whose use and invention of progressively more sophisticated technologies culminated in the creation of professional firefighting. In the 1990s, Targee's portrait again found public exhibition in the headquarters of the St. Louis Fire Department, where it educated St. Louis firemen about their service.

The history of volunteer firefighting has been obscured by decades of storytelling that has caricatured volunteers—as community-oriented leaders, fire-obsessed deviants, or fun-loving pranksters. In actuality, the history of volunteer firefighting is more complex and often contradictory. If volunteer firefighters exhibited a playful culture, they did more than frolic. If they sometimes fought among themselves, they were no more contentious than their neighbors, and they fought fires more frequently than one another. If they advocated public spirit and civic responsibility, they rarely lived up to the democratic ideals identified by Tocqueville in *Democracy in America.* Paradoxically, early firefighters volunteered their labor but nonetheless transformed a communal responsibility into a public service performed by specialists. They placed the actual work for firefighting into fewer and fewer hands, a trend that intensified during the nineteenth century. As volunteers argued for the distinctive importance and nature of their service, they divided the labor of firefighting into increasingly specialized tasks. In the process, volunteer firefighters set the stage for the shift from avocation to a vocation, mirroring broader patterns of change in the industrializing nation.[3]

This story begins in the nineteenth century, when urban residents reorganized fire protection and invented a new approach to putting out fires. They produced a system of labor based on specialized work activities that both preserved order and encouraged urban growth. They removed responsibility for the problem of fire from the entire community, and reorganized it almost exclusively into the body of their organizations (volunteer fire companies). With tacit community approval, fire companies' reimagined danger, civic duty, and invented themselves as embodiments of the body politic. They constructed a culture of manliness that valued physical virility, technological innovation, and commitment to the polity. That culture of manhood contained dramatic contradictions—between competition and cooperation, between self-discipline and expressive personal/team style, between public labor and private interest, between the continued work specialization and the desire to retain traditional communal identity. As firemen struggled to improve their efficiency and embraced social and technological change, their manly culture became less able to contain its contradictions. Increasing demographic and social differences within cities and between volunteer fire companies eroded firemen's shared culture. As such differences grew more pronounced, especially after the 1830s, firefighting became less an avenue to restore order than a theater in which firemen contested for social and political power.

Despite many contradictions, volunteer firefighters helped to make rapid urbanization and industrialization a reality, and as they organized and transformed firefighting, volunteers' beliefs about gender and technology structured urban landscapes, as well as conceptions of the public good and public order. As fire threatened the material viability of nineteenth-century communities, citizens suppressed the danger in a manner consistent with prevailing attitudes about manhood, technological progress and efficiency, ethnicity, and race. Their strategies—embodied in voluntary fire associations—literally allowed the dominant social system to thrive and made urbanization into a reality. However, the organization of fire protection also played an equally important symbolic role in preserving community. As fire threatened to destroy order—and its economic, political, and moral bases—fire companies preserved the social order metaphorically, by representing themselves as bastions of virile male virtue, whose members voluntarily performed an obligation of citizenship. More importantly, as volunteer firefighters embraced both the specialization of their labor and technological innovation, they offered support to industrialists' vision of the future. They championed a vision of a social order in which each individual and group held a functional niche (which sometimes contradicted their rhetoric). In so doing, volunteer firemen pioneered and reinforced the process of industrial and urban change.[4]

Specializing Civic Obligations

In 1803, several young Philadelphia apprentices established the Philadelphia Hose Company and altered how urban communities confronted the problem of fire. They literally reinvented volunteer firefighting by adapting hoses for use in firefighting, and the company transformed volunteer protection into specialized public labor. Of course, hose was not new, but the Philadelphia Hose Company became the first fire company in North America to adapt it to fire protection. Hose, especially the high-quality variant developed in Philadelphia's workshops by artisan firemen, spawned a new division of labor that allowed firemen to remove responsibility for fire extinction from the community and to locate it in their organizations. Furthermore, volunteers recognized the significance of the new mode of organizing, and they established a new rationale to govern their service. They argued that firefighting was a special calling, through which volunteers demonstrated their capabilities as men, innovators, and citizens.

The formation of the Philadelphia Hose Company transformed what in the eighteenth century had been understood as the responsibility of the entire community. Based upon European (especially English) methods of fire protection that were modified by North American colonists, eighteenth-century Americans relied on community organization, buckets, and occasionally explosives (gunpowder) to check the spread of fire. Additionally, many municipalities created ordinances that mandated fire preventive practices, such as laws that asked property owners to sweep their chimneys regularly. Extinguishing a blaze often required hundreds of

Thomas Targee (Captain, Missouri Fire Company Number 5), 1902, Matthew Hastings, oil on canvas. Thomas Targee died in 1849 trying to control the Great Fire in St. Louis. In 1902, former volunteer firefighters celebrated their nineteenth-century service by commissioning a portrait of Targee, which hung at the city hall before making its way into the collection of the Missouri Historical Society.
Courtesy, Missouri Historical Society, St. Louis

urban residents organized into bucket brigades. Additionally, in many places, municipal law allowed private citizens to form voluntary associations to help in this struggle. These associations often purchased, maintained, and operated primitive fire engines. When available, such engines would stand at the head of the bucket brigade, intermittently squirting water onto fires. Although such associations frequently played a central role in fire extinction, the civic responsibility for fire protection nonetheless fell to all urban residents, especially property holders. Every member of a community—including perhaps women and slaves—donated physical labor and/or buckets in the struggle to preserve community.[5]

Against this backdrop, Rueben Haines, Roberts Vaux, and the other young men who founded the Philadelphia Hose Company envisioned a more efficient and specialized method of fire protection. The company used hose, made of leather and joined by copper rivets, to conduct water from the newly constructed system of water pipes (built in 1801) to fire engines. With exacting calculations, an early company member and printer, James P. Parke, quantified the efficiency of hose: "A hose of 600 feet in length, would, perhaps, deliver as much water . . . as 3600 men, with common fire buckets. . . . The benefit of hose may perhaps be seen in a clearer light, when we say, that 120 [fire company] members can deliver as much water at 1800 feet distance as 11,000 men with buckets from each bucket." Hose companies reduced the labor needed to extinguish fires by as much as 99 percent. If the innovation promised to reduce the amount of labor needed to put out a fire, it also promised to increase the speed in which water was applied to a blaze. Parke reported that "an accurate observer, long accustomed to attend at fires" had informed him that hose reduced the time needed to fill an engine from fifteen minutes to ninety seconds. With this argument, firefighters established the standard for measuring the effectiveness of their work for decades (and also created a norm that what would return to haunt them in the 1850s).[6]

Hose companies had reorganized a formerly communal responsibility into a job for a small cadre of men. According to Parke, as few as 120 men could handle the work that had once been performed by 11,000 urban residents. Although it seems certain that Parke overstated the efficiency of hose companies, his claims nonetheless are stunning. According to his calculations, as much as 25 percent of Philadelphia's population (which at the time was just over forty thousand residents) could be called upon to fight fires. Even if he exaggerated by a factor of ten, there can be little doubt that by adapting hose to firefighting, the Philadelphia Hose Company dramatically shifted how the city's residents dealt with the problem of fire. This shift signaled the start of a long, gradual transformation in how the responsibility for restoring order and providing fire safety would be constituted. Although forms of collective organization would continue to play an important role, the associations responsible for fire protection became ever more particularized.[7]

In Boston, where hose technology was adopted in 1823, citizens' reluctance to cede responsibility for fire protection proves instructive about the stakes involved in embracing this new division of labor. Bostonians feared the consequences of placing firefighting in the hands of just a few men. As one fire company leader stated, "Do you think sir, that the citizens of Boston will ever be prohibited from assisting a fellow townsmen in distress?" Yet such complaints ring hollow to some degree because hose was introduced at a time when Bostonians lamented the lack of adherence to community obligation in time of crisis. There had been, observers noted, poor community turnout at fires, which had increased in severity. Nonetheless, Bostonians—even some firemen—appeared to have been hesitant to position such a crucial public responsibility in so few hands. The manner of reorganizing fire protection proposed by the Philadelphia Hose Company altered how urban residents throughout the nation would approach the risk of fire in the nineteenth century.[8]

As firemen connected hoses to the city's burgeoning water systems, the Philadelphia Hose Company discerned that it had altered significantly the division of public responsibility for the problem of fire. The company recognized that fire literally and figuratively symbolized disorder; it provoked bewilderment, confusion, and chaos, in which "stentorian strength of lungs" too often replaced "cool and accurate judgment." As a result, prior to going into service, the company considered a number of questions. How could the new arrangement obtain support from proponents of the previous method? How could it manage operations at fires in a fashion that inspired public confidence? How could the company justify its service? How could it convince neighbors and peers that it was capable of handling the severe crisis of firefighting? Sensitive to these social implications, the company devised a plan to legitimize the repositioning of firefighting authority within a body of specialized workers. The company adopted a strategy rhetorically based in the democratic principles of the young nation and a faith in rationality and collectivity; the company proclaimed that it would discipline itself under "the dominion of *government* and *order* in time of fire."[9]

The Philadelphia Hose Company constituted itself as a democratic body by establishing an organizational structure that balanced authoritative leadership with the company's communal sensibility. Company members elected their own officers, to short terms. At fires the association appointed "directors," vested with sufficient authority to require "obedience." At the same time, the company also sought "to prevent a monopoly of the office [of director]" by holding elections every six months, which limited leaders' power. Through democratic election, as well as rotating leadership, the company further hoped to prevent private interest from overwhelming its and the public's goodwill. In addition, the company believed that rotating leadership posts fostered collectivity of action—a principle that carried over into the methods that fire companies used to organize work activities.[10]

The men of the Philadelphia Hose Company believed that the order provided by firefighters transcended the problem of fire, and went to the very core of the polity. As firefighters inscribed the nation's ideals into the organizational structure of the company, they became exemplars of democracy. Likewise when firefighters protected the built environment, they preserved the core values of American society. At moments of extreme environmental turmoil—when Philadelphia stood on the brink of chaos—firemen brought discipline to the city through their rational and democratic action. The company removed power from the hands of sometimes unruly mobs and placed it exclusively into the collective rationality of a male society. James Parke recalled, "It was their duty to preside with a cool, deliberate judgment . . . when tumult would suspend their faculties, and when the prejudice of a mob—a vulgar and motley herd, among whom, all are, or wish to be, masters, and none scarcely will be servants—would prove a strong and almost insurmountable barrier to their success." When it fought its first fire on March 3, 1804, the company not only preserved the built environment but it reordered civil society. The Philadelphia Hose Company had taken the first tentative steps toward recreating the city's and the nation's methods of fire extinction.[11]

Public Duty and Community Support

Across the nation, firefighters adopted the Philadelphia Hose Company's innovative new mode of organizing at the precise moment when cities and industries began to grow intensely in America and when fire became an increasingly significant threat to societal stability. Structures built of flammable materials gradually clustered more densely within city limits. Ironically, in mercantile cities like Philadelphia and St. Louis, the very lifeblood of the widening economic networks—riverboats—became a source of great danger, as in the 1849 conflagration in St. Louis. Indeed, during much of the century, nearly every fire threatened to erupt into a conflagration. Though firemen often did manage to forestall conflagrations, judging from the ledgers of the Missouri Fire Company, they were far less successful at confining flames to a single structure. Most cities recognized the severity of this threat and codified their respect for the flammability of their built landscape into city ordinances. In the 1830s, for example, Boston authorized the fire department to destroy buildings adjacent to a fire to prevent its spread.[12]

Of course, the large number of well-documented nineteenth-century conflagrations, which affected most industrializing cities, attest to how often severe fires became reality. Philadelphia experienced several sweeping fires, including those in 1839 and again in 1850. In the latter blaze over one hundred structures burned in the city's commercial district, along the Delaware River. Yet Philadelphia was spared a conflagration of the magnitude of the one that struck St. Louis in 1849. Fire raged out of control for nearly two days and destroyed over $3 million in property and goods—approximately 75 percent of the city's accumulated capital. More broadly, inadequate insurance coverage by many merchants, small manufacturers, and homeowners—not to mention the limited financial reserves of insurers that forced firms into bankruptcy—exacerbated the economic consequences of fire in the antebellum United States. Moreover, fire did not just threaten a city's economic future, but it also menaced the financial stability of local governments. For instance, the great fire in St. Louis severely impaired the city's financial stability—approximately two-thirds of city revenues had been collected from property taxes.[13]

In this volatile context urban residents endorsed the technological and social innovation of the Philadelphia Hose Company. Over the course of the next several years, men in Philadelphia and other cities formed hose companies and reorganized engine companies, meanwhile remaking the engines to operate with more efficient flows of water. Most importantly, though, urban dwellers legitimized the new division of labor through moral and financial support. When the Philadelphia Hose Company formed, its members took out a "subscription." That is, they appealed to their neighbors for financial support to purchase equipment and cover operating expenses. Through a circular letter, the company collected over $700 from "the insurance companies and citizens generally." During the first several decades of the nineteenth century, such contributions were commonplace. In 1809, for instance, Phoenix Assurance, a London insurance company, awarded Philadelphia's nine hose companies $100. Shortly thereafter the Insurance Company of North America regularly donated $25 to each of Philadelphia's fifteen engine and hose companies. Besides pecuniary support, communities offered other intangible forms of assistance. In addition to providing a steady stream of men to serve in fire companies, communities also dispensed food, coffee, and alcohol to hungry volunteers at fires, or they spontaneously made donations—especially when their property had been saved.[14]

As the nineteenth century progressed, companies more frequently used subscriptions and/or held balls to collect money to finance much-needed projects—to procure new engines or hose, to improve engine houses, or to purchase other accoutrements of firefighting, such as alarm bells. Sometimes these requests were tainted by hints of extortion. In 1828, for instance, Philadelphia's Hibernia Engine Company told its neighbors, "Never having but once before asked for assistance, they [the company] trust this call will be a successful one; should it fail, the engine must of necessity be abandoned, and the public deprived of the services of the company." However, more often than not, local residents and businesses willingly filled the coffers of their favorite companies. This further cemented the bonds between companies and communities and encouraged a redefinition of fire's dangers. Through such regular donations, urban residents continued to transfer their civic obligations onto the shoulders of firemen; fire buckets became relics of a bygone age.[15]

Municipalities, too, supported the new distribution of risk. They sustained fire companies in many ways, including cash donations—a phenomenon that became more pronounced over the course of the nineteenth century. Though they did not bear the entire cost of fire protection, or its risks to life and limb, cities routinely contributed to companies. For instance, public funds accounted for half of the St. Louis fire department's operating revenue in the years before 1850, totaling over $110,000. The Philadelphia Council established a city waterworks in 1801 and began supporting volunteer fire companies in 1811—a practice that was formalized in municipal laws throughout the region by the 1830s. In addition to sometimes exempting volunteers from taxation, state and local governments often excused volunteers from various responsibilities of citizenship, including jury duty and military service. A generous symbol of support, the exemption for firemen was predicated on several years of service to a company. In 1844, for instance, the Perseverance Hose Company of Philadelphia presented William Thorp a membership certificate that announced his exemption from military service as a result of a state law passed in 1842. During the Civil War New York volunteer firemen protested the draft precisely because this exempt status had been lifted.[16]

As urban residents surrendered participation in favor of efficiency in the battle with fire, fire companies construed service in a multitude of ways that emphasized their connections to their neighbors and/or obscured the increasingly specialized division of responsibility. For instance, many firefighters clothed their service in the language of republicanism—the idiom of reciprocity and public good so pronounced among urban artisans and journeymen and that seemingly existed everywhere in the early republic. Companies expressed this affinity by choosing slogans or images, such as the goddess liberty. Other companies, though, selected symbols that could be identified with nineteenth-century liberalism, such as the metaphoric image of the beehive. Yet others chose symbols that resonated exclusively within their local community, as was the case of the Shiffler Hose Company in Philadelphia, which chose an anti-immigrant rioter to express its values. Whatever their costumes, firefighters clothed themselves in a variety of political and social symbols, especially those that appealed to their neighbors and supporters. Somewhat paradoxically, these symbols concealed the degree to which firefighting was becoming less community based, as increasingly it was carried out by specialists. Firefighters began to forge a collective interest that superseded their connections to their neighbors.[17]

More broadly, firefighters obscured the shifting responsibilities of fire protection by cloaking their service in the rhetoric of democratic citizenship and public spirit. For instance, when companies incorporated, their state-issued charters of-

The American Fireman. Always Ready, 1858, Louis Maurer (artist), Currier & Ives (lithographer/publisher). This 1858 portrait of a volunteer firefighter pulling a small hose reel captures firefighters' romantic perception of their public service. Courtesy, CIGNA Museum and Art Collection

ten likened the organization of firefighting associations to the "body politic." This sentiment captured the full dimensions and complexity of firefighters' claims on the public imagination—especially the degree to which democratically organized fire associations protected the nation's physical, as well as political, order. Conflagrations threatened many different types of bodies, from the general to the particular: the state, urban communities, fire companies, urban residents, and individual firemen. Each body perceived and responded to the threat of fire differently. Fire threatened municipal governments' solvency, vitality, and authority, and through legislation the state devolved that responsibility onto all citizens. Urban residents also feared fire, and they contributed to specialized organizations of firemen through financial donations and moral support. Fire companies, collectives of individual men, shouldered the community's and the state's collective risk by constituting an image of the body politic. The labor of firemen and fire companies symbolized a broader conflict between the nation's democratic collective and disorder, even as it offered a model for how society should be organized.[18]

Firefighting Work: Physical Prowess, Danger, and Technological Competence

Although their efforts had far-reaching effects on the social order, firefighters engaged in very immediate and very taxing labor. Indeed, firefighters' material labor stood at the center of their relationship with their communities. Firefighters performed a demanding physical enterprise that required an array of technology and skill; they also placed themselves at personal risk while working in difficult, if not exceptionally dangerous conditions. Volunteers divided their work into two components: work involving engines and work with hose. Typically, they split these functions into separate work groups with different apparatus, responsibilities, and traditions. Engine companies drayed apparatus, which sometimes weighed several thousand pounds, to fires and operated it by pumping brakes attached to the sides. Meanwhile hose companies brought hose and used spanner wrenches to connect it to nearby plugs and engines.

Indeed, firefighters organized their labor in a manner that reflected widely held understandings of manhood and work in the early nineteenth century. They placed a special emphasis on the physical power needed to operate engines and the technical skills needed both to keep their apparatus working properly and to improve equipment. Moreover, firefighters balanced individual skills, strength, and commitment to community against the demands of the need to be part of the larger collective, to work in tandem with other men. If volunteers acquired their competence primarily while fighting fires, they also learned and exercised them publicly in informal settings. Firefighters argued that spontaneous competitions improved efficiency and provided a forum to practice skills. These contests also served as tests of firefighters' manhood. Gradually, over time, competitions and work at actual fires took on the air of sporting events, although the gravity of the work remained a sobering and persistent undercurrent. Volunteer firefighters believed that their labor, skill, teamwork, and efficiency, not to mention the dangers they faced willingly, made them into firemen, and marked them as especially able and valuable citizens.[19]

Successfully arresting fire necessitated that firefighters arrive at blazes quickly—a pressure they felt keenly because fire spread quickly through densely clustered wooden cities. Promptly receiving word of an alarm was essential. Before cities began to adopt fire alarm telegraph systems in the 1850s, alarms spread via word of mouth and/or by tolling church and fire bells. When firemen "repaired" to their engine houses and headed toward the fire, they often had difficulty in locating the fire. They relied on directions from bystanders or followed trails of smoke—sometimes a difficult prospect, given widespread use of fire in industrial shops. Unpaved, crowded, and narrow streets or alleys impeded their speed. Pulling engines proved hazardous under the best conditions, and it grew more harrowing at night or in inclement weather.[20]

Upon the alarm of a fire, hose companies whisked two-wheeled hose tenders out of their houses first. Next they pulled larger, four-wheeled hose carriages—loaded with as much as several hundred feet of leather hose—to the blaze. Using the two-wheeled tender, hosemen established the initial plug connection, which became easier as urban infrastructures developed. Prior to the construction of efficient water systems, some cities established networks of cisterns, filled with water, for use in firefighting. Hosemen sought the nearest plug connections, and they connected hard leather suction hose—joined by rivets on the seams—to the hydrant and waited for the arrival of an engine. Hardened hose allowed the suction engines, commonly used after the 1830s, to draw water from the hydrants into the engine box. Hose carriages conveyed leading hose—constructed of riveted leather, and later on of soft hemp. Firemen connected this hose to a fire engine's external output valve. Once linked to a water source, firefighters began to operate engines, expelling water onto the fire.[21]

If getting to fires required great effort, the work needed to operate hand-pumped engines demanded significant amounts of physical strength and stamina. Firemen pulsed the levers or "brakes" attached to the sides of the engine vigorously in order to generate streams of water. In order to create the vacuum pressure necessary to generate a continuous stream of water, they had to pump the brakes continuously with a steady cadence, and in order to keep water pouring onto the blaze they had to stroke for long periods of time, with little rest. Although the machines used by firefighters varied greatly by the 1830s, most engines, such as those manufactured by John Agnew of Philadelphia, operated optimally with as many as forty men working the brakes at any given time. Companies divided into two or three work groups, each taking a turn on the brakes; sometimes companies accepted or even solicited assistance from bystanders as they struggled to generate consistent water pressure. Even with the concerted effort of scores of men, arresting a blaze sometimes took many hours. An 1839 news report from *Niles Register* about an especially large fire in Philadelphia underscored the intensity of firefighting labor, which could even last for days. The fire, which broke out at 11 P.M., continued to rage at 3 A.M. the next night, and "had spread to such an alarming extent, that . . . not less than 45 houses [had] been laid in ruins."[22]

Recognizing that pumping engines was their fundamental work activity, firefighters typically measured company effectiveness and their individual competency as firemen by regularly and constantly evaluating the performance of their company and engine. At times, firemen assessed their performance in terms of the amount of time they and their company worked at a fire, pumping water onto it. More commonly, though, firemen measured the efficiency, power, and stamina of the team by assessing the length of the engine's stream. During a trial of its newly purchased engine, for instance, the Missouri Company reported that it "threw one hundred and ninety four feet from the end of the nozzle one inch [in diameter] through a ten foot section of hose. My opinion of the engine is that she is a powerful engine, she throws well but she is a perfect mankiller to work, give her as much water as she can use and it would take a company of two hundred men to keep her working steady." The trial attested to both the manliness possessed by individual firemen and the technological skill of the collective.[23]

Besides reinforcing firemen's sense of themselves as virile and robust, working an engine's brakes encouraged mutuality and reciprocity. Companies required their members to rotate continuously on the brakes, since a single man could muster the strength to work for no more than ten or fifteen minutes. When few men turned out, company secretaries gnashed their teeth at the company's lack of commitment. Phillip Branson, the secretary of the Missouri Fire Company, noted that on one cold February night "St. Louis did not get out of the house. Missouri had only 6 members out." On another occasion, Branson sarcastically reported that the company had "seven members out. Company is improving fast, Damn Fast; *next fire she will go out the back door.*" More commonly, though, companies turned out more than enough men to operate the apparatus. Although a St. Louis ordinance limited company membership to seventy-five, company ledgers often reported that many more men than the legal number showed up to battle the flames. In one instance, the Missouri Fire Company recorded "engine on the ground, plaid [played] first water, had a very large turn out; 69 men on engine, 34 hosemen and if ever an engine went over the ground fast the Missouri Engine did this night. We past tenders and Engines the same as throwing your hat."[24]

The reciprocity fostered by teamwork at fires also nurtured firemen's relationships to their communities. Indeed, many men and boys from local neighborhoods helped to man the brakes, dray engines or hose carts. Fire companies in St. Louis encouraged these "runners" to attend fires, perhaps because they added substantial manpower to a company—sometimes propelling the numbers of workers at fires above legal limits on membership. Occasionally these unofficial "volunteers"—sometimes boys as young as twelve or thirteen—created hose companies associated with officially sanctioned volunteer fire companies. Naming themselves after swift ferocious hunting animals, like "Tigers" and "Hounds," these young men dragged the crucial two-wheeled hose tender to fires. On one occasion, the Missouri Company reported, "Missouri on hand; our tender past [*sic*] the *Tigers*, *Hounds*, and *Grey Eagles*, and got a first rate plug." Such informal understandings between fire companies and their communities cemented the bonds between them, and also trained the young men of the community about male civic responsibility and about the techniques of firefighting.[25]

That antebellum volunteer firefighters so emphasized pumping engines and dragging apparatus as key work activities reveals much about their priorities; in particular, it strongly suggests that these volunteers did not attempt to save their neighbors' lives as much as they protected property. In contrast to the lifesaving narratives of heroism so common later in the nineteenth century, antebellum volunteer firefighters rarely described their work in terms of rescues. Indeed, of the hundreds, if not thousands, of extant images commissioned by fire companies in American cities of the time, very few depicted firemen actively saving lives. An equally small number displayed firefighters entering or working inside burning buildings. Rather, the art commissioned by volunteers prior to the Civil War emphasized the mundane physical work of pumping and pulling engines. Membership certificates, for instance, portrayed firefighting through allegorical images of service, representations of company engines or speaking trumpets, illustrations of men pulling hose carts, or views of apparatus positioned outside of blazing buildings dumping water into upper-story windows and onto roofs. Images of the time sometimes showed hooks, ladders, or axes but rarely depicted them being used by firefighters. Additionally, the term *ventilate*, which denotes action that involves climbing onto a building's roof, was not in volunteers' lexicon of work techniques, at least as described in their ledgers. Moreover, antebellum volunteer firefighters rarely formed and sustained active hook and ladder companies, which further underscores the primacy of hose and engines before the Civil War. Of course the character of the built landscape prior to the Civil War—in which buildings rarely exceeded three or four stories—and limitations in hose and engine technology contributed to the strategies used by volunteers. Thus the indication is that antebellum volunteers typically attacked blazes from outside of structures, rarely venturing into building interiors to face fire at close range.[26]

Even so, the amount of physical risk to firefighters intensified, reinforcing their seriousness of purpose. Typically, volunteers set up their apparatus in close proximity to buildings, for maximum effectiveness. At such nearby positions, especially when the burning structure was large, firefighters found themselves exposed to death. In Philadelphia, for instance, a member of the Diligent Fire Company died when the wall from a downtown building fell onto the company's apparatus. All in all, however, such cases were highly unusual. Few volunteer firefighters died in the line of duty, and those who did typically faced exceptional circumstances, as the story of Thomas Targee indicates. Actually, volunteer firemen faced the greatest physical danger when racing to fires, pumping engines, or sparring with one another. In St. Louis all of the injuries reported by the Missouri Fire Company occurred when firemen rotated on the engine's brakes or fell under the wheels of a moving apparatus. Minute books and "official" department histories in New York, Pittsburgh, Philadelphia, and Cincinnati contain similar reports. Of course, in antebellum America the possibility of serious injury could not be taken lightly. Although volunteers faced fewer hazards, relative to those who later fought fires, they made much of the dangers. Physical risk defined their relationship to other firemen, bonding them as "brothers" facing a common peril. More broadly, volunteers emphasized this exposure frequently; they claimed to be noble and public spirited because of the hazards they confronted.[27]

In order to mitigate the impact of workplace hazards, volunteer firemen created and sustained disability pensions beginning in the 1830s—often with the support of local municipalities and fellow citizens. Precursors to the relief associations formed by professional firefighters and to disability pensions, these funds received warm support from local political leaders and the community. They represented popular appreciation for firefighters' specialized labor and acknowledged the particular dangers faced by firemen in protecting their neighbors' property. In 1835, for instance, firemen in Philadelphia established the Association for the Relief of Disabled Firemen in Philadelphia, which provided a pension to men injured or disabled while serving the city. The mayor spoke at the organization's inaugural benefit, and community members, firemen, and merchants subscribed to the fund. Although the extent of the membership in such associations is not entirely clear, their development both underscored that firefighters faced very real danger and that communities were aware of those risks. Importantly, by demanding and receiving such special recognition, firefighters separated themselves as men whose service to their communities deserved special approbation. Danger—or "exposure" as one observer put it—gave firefighters a common rallying point and a basis for a shared culture as firemen.[28]

In addition to emphasizing the physicality and danger of their service, firefighters also celebrated its links to the culture of urban craftsmen, especially their

Neptune Hose Company Banner, 1830, John Archibald Woodside Sr. The motto of the Neptune Hose Company—TO SAVE OUR FELLOW CITIZENS WE HAZARD OURSELVES—reflected one fire company's perspective on its service. Courtesy, CIGNA Museum and Art Collection

ability to deploy complex equipment. Firefighters placed a particularly high premium on possessing technological acumen. Like other workers of the early nineteenth century—especially those laboring in artisan shops—firefighters constantly improved upon and tinkered with their equipment. Not surprisingly, many significant innovations in equipment drew directly from the work and experience of the volunteers. In Philadelphia firemen like George Mason, Patrick Lyon, John Agnew, and John Parry built or designed apparatus. Other firefighters experimented with new hose design. Led by the firm Sellers & Pennock, Philadelphia volunteer firemen introduced leather hose held together by metal rivets. This new hose stood up better to the higher pressures of city water systems and was indispensable to ongoing attempts to develop and use of more powerful fire engines during the first decades of the nineteenth century.[29]

Firefighters invented and built many of their work tools, and they constructed a connection between artisanal competency, innovation, and public service. They developed and expressed such values when companies of firemen met informally at local workshops to build their apparatus. Some companies in particular took special pride in the connection between the technical labor of firefighters and their service. For instance, the Mechanics Company of Philadelphia celebrated members' occupations and the role artisans played in constructing public safety by representing itself using the well-known 1830s painting *Pat Lyon at the Forge*, which showed a muscular man, working a steel forge, controlling fire to create machinery. Not only was Pat Lyon a prominent artisan, he also pioneered the manufacture of hand-operated fire engines in the United States. Consequently, when the Mechanics Fire Company adopted his image, it honored the involvement of volunteers in conceptualizing and manufacturing fire suppression technologies—including objects as seemingly mundane as wrenches or hose.[30]

Of course, not all firemen possessed close connections to manufacturing shops, but they nonetheless valued technical competence and a culture in which skills and technologies were shared. For example, fire companies frequently corresponded with other associations located throughout the nation, especially exchanging information about improvements in firefighting apparatus. Through such communication, fire companies—especially those in small towns that might not have been able to afford the latest and most expensive technology—learned where to buy an excellent or inexpensive machine. Other companies published newspapers and disseminated a wide variety of news. For instance, a circular sent by a Baltimore company included reviews of the latest firefighting technology on display at New York's Crystal Palace. Such discussions make clear a process through which companies emulated, often with great swiftness, the successful innovations of their peers. Even after the construction of engines became more specialized, fire companies continued to specify their designs. St. Louis fire companies, for instance, corresponded regularly with manufacturers like John Agnew over specifications and requested special designs, colors, and features on their apparatus. When such engines needed repair or modification, companies took them to a local manufactory or fixed the machines themselves.[31]

Firefighters further demonstrated technical competence by operating and maintaining their apparatus. Even what seemed to be the most basic and dreary of tasks, such as pumping an engine's brakes in unison, required technique learned over time and honed through contests and in battle with fire. One firefighter recalled, "The easy and successful working of the machine was of great consideration." Rotating in and out of the team operating an engine's brakes required adroit timing and skill. If firefighters did not carefully coordinate this switch, the strength of an engine company's hose stream could be diminished or a hapless volunteer could sustain serious injury. Firefighters drilled regularly, and especially prior to contests, to assure their fitness for this aspect of their labor.[32]

Additionally, firefighters kept apparatus operational with constant care. They developed an intimate mechanical knowledge of their tools and work by regularly tinkering with and cleaning their equipment. Company by-laws sometimes codified the routine maintenance activities, and working with the engines kept them running smoothly; but more often than not firemen tinkered with their apparatus to prove their "inventive genius." Especially before contests, firefighters fiddled with engines to gain advantage. As a New York firefighter remembered, "The works of an engine would be all taken apart, each screw, joint and valve thoroughly examined and studied to see if some arrangement could not be adjusted whereby greater power could be achieved." Reconfiguring apparatus took on an importance apart from the functional need to keep equipment running well. The words of the fictional fireman Mose—a well-known character that appeared in New York theaters in the 1840s and 1850s—captured firefighters' intimate connections to their technology. In one drama, Mose effused, "I love that ingine better than my dinner; last time she was at de corporation-yard, we plated de brakes, and put in new condensil pipes; and de way she works is about right, I tell you. She throws a three-inch stream de prettiest in town."[33]

Firemen identified with their equipment in a manner that transcended even the pride artisans felt for items they produced. For many firemen, their apparatus and more basic tools, such as spanners, hose, axes, and speaking trumpets, embodied company identity, linking their beliefs about technical competence, physical power, and service. Fire engines became works of art as firemen maintained and decorated them with an obsessive devotion. Firefighters zealously burnished, repaired, and cleaned their engines. They commissioned paintings of heavy wood to adorn apparatus, primarily during parades, but sometimes art remained on apparatus at fires. Volunteers also transformed certain work tools—especially speaking trumpets—into ceremonial gifts presented to firefighters for notable service or meritorious actions. Made of bronze or sometimes silver, these ordinary items became significant markers of manliness and firefighting prowess. In a similar vein, firefighters decorated themselves, their membership certificates, and engine houses with images of their apparatus, as well as other equipment. For instance, on the ornate borders of certificates, firefighters depicted their machines and tools, thereby connecting them to their public service. These certificates offer a pictorial history of technological progression, as firefighters adopted the latest and most efficient equipment.[34]

The effort to master and to improve upon firefighting technologies also was intertwined with the larger ethos of competitiveness that drove firefighters, relentlessly pushing them to perform effective service and to outdo each other. At every fire and in frequent contests, firemen measured themselves, their companies, and apparatus. Informal and formal equipment trials transformed daily work activities into rituals that underscored firefighters' commitment to their task. For instance, the imperative to reach a fire quickly became an informal expression of firefighters' competence and a test of collective manhood. Reaching a fire first—usually with the specially designed two-wheel hose tender—signaled that a company possessed discipline, team spirit, and physical prowess. In ledgers, companies regularly expressed pride at arriving first or dismay at being outrun. In St. Louis the Missouri Fire Company reveled in besting a rival: "Union suction tried to pass us but we ran away from them and had water on the fire before they had their lead hose off." Alternately, for firefighters across the nation, being "passed in the engine house" was a great ignominy. The Missouri Fire Company's minutes suggest how seriously it took such informal competitions when it related the company's dismay at being passed: "The St. Louis, Union, & Liberty all past [*sic*] the Missouri in the house. This is the first time the Missouri has been past [*sic*] for 2 years & 3 months and I hope it will be the last." Firefighters had transformed daily work activities into ritualized competitions that carried great significance. Winning such contests meant more than getting to fires quickly; it indicated a company's fitness as men and as public servants.[35]

The most significant competitions involved "throwing" water high into the air, in vertical streams. Firefighters judged—usually against a point on a tall building—the length of the hose streams, and elaborate rules specified the length and size of hose and contest duration. The company whose stream went the greatest distance (usually measured vertically, which most mirrored work at a fire scene) was deemed the victor. In Philadelphia, for instance, after putting out a large fire on Market Street in 1850, the city's volunteers stayed on the scene and tried to outdo one another by arching their streams high over the burned-out building. Two years later, the most powerful hand engine built and operated by Philadelphia firemen, the Diligent Engine, competed against the pride of Baltimore's fire department. The Diligent triumphed when it threw water more than 170 feet into the air. As with informal races, fire engine contests developed as the number of fire companies increased and as firefighters both improved upon and became more invested in their tools and apparatus. By the 1840s in St. Louis, there was a tradition of challenging rival companies to assess which company possessed the best equipment and members; of course, challenges rarely went unanswered. Companies sometimes met for weeks prior to competitions to debate and clarify arcane rules. Over the course of several meetings rivals decided on judges, stakes, and standards for victory.[36]

Losing these contests could be devastating. For nearly a decade, St. Louis's Union Fire Company No. 2 (using an Agnew engine) fended off all challengers; the company dubbed its engine "Emperor" because it won one contest after another. Before the Emperor vanquished the Missouri Fire Company's "Old Bull" engine in 1847, the Missouri Fire Company had held the honor of having the city's best engine—and the reputation as its most effective and powerful company. The Missouri company so coveted the honor it had lost that, for the next decade, it purchased a series of engines, even asking Philadelphia's John Agnew to build a new pumping mechanism for the Old Bull. Left gnashing their teeth by repeated failure, the disconsolate company questioned not only its technology, but also the potency of its members.[37]

Firefighters transformed competitiveness into a shared male virtue by rhetorically casting competitions with one another (and with nature itself) in terms of sexual domination over women. In descriptions of their work, firemen sometimes used sexualized language to describe their control over nature and equipment, and they often described their apparatus as being feminine. Firemen "played on" fires with hose, and company leaders encouraged the volunteers to work on an engine's brakes with phrases like "up and down on her, men." This was especially true of contests. In one form of competition—more typical in New York than either St. Louis or Philadelphia—fire companies connected their engines in a line, as they might to combat a fire a long distance from a hydrant. Joined by hoses, the first company pumped water into the second company's engine box, which then pumped water into the third company's engine, and so on. The company with the least water in its engine compartment was declared the winner. A company was declared "washed" if its engine box had overflowed, and its washed engine was said to have lost its "virginity." Filling virginal engine boxes and spewing copious amounts of water provided tangible evidence of firefighters' manhood.[38]

If engine competitions represented firefighters' most visible and significant battles, firemen's combativeness emerged in a host of more mundane and less tangible ways. In the 1810s, for instance, several Philadelphia fire companies litigated the type and construction of bells attached to engines. During the 1840s and 1850s, the Hibernia and Hand-in-Hand engine companies of Philadelphia battled over which company had the right to the designation "No. 1," which signified the company that had been in continuous service to Philadelphia for the longest time. Eventually settling the issue in court, each company produced a lengthy company history as evidence of its lineage. In the 1850s, Philadelphia's Hibernia Engine Company instituted a competition within its own ranks in regard to attendance at fires. The company divided its members into two groups and recorded each team's appearance at fires. Similarly, the emergence of "bunking" as a tradition likely had its roots in the competitiveness of firefighting culture, not to mention the economic vicissitudes of everyday life. Indeed, this intense competitiveness literally colored firemen's other expressive activities, including especially ornate displays at parades, exorbitant costumes, and ornate member certificates.[39]

Passionate competitiveness frequently led to long-term disputes and exacerbated standing animosities that existed for reasons of class, ethnicity, or politics. Indeed, even at fires, long-standing rivalries often threatened to erupt into conflict. Keen competition sometimes led rival companies to cut towropes or to jam carriage spokes, and may have impeded quick arrival. In addition, the competitive aspect of volunteer firefighters' culture resulted in dramatic street brawls. If the frequency and scale of such fights have been overemphasized in previous studies, physical confrontation did play an integral role in firefighters' culture. Often a response to real or imagined slights, melees between companies typically involved only a few men. Just as frequently, company confrontations resulted in attacks on an engine, which amounted to a blow against each member of the opposing company.[40]

More often than not, however, competitiveness confirmed firefighters' common sense of public duty, especially at fires, when they usually forgot their animosities, at least temporarily. As was true in other cities, firemen in Philadelphia and St. Louis impeded their brothers' work at fires only rarely. In striking contrast to well-publicized brawls, firefighters' cooperation appears to have been pervasive—even between bitter rivals. For instance, when fireplugs were not available nearby to a blaze, companies put aside difference and pumped water from engine to engine. Whether frivolous or calculated to improve the efficiency of service, during the first forty years of the nineteenth century competition colored disparate aspects of firefighting—its physicality, technological skill, and public spiritedness—and unified firefighters. And, indeed, firefighters' intense combativeness matched the rough-hewn character of early-nineteenth-century masculine norms, and fit well a society driven increasingly by individualism and a capitalist economy. Somewhat paradoxically, the ethos of competitiveness served to unify firemen, helping them to define the boundaries of their increasingly specialized work culture. However, by the 1840s, the ability of firefighters' competitive brotherhood to contain differences had begun to diminish, especially in cities that experienced rapid demographic change.[41]

Fire Engine "Dinkey," Union Fire Company No. 2, St. Louis, 1852, daguerreotype by Thomas Easterly. Firefighters built, designed, or assigned the specifications for their equipment. In particular, they invested much of themselves in their engines and viewed them as integral to the company identity. The Union Fire Company observed the acquisition of a new engine in 1852 by commissioning this daguerreotype. Courtesy, Missouri Historical Society, St. Louis

Demographic Change and Volunteer Firefighting

When onlookers witnessed firemen in competition, as they did along Market Street in Philadelphia in 1850, they watched a wide cross-section of the city's male population assert contrasting beliefs about order, public service, and manhood. Volunteer firefighters purposefully wore different costumes, employed varying symbols, and shouted orders in the unique patois of their communities. The contrasting apparel and dialects revealed firefighters' vastly different ethnic, class, and political backgrounds. In Philadelphia and other American cities, volunteer fire companies represented a cross-section of increasingly diverse urban communities. As the century progressed, these social and demographic fissures grew more profound, and by the 1840s and 1850s differences between companies within the same city had become quite significant. In the decades after they had created firefighting as a specialized communal public service, firefighters began to imagine their service in contradictory ways. Paradoxically, as firefighters ordered their work around a common ideal of male duty, they also organized themselves in a manner that undermined the unity of that vision of service.

Whatever their differences, volunteer firefighters uniformly agreed that firefighting was the purview of white men, and their beliefs about race explicitly structured how they provided service. With few exceptions, firefighters denied African Americans the opportunity to join fire companies or to establish their own companies. As early as 1818 in Philadelphia the city's free black community attempted to form a fire company—the African Fire Association—in the same manner that other communities formed fire companies. Through a series of published circulars, the company sought subscriptions to finance its organization, which brought an immediate response from nearly all of the city's twenty-five fire and hose companies. If not all firefighters rejected the claims of their African American neighbors, a majority of the city's companies opposed the formation of the association, under the guise of preserving the public good. Opponents of the organization circulated a statement that argued against "the formation of fire-engine and hose companies by persons of color" because it would produce a "serious injury to the peace and safety of citizens in time of fire." They further urged "the citizens of Philadelphia to give them no support or aid, or encouragement in the formation of their companies, as there are as many, if not more, companies already existing than are necessary at fires or are properly supported."[42]

Philadelphia firemen ultimately sought legal sanction against the African Fire Company, asking the "proper authorities" to prevent the new company from taking water at the city's fireplugs. Much to their chagrin, the city council reported that it did not have discretion in the matter; any fire company could use city hydrants. The municipality did not reveal hidden support for the African Fire Company or abrogate its interest in fire protection, but it underscored a broader civic principle of the time: authority over and responsibility for fire protection lay within the community more broadly. Volunteer firefighters thus were obliged to achieve their goal through the power of social sanction. Eventually, strong community pressure swayed—perhaps intimidated—African American community leaders, who urged the company to disband. They feared public reprisals or worried that fire companies would not help to extinguish fires in the homes of black Philadelphians. Just weeks after its formation, heeding the remonstrance of their community leaders, the men who had formed the African Fire Association disbanded the company.[43]

Although firemen in Philadelphia, and northern cities more broadly, made systematic racial exclusion an integral element of firefighting service, firefighters in some southern cities had a more complex relation to the labor of African Americans. For instance, firemen in Charleston and Savannah relied on African American labor—both that of free blacks and slaves—to help restore order at times of fire. In both cities, black workers pumped engines under the supervision of white officers. For instance, in Savannah, nearly six hundred African Americans performed firefighting service. "Free persons of color" earned exemption from the poll tax for their service, but slaves also carried out firefighting tasks, earning a small wage for their owners. Given the high premium that firefighters in most American cities placed on the skill and strength required to operate engines, it perhaps comes as a surprise that some white southerners gave this responsibility to African Americans. However, in neither Charleston nor Savannah did black firefighters operate independently of the authority of white firefighters. By placing the labor of African Americans under the watchful eyes of white overseers, southern volunteers replicated the division of labor in the American South more broadly.[44]

Variations in the ways that firefighters of different regions excluded people from service underscores the complex demographics of volunteer firefighting in the decades prior to the Civil War. In that records are incomplete and membership patterns varied from town to town, generalization is difficult. Yet we can see that firemen did not simply use community pressure to exclude others from participating in the fire service, but created membership practices that included rules about sponsorship, lengthy discussions of prospective candidates, and arcane voting paraphernalia that allowed one voter to block someone's membership. Through this structure, members of individual fire companies very consciously chose those people with whom they served. Not only do firefighters appear to have excluded African Americans but they also sought to restrict membership along class lines, with artisans, journeymen, and the middle class being the most well-represented groups in volunteer fire departments. Attempts to limit the participation of unskilled laborers may have achieved a boost from the fact that these workers found it difficult to leave their jobs to fight fires or that they were more likely to be transient. Likewise, firefighters in different cities and regions appear to have attempted to restrict membership based on ethnicity or politics, though this strategy appears to have been relatively unsuccessful.[45]

The tensions over membership in fire companies were exacerbated when the demographic composition of American cities began to change dramatically in the 1830s. Indeed, the fracturing of American society along class, ethnic, and political lines resulted in the emergence of significant diversity in volunteer firefighting, the particulars of which varied by city and region. For instance, by 1850, volunteer firefighters in Philadelphia and St. Louis represented nearly every neighborhood, ethnic group, and class of workers, not to mention the different political constituencies that may have crossed such divides. In St. Louis, according to departmental historian Edward Edwards, the Liberty Company consisted "almost entirely of mechanics, the nucleus being formed of foundry employees of Gaty, McCune & Company." Likewise, French and German men in south St. Louis formed the Washington Engine Company and "well-to-do citizens of the extreme Northern part of the city" incorporated the Mound Fire Company. Additionally, membership in the Missouri Fire Company included a variety of merchants, clerks, and "first-class" mechanics. Approximately three-quarters of the men serving in the Franklin Fire Company worked as artisans, although the remainder split between laborers and white-collar occupations. Similar differences partitioned the fire service in Philadelphia. The Hibernia Fire Company, for instance, consisted of Irish Protestants who had emigrated in the eighteenth century, while the Mechanic's company was dominated by artisans. The Pennsylvania Fire Company was composed of non-manual laborers in the 1840s, and judging from the Philadelphia Hose Company's membership roster, artisans and white-collar workers dominated the organization from its founding in 1803. Likewise, in the Southwark neighborhood diverse groups of men banded together in particularized, homogeneous firefighting associations to enact their obligations as citizens. For example, located just blocks apart, the Weccacoe Engine Company and the Weccacoe Hose Company espoused different political beliefs, the former advocating nativist politics and the latter Democratic, antitemperance values.[46]

Ethnic, political, and social divisions developed or were accentuated by rapid industrial and demographic changes associated with immigration and urban growth, which occurred at different times in various cities. The 1830s and 1840s, respectively, were decades of such change in Philadelphia and St. Louis, and the composition of the cities' fire protection changed during these chaotic decades; as each locale grew in population, the number of fire companies increased substantially. For instance, in 1817 a core of approximately twenty fire companies served Philadelphia, but by the 1840s, the department had more than doubled in size; in 1855, seventy-five fire companies operated in the expanding region. St. Louis experienced a more dramatic population explosion and corresponding shift in the number of fire companies. In 1845, eight fire companies protected St. Louis, up from two such associations before 1832.[47]

The formation of new fire companies and the expansion of firefighting services generally followed a distinct spatial logic. Not surprisingly, in both St. Louis and

Philadelphia, companies formed in areas of rapid population growth and substantive development. For instance, comparing company locations with their dates of formation in Philadelphia between 1817 and 1856 suggests much about the social and political geography of the fire department. Of the twelve companies located in the central portion of the city, ten were created prior to 1817 (nine prior to 1811). The other two, the Schuylkill Hose Company and the Warren Hose Company, formed in 1833 and 1838 and located themselves very near to the center point of the city at that time, what is now the intersection of Broad and Market Streets. Other older companies moved away from the city's center, though they remained proximate to the Delaware River (the city's eastern edge). This allowed them to continue to play an important role in defending the large concentrations of merchant capital amassed along the river. And, generally speaking, they remained adjacent to the locations in which they were founded. This pattern fits the region more broadly, as well. During and after the depression years of the 1840s, much of the expansion of Philadelphia's volunteer fire department occurred along the edges of the city, where the population was exploding. Companies tended to form in recently settled, populous districts such as Spring Garden, Southwark, Kensington, and Moyamensing—precisely where the city's working-class and immigrant population was settling.[48]

The increasing ethnic and class diversity so evident in the growing urban population did not manifest itself within the ranks of particular companies. Rather, it appeared across the spectrum of *all* the city's companies. Put another way, membership in individual companies remained relatively similar, but the body of firemen came to include a more diverse mixture of fire companies, each of which included homogeneous groups of men from varied political, ethnic, or class backgrounds. Once again, it is difficult to make a simple generalization about how fire companies came into being and developed because every company produced a distinctive identity based on where it was founded, when it was organized, the composition of its initial membership, and its particular historical trajectory. The history of the Philadelphia Hose Company, which remained relatively homogeneous between 1810 and 1850, argues for continuity of membership, rather than change. In this instance, the company remained demographically homogeneous from its founding, having been created in a particular neighborhood and developing a rich tradition. A longitudinal examination of the Philadelphia Hose Company's membership records shows that professionals, merchants, manufacturers, clerks, or artisans accounted for at least 90 percent of all members throughout the period. By contrast, other companies experienced membership transitions, in which men of one background replaced those of quite different circumstances. For instance, the

Delaware Fire Company had a relatively diverse membership in 1824, but by 1858 working-class men dominated the company, reflecting the demographic composition on the city's western edge, where the company had relocated its engine house. Similarly, middling Philadelphians dominated the United States Hose Company at is formation, but manual laborers controlled its membership at midcentury. Although the histories of individual companies varied, evidence suggests that most firefighters served in companies composed of men very much like themselves in the 1850s. Firefighting was enacted in the context of small groups whose membership was bounded by occupation, ethnicity, or geography.[49]

Although no single pattern easily describes changes in urban fire service, by the 1850s, firefighting had become an increasingly mixed-class and multiethnic endeavor in cities across the nation. Whether firefighters served in New York, San Francisco, Cincinnati, Chicago, New Orleans, or St. Louis, even though they worked in companies populated by men much like themselves, they battled flames alongside men from diverse groups of fire companies. This diversity certainly holds for Philadelphia, which enumerated its volunteer firefighters in 1868. The majority of firemen labored as artisans or journeymen, and approximately 30 percent of firemen labored in white-collar occupations—clerks, professionals, or prosperous merchants. Fewer than 10 percent of the city's firefighters worked as wage laborers. The enumerators did not collect data on ethnicity, but there is no reason to think that Philadelphia's fire department did not mirror the city in this regard, which is confirmed by anecdotal evidence that points toward significant ethnic diversity. More broadly, although firefighters engaged in a common labor, they increasingly interpreted that duty differently, depending upon their class, ethnicity, or political beliefs. Clothing themselves in the idioms of their communities, volunteers struggled to balance their particular identities against their shared brotherhood.[50]

Symbols and Streets: Balancing Brotherhood and Neighborhood

As the century wore on, volunteer firefighters saturated the public sphere with richly colored symbols that articulated a contradictory message—of difference and commonality. On one hand, firefighters sharply distinguished themselves from one another, in a variety of ways. They commissioned or composed playful musical scores and lyrics about their companies; they selected costumes and colors for their associations, and debated political and social issues through their choices of company symbol. On the other hand, volunteers developed more common activities

Central Fire Company No. 1 (n.d.), Mat Hastings, watercolor on paper. Prior to the Civil War, firefighters controlled blazes by taking positions outside buildings. Volunteers pumped levers, or "brakes," located along the sides of an apparatus to produce enough pressure to force water onto a fire. They rotated onto and off the brakes in short bursts of five to fifteen minutes and worked continuously for long periods of time. Courtesy, Missouri Historical Society, St. Louis

for their fire companies. They marched together at parades, transformed informal competition into formal contests, visited and exchanged gifts with other associations, and held balls to raise money. Although firefighters had a difficult time balancing company identity against the interests of their emerging brotherhood, their expressive public performances nonetheless served several important purposes. Firefighters continued to assert, more boldly than ever before, the critical importance of their increasingly specialized public service, and volunteers reassured property holders, politicians, and neighbors that the urban environment was protected from fire. Finally, and perhaps most critically, when they began to expand gift exchanges beyond their home cities in the 1840s, firefighters began to develop a notion of brotherhood that transcended local and regional peculiarities, establishing a framework for a national community of firemen.[51]

Firemen invested an unusual amount of time decorating themselves, their engines, and their engine houses. At monthly company meetings, firemen discussed all aspects of company business from membership to finances to upcoming drills. In particular, they spent a significant portion of their time deliberating on issues involved with their dress. Companies held protracted debates about which hats members should wear to upcoming parades, what symbols should be painted on those hats, what types of costume should be worn. Quite often, volunteers adjourned special meetings just to consider issues of costume left unresolved at regular meetings. Perhaps companies spent so much time discussing costume because they had such a wide variety of choices in their extent and substance. For instance, they could choose between hats of pressed velvet and helmets of molded leather with high tops, short bills, and sloping projections off the rear of the cap. Companies also coordinated shirts with pants, belts, and other accoutrements that varied in color and style. In addition to these items, firemen sometimes wore oilskin capes, specially purchased coats, and leather belts. They further emblazoned their costumes with the company's name, motto, and/or number. Firefighters selected images that reflected their differing interpretations of firefighting service: political figures, patriotic icons, and more particularized, local emblems.[52]

Although the choice of company symbols varied widely, they typically placed firefighters' local connections in the context of a broader symbolic universe. For instance, many companies—such as Philadelphia's Northern Liberties Hose Company and United States Hose Company, and St. Louis's Liberty Fire Company—chose the goddess Liberty as their symbol to underscore the importance of firefighting within the polity. These images suggested that when firefighters protected property and community they also preserved democratic values. In Philadelphia, the "Young Ladies of Spring Garden" commissioned a five-by-eight-foot banner of Liberty for the Western Fire Company. On it, Liberty treads upon a crown and shackles, symbols of tyranny and oppression, as she gestures heavenward to a bright break in the darkened clouds. A sword in her right hand points to the Declaration of Independence; to her left rests a shield painted in the colors of the American flag, which is surrounded by the motto "E Pluribus Unum." The banner connected the members of the Western Engine Company to the politics of artisanal republicanism so prevalent in Jacksonian America. St. Louis's Liberty Fire Company varied its portrait of Liberty on hats acquired in 1851. It placed Liberty on a rocky crag overlooking a steamboat floating down the Mississippi River, which signified both the social and economic link between St. Louis and the rest of the nation and represented city boosters' beliefs about the role of commerce in the city. By depicting Liberty guarding the city and its economic fortunes, the Liberty Fire Company articulated the loftiness of its responsibilities, and at the same time expressed its commitment to the political principles of the young nation.[53]

Likewise, company mottoes and names reflected the diversity of fire companies, often expressed a commitment to unified service, and frequently implied that firefighters' service was a distinctly male endeavor. Some mottoes expressed national pride and others captured the public-spirited sentiments that drove firefighters' culture. In Philadelphia, the Neptune Hose adopted "TO SAVE OUR FELLOW CITIZENS WE HAZARD OURSELVES," and in St. Louis the Washington Hose Company chose "ALL PRIVATE DUTIES ARE SUBORDINATE TO THOSE WHICH WE OWE TO THE PUBLIC." In addition, other companies chose symbols with multiple meanings. The Western Engine Company's slogan, "E PLURIBUS UNUM," suggested the unity of the nation and the ideal of brotherhood so important to firefighters. The St. Louis's Union Fire Company also chose a phrase of double intent—"IN UNION THERE IS STRENGTH." It eloquently underscored the connection between masculine physical prowess and firefighters' specialized responsibility to protect property and the state. For the most part, companies chose names, mottoes, and symbols that articulated a commitment to a broader brotherhood of public service.[54]

In much the same way, balls and social events confirmed the unity of a city's fire department, reinforced common principles, and also served as a vehicle for establishing a national culture among firemen. For instance, a poem written by Washington Fire Company foreman Charles Kumle and read at the St. Louis firemen's ball in 1844 shows how ritual and symbol interacted to affirm volunteer firemen's understanding of their culture. Performed at a fundraiser and attended by city luminaries, and friends and subscribers to the department, the poem followed the public presentation of a ceremonial trumpet to a company officer.

So gentlemen, I beg you stop the ball,
For I don't mean to make a speech at all
Brothers fare well! Had you but notice spread
I might have said, Long may the **Central** stand
The nucleus of a protecting band.
Rouse every heart! Let every arm be nerved,
and Let the **Union** ever be preserved.
Alert and pure as were the Thessalonians;
The **Washington** may boast for Washingtonians;
Still may St. Louis City laugh at flaws

While the machine **St. Louis** holds her name
When Mount Aetna's burning comes again about
Let Old **Missouri** on, She'll put it out.
From tyranny, destruction, misery,
We find our guard in glorious **Liberty.**
And last not least among the honored names
The **Phoenix** stands triumphant o'er the flames.
All this I might have said but now, I lump it
Into few words: Friend Wilgus have your trumpet,
Another section there; now man the brakes
Down on her men: hurrah she takes, she takes;
Hosemen, real up! now man the rope my men
The fire is out; and hie for home again.

Kumle celebrated firefighters' prowess and bravery, their place at the center of the system of fire protection, and their position as guardians of the polity. His declaration also united the city's diverse fire companies into a single cause, bound by a common brotherhood. Such poetry, as well as the social life of the engine house, preserved harmony among companies, and intense physical competition subsided at social events. Dinners, balls, saloon gatherings, or picnics helped to cement ideals of brotherhood. Firemen exchanged stories and created myths; they assigned reputations—to men, companies, and equipment. Of equal import, such gatherings served an important community and public relations purpose. They confirmed firemen's readiness to confront the danger of fire and validated firefighters' importance.[55]

Perhaps more significantly, ritualistic gift exchange bound companies across geography and underscored a developing vocational identity that transcended locale. Firemen frequently contacted firemen in other places simply because they shared the same company number, the same name, or the same engine make and model. As early as the 1830s, fire companies invited other units from nearby cities to parade or drill with them. In addition, many companies traveled great distances to "visit" firemen in other cities. In 1856, for instance, the Hibernia Fire Company of Philadelphia spent several weeks visiting cities on the eastern seaboard from Boston to Charleston. In their book commemorating the occasion, the company secretary related anecdotes of the various dinners, drills, and balls thrown in honor of the visit. More common were yearly visits between cities; every year during the 1840s and 1850s, before the dissolution of their volunteer departments, Union No. 2 of St. Louis and Relief No. 2 of Cincinnati visited each other's home city. In these exchanges of greetings and warm wishes, many companies exchanged daguerreotypes, plaques, and fire hats. Gifts and messages would also accompany a tragic fire or the death of a fireman. Firemen gave many gifts, but ceremonial speaking trumpets were especially popular. Made of silver, these firemen's trumpets, duplicates of the speaking horns used by company officers at fires, were usually engraved with gilt or ornate images of flowers and/or firefighting. In a more practical vein, such visits, correspondence, and gift exchange served as a conduit through which firemen often shared notes about new technologies and departmental organization, and helped to establish a market for secondhand equipment.[56]

Despite startling displays of unity, the romantic idiom created by firefighters should not be overstated to suggest that firefighters' identities as firemen superseded their particular connections to local communities. From the very first firemen's parades held in the 1820s and 1830s—usually as part of celebrations of national identity—firefighters' symbolic culture became a prime locale for interpreting the political and social order. More broadly, firefighting symbolism grew more expressive as American urban life became increasingly fractured, and "private," to use Sam Bass Warner's phrasing. The pace of urbanization, immigration, and industrialization divided the city into increasingly atomized communities, and the expression of particularistic identities undermined firefighters' unity. Asserting differences increasingly moved from the realm of ritualistic debate into physical confrontations between firemen. On the streets of Philadelphia, strife between fire companies grew as ethnic and class competition became more pronounced and levels of urban violence rose. This was especially true during the 1830s and 1840s. Perhaps the most exceptional instances involved conflicts between the Shiffler Hose Company and Moyamensing Engine. Composed of nativists and emulating its namesake George Shiffler—a "Native American" killed while burning Irish Catholic churches during the 1840s—Shiffler Hose frequently skirmished with the Irish immigrants of Moyamensing Engine, whose famed leader William "Bull" McMullen was notorious on the streets of Philadelphia. Symbolic competition boiled over into street melees, and despite calls for brotherhood, firefighters could not contain the increasing social and cultural friction between companies.[57]

Yet physical confrontations between firemen do not appear to have been any more frequent than other neighborhood disturbances. For instance, in Philadelphia, fighting between firemen appears to have increased in proportion to the intensification of violence on the city's streets. Not surprisingly, fire companies participated in some of the region's most acute crises, but more often than not firemen played a crucial role in restoring public order. With a few notable exceptions, firefighters helped to quell riots, protect property, and restore the peace. For instance, in 1838, when antiabolitionists burned Pennsylvania Hall, a mob tried to prevent firemen from reaching the building. Several of the city's companies battled their way to the blaze and received commendations from the community for their service. One local church presented a silver ceremonial trumpet to a member of the Good Will Fire Company with the inscription, "Reward of Merit, Presented by a Portion of His Fellow Citizens as a Mark of Their Approbation of his Gallant and Meritorious Service During the Recent Convulsions in this City and in which he periled life itself for the protection of private property amidst the threat and intimidation of an infuriated and exasperated mob." Likewise, in 1849, a notorious street gang, and at least one fire company, instigated the "California House riots," but the department performed with élan in the incident. Most firefighters stood against the rioters. On balance, even if they sometimes gave their critics ammunition, firefighters played an active role in preserving order. In fact, during the tumultuous 1830s and 1840s, some of the most ardent opponents of volunteer firefighters defended them, reporting that firemen protected property with great zeal. In 1841, for instance, businessman Sidney George Fisher effused, "The fire department here is very energetic and efficient."[58]

Volunteer firefighters struggled to balance brotherhood against the particularistic expressions and beliefs of individual companies and their communities. These tensions seemed natural to firefighters and their neighbors, but have been misunderstood by historians, who have emphasized to the point of caricature either the differences that firefighters expressed or the mythic brotherhood to which volunteers claimed membership. When viewed in the context of the *work* performed by firefighters, and their functional and symbolic role in shaping and restoring the social order, the roots of these contradictions become apparent. In the context of battling the problem of fire, firefighters shared and expressed a common understanding of what constituted manhood, service, and order. Take, for instance, the New York firemen who challenged the draft in the 1860s. Volunteers began the day protesting the draft, but, once rioters set fires, they literally clothed themselves in the garb of firefighters and protected the social order. At the same time that firefighters protected and maintained society, they expressed widely divergent understandings of how order should be constituted and where power should be located. Indeed, firefighters were powerful and public advocates of diverse political and social opinions. Constraining these seemingly competing impulses was not an easy task for firefighters, and it became increasingly difficult as immigration and industrialization fueled the growth of American cities.[59]

The Old Philadelphia Fire Department of 1850. The Great Engine Contest on Sunday Evening July 7th 1850 at 5th & Market Streets, 1882, H. Sched (artist); Charles H. Spieler (engraver); Theo. Leonhardt & Son (publisher). Intense competitiveness drove the service of volunteer firefighters, as suggested by this 1882 engraving of an engine contest that spontaneously erupted among Philadelphia's fire companies after a blaze in 1850. Firefighters dressed in diverse costume struggled to throw water to the greatest height, thereby proving the company's competence and virility. Courtesy, CIGNA Museum and Art Collection

Maintaining Order: Management Associations and the Problem of Unity

Balancing public service against particular company interests remained an issue for volunteers, whose identities continued to be circumscribed by social, cultural, and political realities. Indeed, volunteers' sense that they belonged to a distinctive "fire department" was emerging only slowly at midcentury, and it often was subordinate to their affiliation with a company, community, or political unit. The clearest sense of how volunteers gradually unified into a well-defined political collective—and became part of a volunteer fire department—appears in efforts to manage service at fires and conflicts between companies. Early in the nineteenth century, companies organized leadership associations for practical reasons, especially to direct firefighting work. These administrative bodies forged harmony between companies and represented the needs of all firefighters before elected political bodies. However, except for some limited successes in the first decades of the nineteenth century, these associations largely failed. They could not transfer—at least regularly—the unity of firemen's brotherhood to the political arena.

In Philadelphia, as early as 1805, volunteer firefighters formed management associations to prosecute the common interests of firefighters, to settle disputes, and to resolve jurisdictional issues that arose in the region's fractured political structure. The city's hose companies formed a Fire Hose Association to represent members' interests before the municipal government and city merchants. Just a decade later, in 1817, firefighters established a more powerful but specialized management institution designed to coordinate their labor, govern conflicts, and represent their common interests to the polity. An outgrowth of the Fire Hose Association, the Fire Association (FA) renamed, reorganized, and restyled itself as the leadership body of the entire department. The FA represented fire companies before the municipal government; mediated disputes between companies; encouraged parading; and eventually fostered the formation of an association for the relief of disabled firemen and a board of control for the department. It also planned for company self-sufficiency and independence. Indeed, the FA hoped to finance companies through dividends earned from its business as an insurance company. Although it is unclear just how significant the FA's donations were, by 1859 the organization distributed $738 to each of forty-nine member companies.[60]

The Fire Association succeeded less well in its other ambitions. As a mediator of disputes between companies it failed. In the 1810s, for instance, the Philadelphia Hose Company and the Good Intent Company became embroiled in a protracted dispute centering on the patent of an alarm bell. Philadelphia Hose asked the FA to determine if Good Intent's bell infringed upon its design. After the management association decided in favor of Philadelphia Hose, the Good Intent Company withdrew from the FA (but remained in active service). Only after the Court of Quarter Sessions ruled against it did Good Intent reconcile with Philadelphia Hose and petition to rejoin the FA. The inability of the FA to settle disputes demonstrated the limitations of its authority—a recurring problem faced by later management associations. Other factors limited the organization's influence as well. Over time, the FA included proportionally fewer of the region's fire companies. And, if few of its own members recognized the FA's authority, nonmembers felt even less compelled to abide by its decisions. In addition, the region's various municipalities did *not* recognize the FA's authority over firefighting. Thus the FA became less and less relevant.[61]

In 1839, firemen renewed efforts at self-discipline, forming a more ambitious board of control. Over twenty companies formed the Board of Control of the Philadelphia Fire Department according to the same plan as member fire companies. Adopting constitutional language and procedures, the board elected officers and "directors" whose duty it was to "superintend the operations of the Controllers in times of fires [and] place the members in whatever situation they may deem necessary." In addition, the board created a system of fines that it meted out for not attending meetings or fires. Like its member companies, the board expected such penalties to foster unity by establishing common institutional responsibilities. In addition, democratic elections and rotating committee leadership encouraged cooperation without obliterating the individual identities of member companies. The board further codified informal intercompany relationships into formal processes, including printed work rules that required members to cooperate. The constitution also demanded that controllers report any "violation of the peace or principles contained (in the constitution) . . . or infraction upon the rights of others." Abrogation of communal interest—such as refusing to share a plug connection, racing, or failing to recognize a leaders' authority—met with stiff penalties.[62]

The Board of Control, like the FA, never fully realized its goals. Within several years, it disbanded. A number of factors impeded its success: intense local and regional political divisions, firefighters' reluctance to accede to centralized authority over their division of labor, and the existence of many informal modes of resolving disputes. The Board of Control's authority resided within the narrow corridor of Philadelphia's central city, where twenty-five of its twenty-nine members were located. In this part of the region, the board's power expanded beyond a mere advisory role. Over time, Philadelphia's councils gradually invested the Board of Control with increased power to shape the department—especially at fire scenes. It provided the board with influence over municipal appropriations made to fire companies. In addition, the state gave the board limited power to act as police at fires and to "protect property and quell riotous behavior." Outside Philadelphia, where over thirty-five of the region's seventy-plus companies were located, geography, demographics, and political divisions weakened the board. Fire companies and communities in those districts, such as Southwark, Kensington, and Northern Liberties, jealously guarded their political prerogatives against encroachments by interests associated with the larger and more economically powerful city of Philadelphia. After one company joined the board, it later resigned because its members "found [themselves] placed so far from the city" that they were "deprived of the benefits of the Board and for the same reason our Controller is of little or no service to you." In 1840, the Pennsylvania Hose Company remonstrated against a plan by the Kensington Board of Commissioners to recognize the board's authority. The board, which craved such legitimacy, sent a delegation to "refute the charges." Its lack of control beyond Philadelphia underscores the extraordinary demographic and political differences between the districts and the central city.[63]

Informal modes of dispute resolution also impeded the Board of Control's efforts to govern fire companies. When companies found themselves at odds, they followed several routes to resolving their problems: formal contact between company leaders, physical retaliation, public appeal, and informal legal action. For instance, sometimes companies appealed to local municipalities and neighborhoods to redress grievances and force foes into submission. Only rarely did firefighters turn to formal organizational bodies to resolve disputes. In September 1839, for instance, the America Hose Company sent a letter of complaint about the Philadelphia Hose Company to the Board of Control, which subsequently enjoined both parties to testify before it. After failing to appear on several occasions, Philadelphia Hose informed the board that it "declined complying with the requests as this company is not represented in that Body and cannot therefore acknowledge its jurisdiction." Even so, the Philadelphia Hose Company agreed to meet with members of America Hose to "assuage the misunderstanding satisfactorily."[64]

Fire companies, like ordinary Philadelphia residents, made little distinction between legal modes of settling conflicts and other methods. In practice, Philadelphia courts provided litigants with extraordinary control over a system ostensibly designed to discipline them. In the hurly-burly of the courtroom, disputants jockeyed for favorable negotiating positions using a host of techniques, such as counter-suits, or "cross-bills." As a result, few cases ever went to trial. Such a seemingly rag-tag system of justice worked effectively for residents in the politically and socially divided region. It also worked for fire companies. As Allen Steinberg argues in his study of Philadelphia criminal justice, "In gambling and fire riot cases most private prosecutors were not interested in convictions or even punishment. Instead, they sought the restitution of a desired balance in a private relationship." The city's courts became another arena in which firemen competed. Companies used the legal system to avoid facing contradictions inherent in their relationship to the collective body of firemen.[65]

If firefighters did not redress grievances through a centralized authority, they nonetheless continued to establish management associations to prosecute their political and social interests. In fact, fire companies in Kensington, Northern Liberties, and Southwark established management associations akin to the Board of Control. In 1842 fifteen fire companies in Northern Liberties and Kensington established the "Firemen's Convention for the Incorporated District of the Northern Liberties and Kensington" to nurture firefighters' claims. Firefighters organized such associations much like the Board of Control, right down to its rules of conduct. Yet, many companies remained hesitant to allow these management associations to limit their autonomy. And, indeed, few nineteenth-century volunteer firefighters conceived of themselves as members of a "volunteer fire department" even though they had developed a sense of brotherhood around their specialized service. A formal bureaucratic vision of firefighting had not yet emerged. In the years before the 1850s, as firefighters struggled to balance their interests as firemen with those of their local communities, they experimented with, but did not fully embrace, more formalized management bureaucracies.[66]

Conclusion: The Atomization of Service

As firefighters struggled to discipline themselves, they continued the trend toward specializing their civic obligations. Beginning in the 1830s, more than half of Philadelphia's fire companies introduced innovative new membership strategies that distinguished between types of service. Hose and engine companies alike, including such departmental stalwarts as the Diligent Engine Company and the Hope Hose Company, diversified membership into several grades—active, honorary, and contributing. Active members extinguished fires (and sometimes paid minimal dues). Honorary membership was given to men who performed notable active service; such firemen had the right (but were not obligated) to attend fires, and also paid a small amount of dues. However, when companies allowed contributing members, they radically changed the conditions of service. Unlike active and honorary members, contributors did not have to fight fires. Rather, contributors paid heavy dues, and they were allowed to participate in parades and company functions, and sometimes received voting privileges. Men holding this membership grade did not have to perform physical labor, which transformed the very fabric of firefighting service. Some men worked while others served as firefighters in name only, as a reward for their financial contributions. A new set of beliefs—based on the amount of dues a member could pay and/or his predilection toward dangerous physical work—began to define firefighters' brotherhood.[67]

The roster of the Philadelphia Hose Company illustrates this trend. The company began accepting contributing members in 1837, and by 1841 contributing members accounted for over half of the membership; by 1853, contributing members comprised nearly 75 percent of the company's strength. The experiences of the Philadelphia Hose Company mirrored trends in firefighting more broadly, and the proportion of *nonactive* firefighters in Philadelphia fire companies increased steadily through the 1860s. In 1842, 60 percent of all fire companies in Philadelphia had accepted contributing members into their ranks. And, there were as many contributing and honorary members in Philadelphia's fire department as there were active firemen. In other words, in the early 1840s, only about half of the men serving as volunteer firemen in Philadelphia expected to perform firefighting labor.[68]

The gradation of service levels corresponded to the intensification of firefighters' cultural differences as well as the heightening of differences within the department, which also began in the 1830s. The coincidental timing of these shifts is revealing. By increasing income, a company would have been able to afford to commission expensive engine engravings, new costumes, banners, or other symbolic items like trumpets. Likewise, a steady revenue stream from contributing members could help to defray the constant expenses of new hoses, tools, and apparatus. In addition firefighters could make bolder statements of company identity—making it possible for active members to express the group's male bravado and technological acumen. Moreover, by adding contributing members, companies increased the available labor pool as well as the representation of their communities.

The reorganization of firefighters' labor illuminates the broader manner in which Americans ordered their lives and their cities. As the problem of fire grew, firefighters argued that controlling the problem was an important public function that should be performed by particular groups of skilled men—whose service was characterized by technological competence, physical strength, and selflessness. Rhetorically, firefighters and their neighbors advocated discipline and unity in the face of danger. However, increasing demographic differences and intense competition confounded the efforts of firemen to create a single brotherhood of firemen. Firefighters increasingly interpreted their service in the many different idioms of urbanizing America. Furthermore, beginning as early as the 1840s, the many symbols and rituals that allowed firefighters to negotiate the contradictions in their manhood and their culture—between competition and cooperation, between communal orientation and individualistic expression—lost their effectiveness under the strain of industrialization and immigration. Well before urban America began regularly to erupt in riots, resulting from the dislocations of industrial conflict, cities faced a crisis of order in how they organized against the problem of fire. Firefighters adopted an approach to danger and safety that predicted the broader societal trend. Specialized niches and particularistic loyalties—to neighborhood, class, ethnicity, or politics—governed the ways that Americans disciplined themselves, arranged their economic activities, and structured their communal life.

CHAPTER TWO

The Business of Safety

The American Fire Insurance Industry, 1800–1850

Early in the nineteenth century, fire insurance companies established an approach to the problem of fire that was very different from the tactics and tools used by volunteer fire companies. Whereas firemen endangered their physical bodies in the performance of public service, fire underwriters imperiled financial assets in pursuit of profit. Not surprisingly, the returns on their work also were dissimilar. Volunteer firefighters received social, cultural, and political esteem from their neighbors. By contrast, fire insurers earned dollars, and in time they acquired increasing amounts of political and social capital as well.

If underwriters knew well that fire had a physical dimension, they nonetheless viewed it primarily as an economic problem. To entrepreneurial merchants, fire represented a commercial opportunity, even as it threatened their mercantile endeavors. However, turning a profit—much less building a durable industry—proved exceptionally difficult, even though conceptually the business of fire insurance was relatively simple: underwriters sold contracts that furnished indemnity against economic loss caused by fire. Profits were achieved by setting the price of insurance high enough to cover losses without driving business away and/or by preventing losses from occurring altogether. Over time, accomplishing this goal generated a host of different management strategies, organizational forms, and systems of practice. Although these tactics would change over time, the same basic conundrum confronted every insurance company and every generation of insurance worker: How was it possible to make income (from premiums and investments) exceed liabilities (losses and expenses)?

Underwriters' approach to fire—as a business and societal problem more broadly—is reflected in the language with which they described their endeavor. When insurers underwrote policies, they "took risks." That is, fire underwriters transformed a highly specific environmental hazard—that a blaze would strike a property located in a concrete and particular physical and cultural context—into an economic abstraction. Represented as a rate, this premium reflected, in theory at least, the likelihood that the insured property would be destroyed by fire in a given time period. However, firms were less concerned about the effect that fires had on individual policies than they were about their effect on the thousands of risks that they had accumulated. Insurers sought to minimize the risk in their portfolios with a variety of management techniques and information tools—some of which were utilized in other financial industries of the era—including account books, administrative guidelines, behavioral prescriptions, and later on statistics and maps. As they refined the use of such technologies over the course of the nineteenth century, underwriters sought not only to better comprehend the problem of fire but to make out of it an abstract commodity, easily salable in the nation's expanding capitalist economy.[1]

The development of the fire insurance industry paralleled the shifting division of labor in firefighting organizations. In much the same way that the Philadelphia Hose Company's formation exemplified a movement away from the communal arrangement of firefighting, the establishment and expansion of "joint-stock" insurance companies represented a new mode of organizing to combat the problem of fire. By turning to these incipient corporations, a limited number of urban residents embraced a more specialized form of social organization—as compared to the relatively community-focused strategies of dispersing risk that were championed by mutual insurance companies. When the Insurance Company of North America (INA) began selling fire insurance policies in 1794 and wrote its first fire insurance policy in 1792, it offered an alternative to subscribing to mutual societies. Later, when INA expanded its business beyond Philadelphia's boundaries, it further signaled that insurers had begun to sever their connection to local community. The appearance of joint-stock firms also coincided with a broader shift in how underwriters organized their business to confront the problem of fire. Insurers began to create bureaucratic hierarchies, establish formal management procedures to guide their labor, and experiment with new ways of representing danger.

The expansion of the Aetna Fire Insurance Company—an exceptional firm located in Hartford, Connecticut—also offers insight into the development and history of the industry. Aetna's approach to the problem of fire distinguished it from the majority of the industry, in which individual firms cultivated markets in tightly circumscribed geographic areas. Aetna bucked this trend by expanding across the nation almost from its inception in 1819. Yet, like other firms, Aetna's practice centered on gaining and manipulating knowledge of risks. However, because they could not scrutinize every risk personally, Aetna's managers created a network of surrogates to transact business in distant locations. In order to control a rapidly expanding network of agents, the company also developed relatively sophisticated administrative procedures to manage its daily operations. Although Aetna's early innovations in management and strategy made it unusual, other companies were emulating its practices by midcentury. In fact, many of the industry's future leaders learned the business at Aetna, and many practices peculiar to Aetna—such as drawing diagrams of risk—would develop into standards used across the industry. By the end of the century, Aetna became one of the largest and most influential companies, and along with a handful of Hartford, New York, and Philadelphia firms, including INA, dominated the fire insurance industry.

Despite significant organizational innovations, achieving solvency—much less profitability—eluded the fire insurance industry during its initial stages and preoccupied it for much of the nineteenth century. Developing business practices that would help to turn a regular profit proved difficult because underwriters had little skill in predicting how fire would manifest itself in the environment. Insurers lacked a definitive sense of how the problem affected both the policies they underwrote and their portfolios of risk. Of course, from common sense and on-the-job experience, underwriters knew that some buildings were more likely to catch fire than others. For instance, they recognized that wood structures burned more often than stone buildings. However, insurers had no sense of how much more often frame construction burned. Not surprisingly, insurance rates fluctuated wildly. More significantly, perhaps, underwriters did not possess adequate means to assess how fire affected their portfolios more broadly. In other words, they could not precisely determine how a single fire, groups of fires, or conflagrations would affect their financial standing. As a result, insurers faced an uncertain and arbitrary financial future, and writing policies became more akin to gambling than many would have liked. Although ambivalent about the uncertainty of early underwriting methods, one insurer underwriting fire and marine risks along the Mississippi River nonetheless savored the excitement of his risk-taking behavior. In his memoir, James Waterworth recalled with nostalgia the adventures of his early days in the industry, when ignorance about risk had been nearly complete.[2]

The Fire Insurance Industry's Early History

Philadelphia was the birthplace of the American fire insurance industry. The nation's first fire insurance company, the Philadelphia Contributionship, offered fire insurance beginning in 1752. Organized as a mutual insurance company, the Contributionship was a financial cooperative based in Philadelphia's elite and middling communities. Recognizing the omnipresent danger of fire, well-to-do citizens banded together for their common economic good and created risk communities mediated by their face-to-face relationships. Members shared in the company's losses as well as gains. As the company's charter noted, it provided "for the common Security and Advantage of our Fellow Citizens and Neighbors," and it promoted "so great and public Good as the Insurance of Houses from Loss by Fire." However, the Contributionship's vision of the public good was limited. In the eighteenth century, mutual insurance companies placed strict requirements on what constituted insurable property. They extended protection to a narrow segment of property owners—mostly those who owned homes constructed of brick or stone.[3]

Discussions about what type of property the Contributionship should insure reveal the limitations of eighteenth-century fire insurances protection. The company's organizing document, its "Deed of Settlement," specified that it could insure no frame dwelling constructed after 1752. In 1769, the company prohibited insurance on wooden dwellings altogether, concentrating exclusively on property built of brick or stone. In 1781, the Contributionship precluded insuring houses with nearby trees, even refusing to renew existing policies. More than a mere extension of fears about the flammability of wood, the exclusion was a response to a treatise on lightning danger and complaints that trees obstructed fire companies' ability to effectively fight fire. However, the policy of prohibiting insurance on houses with trees did not receive universal welcome from the Contributionship's members, and generated dramatic controversy. About forty members of the Contributionship expressed their dissatisfaction in the *Pennsylvania Gazette;* in particular, they questioned whether trees posed an additional danger. The disaffected members were dismayed that the Contributionship refused to repeal its policy, and were equally distressed that the organization was unwilling to assess them an additional premium "for the supposed risk attending trees in cases of fire." In 1774, after securing enough capital, they formed a rival association: The Mutual Assurance Company for Insuring Houses from Loss by Fire. The new organization adopted a tree as its symbol, and provided insurance to those with shade trees for only a small additional premium.[4]

Strikingly, despite such generous exemptions, the "Green Tree" mirrored the practices of the Contributionship. Neither company provided coverage for goods and personal items located within buildings, nor could businessmen purchase insurance for commercial risks. Indeed, debates about coverage, and insurance policies themselves, occurred among a narrow segment of the population that could afford the relative expense of such restrictive conditions of purchase. Perhaps the most revealing insight about the boundaries of insurance protection can be found among the large numbers of well-to-do property owners who joined mutual societies as an investment strategy because policies sometimes returned a small dividend. Ultimately, mutual fire insurance companies provided only limited protection to a narrow range of demographic and geographic communities.[5]

By the last decade of the eighteenth century the boundaries of fire insurance protection expanded as more companies entered the field and as fire insurance was transformed into a purely commercial endeavor. In 1792 local merchants initiated the first "joint-stock" fire insurance company, INA, and established the Insurance Company of Pennsylvania in 1794. These firms offered a wider range of protection, including policies for commercial properties. Moreover, they differed from mutual companies in that they organized to benefit private interests; these were not communities of property owners banding together to share their risks. Yet, INA and the other new firms entered the insurance business cautiously, reflecting uncertain markets and the relative novelty of insurance in the late eighteenth century. For instance, although conceived as a general insurance company able to write life, fire, or marine contracts, INA concentrated most of its resources in the marine insurance business. Life insurance was not well understood in North America, and mutual insurers dominated Philadelphia's fire insurance market. By contrast, underwriting marine cargoes and hulls offered vast economic potential. The business of marine insurance was relatively better understood than fire or life. Additionally, the marine insurance business played a significant role in international commerce and Philadelphia was one of the nation's principal ports. Moreover, over $600,000 in permanent capital made INA the city's largest marine underwriter. Not only could INA underwrite greater risks than regional firms, but it could competed on equal footing with international companies.[6]

Insurance companies drew upon the business practices and culture of marine underwriters as they prepared to enter the fire insurance field. In particular, the development of INA illustrates how the commercial culture of the coffeehouses—

where marine insurance was commonly underwritten—shaped the early fire insurance industry. Conducted by generalist merchants within demographically circumscribed communities, early marine insurance was a collaborative endeavor in which many different merchants—often familiar with one another, related by family ties, or tied together by significant financial exposure—shared risk. As specific firms formed, they capitalized on family and personal relationships with shippers, builders, and other merchants. Moreover, though marine underwriters insured international and regional cargoes, they usually operated within the context of local ports like Philadelphia or Boston. When INA opened for business, it tapped into the experience of the Philadelphia insurance community, which transacted business in the City Tavern. INA named to its board at least six of the city's most prominent marine underwriters and commercial leaders. However, day-to-day authority resided in the hands of company secretary Ebenezer Hazard, who made business decisions about risks and wrote policies. Hazard and the secretaries of the other upstarts in the fledgling industry struggled to manage a steady flow of information about risk and to incorporate that knowledge into increasingly complex business transactions. Although later firms no longer worked out of the coffeehouse, gathering and evaluating large quantities of information dominated the concerns of underwriters for many years, mirroring the struggles of the financial industries more broadly and predicting a central dilemma facing managerial capitalists in the postbellum era.[7]

Buoyed by a growing market for fire insurance, increasing fire danger, and an expanding economy in Philadelphia, INA began to offer fire insurance in 1794. Unlike mutual companies, which restricted themselves to underwriting dwellings and were backed only by the premiums of their members, INA's capitalization of $600,000 gave it access to an extraordinary amount of funds. Moreover, the ambitions of the merchants who founded INA extended both beyond the local geographic region and beyond the limitations of the market in selling protection for residential structures. INA became the first insurance company on the eastern seaboard to insure "goods, wares, and merchandize, or other personal property" against loss by fire. This significant innovation followed precedents set decades earlier by London companies and was crucial to the expansion of INA and the insurance industry in North America. More broadly, American firms now could assure both the integrity of capital investment in the built environment and in the commodities that flowed through the nascent industrial economy. INA took advantage of its strong financial position; in its first six months, the firm collected over $60,000 in premiums, along with another $3,000 in interest, and paid about $4,500 in losses.[8]

As they cultivated new business early in the nineteenth century, insurance firms began to focus on developing a better understanding of the problem of fire and on setting guidelines for everyday business practices. During the first three decades of the century, several activities became central to the fire insurance business: surveying a risk, corresponding with field representatives and customers about hazards and rates, and compiling records of surveys and transactions in ledgers, and classifying danger. By the 1810s, companies transformed such informal procedures into formal written guidelines and organizational structures. In particular, the industry diversified its risks, and underwriters established rudimentary distinctions between different sorts of property, manufacturing activities, and construction methods. Initially such divisions resided in the minds of company secretaries—in an expanding qualitative knowledge base about fire danger that they developed from their own experience. During the first two decades of the century, many companies transformed these qualitative understandings into formal written categories, albeit incompletely and in an idiosyncratic fashion. Although such systems of classifying risk remained relatively primitive, they nonetheless helped to guide firms in making decisions about risks—especially in setting rates, which were quantitative expressions of insurers' qualitative understandings of danger. Both rates and categories of risk were in a constant state of flux, as underwriters entered into a dialogue with prospective customers and the landscape.

Both extensive use of property surveys and the evolving manner of their use underscore the intensity with which insurance companies gathered information about the risks they underwrote. Of course surveys had long served as the first step in the insurance transaction, but during the eighteenth and early nineteenth century such surveys provided only the most basic information about a structure, its contents, and value. The Philadelphia Contributionship employed two men to inspect and survey property. Surveyors examined how buildings were constructed and commented on basic construction details or unusual architectural features. They especially examined the thickness of walls, materials used, and what types of activity took place within or nearby the structure. In addition to this basic property description, surveys included the policyholder's name and an estimate of the property's replacement value. Finally, each survey was numbered and associated with a policy, which specified the dates of the contract, its value, and the extent of coverage.[9]

The practice of collecting surveys, as well as the surveys themselves, became increasingly formalized in the early years of the nineteenth century as firms spread beyond the locale of their home offices. In 1807, INA, which already wrote contracts in Philadelphia's hinterland, extended its business into the Ohio River valley. As INA expanded, it created more formal procedures for writing insurance contracts, including provisions for obtaining property surveys. The company's proposals demanded that in places where it had no agent, applications for insurance required a property description "made by a master carpenter." The application, which was also to be signed by the "owner" and "attested before a notary or magistrate," was expected to contain a plethora of specific information about the property: its contiguity to water, the state of local fire defenses and fire engines, materials of construction, adjacent dwellings, and how ashes were disposed. In addition, the board directed company president John Inskeep to appoint "trusted persons . . . whose duty it shall be to Survey and Certify the situation of all Buildings and property on which insurance is required, at the expense of the person applying therefore." INA refused to write any insurance policies "without the return of such survey and Certificate."[10]

Evaluating the information provided by property surveys allowed underwriters to differentiate risks and to set rates—the basic business transaction in the industry. Insurers set rates and issued policies based upon their assessment of the information supplied from the field, and in conjunction with a company's financial position and local market conditions. A competitive market, or lack of it, could drive rates down or keep them artificially high; indeed, rates often fluctuated by company, and from year to year. Although market conditions frequently loomed larger in underwriters' decision-making, the process of evaluating information remained central to the rate-setting endeavor. Indeed, risks deemed more hazardous were underwritten for a higher premium, or not at all. But how were such assessments made?

When evaluating a risk, insurers assessed observable characteristics and gathered those notes into surveys; then they evaluated the information against their knowledge about how often a fire might occur in similar circumstances. Amazingly, even though the act of assigning a numerical value to a risk was a quantitative exercise, the process was not based in any formal statistical tables. Instead, insurers drew upon their experience, based almost entirely on personal recollection and anecdote. Indeed, underwriters did not keep systematic records of losses; instead, they noted them in the most general way, by listing them in ledgers and/or in account books. Material facts describing the loss—data easily obtainable from the survey—were not tied to the financial payments. Companies neither compiled quantitative tables about their losses nor engaged in rudimentary actuarial record-keeping. Even so, underwriters codified their qualitative evaluations of fire danger into written lists. In these guidelines, insurers assigned a financial value to various structural conditions associated with property: information about its use, ownership, construction, and spatial relation to other property. If insurers' methods of organizing risk lacked a systematic quantitative character, they nonetheless categorized their knowledge.[11]

Underwriters' classifications of risks responded dynamically to changing business conditions; they grew more elaborate as the industry insured a wider range of property and continually refined its views of danger. At its inception, for instance, INA divided the perils it would insure against into four categories. The company identified buildings constructed *wholly* of brick or stone as first-class hazards, and those structures built only *partially* of brick or stone were deemed second-class hazards. Those buildings that housed particular commodities or that contained certain types of manufacturing processes were labeled "extra-hazardous." Finally, the company refused to insure wood structures, implicitly identifying a fourth category of uninsurable structures. As the industry developed, insurers refined and tinkered with their classification schemes—underscoring the dynamic nature of these seemingly fixed information technologies. For instance, as INA accepted more diverse risks in the first decades of the nineteenth century, the company categorized danger in a more detailed manner. When INA began to insure wooden structures, it explicitly rewrote its list of rates and dangers to include "slight wooden buildings." If frequent alterations reveal the industry's expansion, they also point to the extreme variability of fire danger and the difficulty of pinning it down in rapidly changing urban and industrial settings. For instance, in the 1807 rate table, "first classes of hazard" remained buildings constructed entirely of brick or stone, but now this category reflected an appreciation of how roofs covered with tile, slate, or metal roofs could impede the spread of fire from surrounding buildings. Like other firms, INA remade its classifications regularly and with subtlety, and it repeatedly refined earlier categorizations of risk. Even though such categories reflected the idiosyncratic styles of particular companies, many companies evaluated risk using similar categories. More broadly, such categories represented underwriters' continued desire to move beyond anecdote and personal recollection and create an objective measure of fire risk.[12]

During the first three decades of the nineteenth century, underwriting firms also began to enumerate many exceptions to their policies and guidelines. These alterations reflected the growing scope and intensity of fire hazard in the expanding economy, underwriters' changing—and limited—knowledge of fire risks, and the pressures of a more competitive marketplace. The industry identified many more risks as extra-hazardous, which meant elevated rates, additional inspections, or policy prohibitions. In particular, buildings that held flammable commodities or included fire-intensive manufacturing processes became subject to greater scrutiny. Many traditional artisanal crafts as well as new manufacturing processes were included in lists of "hazardous occupations": distillers, sugar refiners, bakers, painters, carpenters, and turpentine distillers. In addition, some firms, such as INA, also identified certain goods and merchandise as extra hazardous because of their susceptibility to water damage or theft. Although many companies had once prohibited insuring such properties altogether, by the 1820s many companies underwrote these dangerous risks at a higher premium. For instance, the American Fire Insurance Company (AFIC), which shared many of INA's philosophical underpinnings, routinely wrote policies on cotton manufactories, albeit under a plethora of exceptional conditions.[13]

With the development of more detailed classification systems, surveys and correspondence became more effective conduits through which insurers developed experience without having to suffer financial loss. For instance, AFIC acquired significant expertise at underwriting cotton mills from studying its surveys, and it published that knowledge in an 1823 pamphlet. Unusually detailed for its time, the tract commented on the industry's approach to underwriting manufactories, especially the process of setting premiums on cotton mills. The secretary of the AFIC, William Jones, reported that cotton manufacturers were not a typical situation, but could be underwritten if "more ample investigation of the nature of each particular risk" was acquired. Jones specified that two disinterested observers should compile cotton mill surveys. Those observers should sketch "a ground plot" of the premises and all structures within fifty yards, identify the "situation"—its location more generally, including the town and nearby water sources or landmarks—describe the building, and remark on the heating, manufacturing processes, and lighting used in the structure. AFIC's secretary then explained how surveys could be used to calculate adequate premiums on property, and even to encourage manufacturers to embrace safe practices, which "might abate the premium" and reduce the possibility of a fire.[14]

Despite underwriters' growing sophistication, taking risks continued to confound their ability to produce a profit from the chaos caused by fire. If fire risk was difficult to pin down, underwriters nonetheless created financial strategies to minimize their exposure, especially evident in their approaches to single large risks. Then, as now, the industry recognized that writing an insurance contract of an especially large value exposed them to greater jeopardy from a single fire. In light of this concern, in the early nineteenth century, INA established a sliding scale for premiums that depended upon the amount of insurance sought. For instance, on the first class of hazard (stone or brick construction) the company insured structures and goods valued at less than $8,000 at the rate of $0.30 (per $100 of insurance.) If a property was valued at between $8,000 and $16,000, INA charged $.045; property valued between $16,000 and $25,000 was assessed at $0.60. On the more dangerous second class of risk, the company insured only up to a value of $8,000, at a higher rate of $0.75 cents per hundred dollars. As insurers developed strategies for dealing with very large single contracts, they soon realized that the ever-changing urban landscape posed a new, albeit similar, threat. Conflagration became frighteningly common and represented a more profound danger to property owners and insurers. Not surprisingly, insurance firms became increasingly alarmed by concentrations of risks within a single geographic district, but a strategy to deal with losses from conflagration proved elusive.[15]

As underwriters became more and more aware that social and political approaches to the problem of fire affected their bottom line, they demonstrated consistent support for legislation aimed at improving public safety, but they did not advocate a systematic approach to fire prevention. On a number of occasions underwriters supported the restriction of various types of fire use in certain city districts or sought to prevent certain types of construction. For instance, in Philadelphia at the turn of the nineteenth century, underwriters supported limitations on wood construction, and such "fire limits" became standard in many cities throughout the nation. Likewise, many cities also enacted prohibitions on the storage and production of certain very hazardous materials, such as gunpowder, within city limits. These legal codes appear to have had limited effect. They dealt almost exclusively with the most profound threats, and it is unclear how extensive they were or whether they were enforced. Similarly, other economic and political interests in cities too often frustrated attempts to expand public safety legislation. Such difficulties in battling fire speak volumes about the nascent insurance industry and American society more broadly. Significantly, the insurance industry remained relatively small, poorly organized, and fragmented, and developing a focused cross-industry approach to the problem would have to wait until the insurers forged a common interest and built lasting trade associations. Perhaps more importantly, the challenges of industrial and urban change, including especially shifting and new fire dangers, impeded efforts to promote urban fire safety. Therefore it is hardly surprising that most insurance firms focused on the basics of their business: refining daily practices, managing information, administering policies, and paying losses.[16]

Underwriters may have had a limited ability to shape broader issues of public safety, but they sometimes used market mechanisms to promote fire safety. In the face-to-face communities of the eighteenth century, when the insurance industry was small and close-knit, this tactic appears to have succeeded on several occasions.

In a number of instances insurers forced customers/shareholders to revise their construction plans. This strategy worked within the tightly circumscribed community of a local mutual insurance company whose clients were also company owners—a much different situation than that faced by stock insurers during the early days of industrialization in the nineteenth century. Yet, even into the nineteenth century, such strategies may have continued to work. Certainly insurers used an increasing number of restrictions on policies in an attempt to avoid the most dangerous risks. If not heeded, these prohibitions could negate a policy altogether, thus pressuring property owners to build more safely or to avoid renting their property to certain types of manufacturers. Alternately, such restrictions could force industrialists or commercial interests to avoid dangerous practices. However, such market coercion succeeded only when insurers collectively chose not to insure certain types of property. Increased competition for business and the expanding number of risks would increasingly limit the coercive power of the industry early in the nineteenth century.[17]

By the 1820s, the fire insurance industry had expanded significantly, and the consumer market in fire insurance also continued to grow steadily. In the 1790s, at the inception of INA, there was, according to the historian of Philadelphia's insurance industry, J. A. Fowler, "little readiness on the part of even business men to accept the new [from INA] fire insurance." Just two decades later the market in insurance had grown, but it remained relatively sparse. In a letter to *Poulson's Daily Advertiser*, an "Old Citizen" puzzled that "the practice of insuring against loss by Fire, is far from being a general one" and that "but a small proportion of the houses in Philadelphia are insured, and a much less proportion of the goods in stores and warehouses." By the 1830s, fire insurance coverage had become more common. According to the ledgers of a Philadelphia fire company that attended approximately forty fires in 1830, over $110,000 in property had burned in those fires; only $56,000 was indemnified by insurance. Although such figures likely overstate the pervasiveness of insurance, they reveal how rapidly the industry was expanding early in the nineteenth century as it aggressively raced to keep pace with the problem of fire in the changing American landscape.[18]

Aetna, Fire Insurance Practice, and Industry Expansion

Understanding the changing built landscape, which grew more complex and more vulnerable to fire, increasingly posed a dizzying challenge to the insurance industry. For instance, in Philadelphia in 1810, nearly one hundred thousand people lived in 15,814 houses, of which 6,582 were constructed of wood. Within twenty years, the number of structures in the region had increased to over forty thousand, and housed over 160,000 residents. At the same time, the built environment became crowded with many new dangers. Anthracite coal displaced wood as the most common combustible used in homes, and matches replaced steel and flint as the most common way to ignite a fire. On the other hand, gas lighting was introduced in the United States in the 1810s and fats were used less frequently for illumination. As a result of these and other changes in everyday life, the length and complexity of insurance policies and discussions in company manuals lengthened. The insurance business reflected the growing complexity of urban environments—and underwriters' appreciation of the changing landscape. By the 1850s, at least one company's manual included a section on "exceptions" that was more than twice as long as the rating table itself.[19]

To make sense of the changing environment and the fire risks it possessed, the industry expanded dramatically. By the 1830s, firms were entering the business aggressively, in terms of seeking new business and in organizing themselves. For instance, within a year of its formation in 1829, Philadelphia's Franklin Insurance Company had taken nearly $1,115,000 in risks. This rapid growth could only have occurred with access to a market as robust as Philadelphia's, where the company wrote 90 percent of its policies. Although some firms continued to serve local markets, others expanded the geographic boundaries of their business. In contrast to the Franklin Insurance Company, INA occasionally underwrote risks in the Ohio River valley as early as 1807, and by 1840 it had established agents in cities as far west as St. Louis and Cincinnati. Likewise, underwriters in cities without expansive markets, such as Hartford's Aetna Fire Insurance Company, overcame this disadvantage by writing policies over a wide area. Although these business strategies made such companies relatively unusual in the 1820s, underwriting insurance over a broad region had become common by the 1840s.[20]

Examining how Aetna organized and practiced the insurance trade provides a window into the development of the fire insurance industry between 1820 and 1850, and offers insights into the industry's changing approach to apprehending the problem of fire in America. Indeed, although Aetna often adopted business strategies that distinguished it from the rest of the industry, many of the company's practices became standard by the century's final decades. In particular, Aetna placed the manipulation of data at the center of its business activities, reflecting the growing importance and the intensification of information-gathering practices within the industry. To manage it knowledge, Aetna developed an expansive network of field representatives and created formal, bureaucratic procedures to supervise and to guide those agents, as well as the observations they fed to the company. These management strategies rested on Aetna's belief that local connections and personal inspections of risk formed the cornerstone of its business. Although the firm employed many of the same surveillance methods favored throughout the industry, it had expanded their scope, intensity, and range. Aetna's correspondence with its agents underscores the growing sophistication of its knowledge of the built environment and the problem of fire, as well as its evolving business strategies.[21]

In the 1820s, Aetna expanded according to a novel strategy that hinged upon a relatively simple principle. At a time when most companies advertised by boasting of the size of their capital reserves—signaling that a firm could pay its losses—Aetna, took a different tack. It emphasized making carefully calculated determinations about underwriting risks, an emphasis that became more pronounced after a sweeping fire nearly drove the company to bankruptcy in 1827. For Aetna, exercising good judgment meant dispersing risks widely, thereby reducing the firm's exposure to a calamitous fire. As company leaders argued, firms that had their "risks duly dispersed" were the "safest." Aetna distributed its risks by not writing too many policies within a single urban district. Isaac Perkins, Aetna's first secretary, elaborated on the benefits and strategy for scattering risks geographically. He argued that "if an office with a million of capital should insure to [the] amount on every building in a compact part of Philadelphia, its exposure to ruin would be much greater than . . . insuring $100,000 in each of ten towns." Yet, good practice extended beyond the prohibition on writing too much insurance within a single district or even on contiguous structures. Indeed, as Perkins implied, a company's financial exposure derived not just from the position of its risks in the landscape but also from the arrangement of danger within its entire portfolio of risk. If evaluating the dangers evident in individual policies was a good idea, so too was assessing risk in the sum total of a firm's policies. In practice, this meant minimizing exposure in a single type of manufactory or business endeavor. Aetna, then, sought to manage its entire portfolio, to distribute risks not only across the broad range of the built environment, but also to disburse them across the economic landscape. Scrutinizing individual policies as well as the company portfolio made "assurance double sure" and initiated a new business model in the industry.[22]

Aetna arrived upon this strategy through "experience and reflection," revealing the important role that information had begun to play in helping the firm to remake its practices according to the changing dangers of fire. As Aetna learned from its losses, it became increasingly invested in facilitating the gathering and analysis of data about the fire threat. Organizing and categorizing information—about the landscape, about property, about fire—acquired through surveys and writing policies became the backbone Aetna's strategies. Not only did the company scrutinize its own record, but also it subscribed to New York and Boston newspapers, recording the place, property, and other details of fire losses. In 1827 Aetna even convened a conference between itself and three other Hartford companies to compare experiences, and rates. From its observations, Aetna sought to understand the minutest details of fire risk and hoped to use that information to give itself a competitive advantage. Eventually such practices would percolate throughout the industry and become the basis for actively producing knowledge about the dangers in the built environment. However, in the first half of the nineteenth century few companies emulated Aetna and peer companies like INA.[23]

Aetna did not just seek general information about fire danger, but sought to understand each risk intimately, though this posed a serious administrative problem in the decades before the railroad and telegraph would shrink time and space. Hampered by geography, company officials hired intermediaries to conduct business on the company's behalf, believing that familiarity with the environs and people of a particular locale was crucial in underwriting any risk. Indeed, the company secretary sometimes chastised an agent for insuring a property too far from his home. As James Goodwin explained to an agent in 1834, "One object in appointing agents is to have a personal knowledge of the property insured." The company wanted an agent nearer to the insured property to underwrite the risk. It expected that that agent would be able to report on the specifics more easily and also be able to judge more easily if the property were over-insured. Aetna had created an expansive network of local agents to help it manage its portfolio of risk, and also to acquire face-to-face "knowledge" of the landscape. Thus, much was at stake when hiring agents, because they received such far-reaching and important responsibilities. They were charged with soliciting business, inspecting property, executing surveys, writing policies, adjusting claims, and paying losses. Not surprisingly, the company appointed agents carefully. Typically Aetna appointed representatives with whom it was familiar, either through family or other interpersonal business networks. Likewise, hiring practices offered prospective employees a model of the discipline and prudence that it sought to fuse into its bureaucratic procedures. To identify such men and to incubate this work culture, the firm scrutinized prospective agents with a zeal that it expected agents to emulate when they conducted the company's business.[24]

Like other companies of the period, Aetna became increasingly explicit in its quest to understand and to minimize the cultural component of the fire hazard as a critical component of its expansion and search for better methods of managing information about fire risk. The firm did this by explicitly instructing its agents to scrutinize the personal qualities of potential customers—a practice that would later be described as identifying the moral hazard. Indeed, when the firm urged agents to know the character of a risk, it did not mean simply assessing physical particulars; Aetna wanted agents to evaluate the personal reputation and qualities of potential customers. Already in the 1830s, the company demanded that agents evaluate the "moral character" of proposed policyholders. In 1837, a new agent learned that "in all cases" it was important "to *know* that the person for whom you insure possesses a fair reputation." Several years later, in 1844, the company secretary, Simeon Loomis, wrote to an agent in Burlington, Iowa, "We would have our dealings, as far as possible, with men of integrity." Such agents, the company reasoned, would be able to identify and insure only the "best quality" of men. Loomis wrote a New York agent that "we want a living, thinking, judicious man for our business—one who can know how things are done, and how people are getting on in the world—whether they are daily practicing fraud in their dealings, or leading good honest lives." This dictate also urged agents to investigate the people in buildings that adjoined the property under consideration for insurance. Well into the 1850s, the company advised agents to "first learn the character of the applicant and of those who occupy a part of the same, or adjoining buildings—if of bad repute avoid the risk—guard against over insurance, frauds, enemies, or threats to burn, gross carelessness, habitual intemperance, and persons seriously embarrassed or otherwise in desperate circumstances." Thus the firm sought to transact business with men whose approach to commerce matched the firm's conservative, risk-averse behaviors and vision of manhood. Aetna's attention to prudence stood in stark contrast and as a corrective to the aggressive individualism, risk-taking, and competitiveness encouraged by the expanding market economy.[25]

In the context of the disorder that the middle class so often perceived in the nascent industrial and urban world, Aetna especially expected its representatives to be examples for the agents of other firms. In a letter to a company representative Loomis warned him to be wary of his peer agents: "If they are the agents of reckless, irresponsible companies, they will be reckless too, and their example unsafe." Aetna further charged its representatives with counteracting the negative example that competing (and careless) agents had on local insurance practice. Aetna urged collaboration as a way "to improve such agents, if their principal has any reputation. It is desirable that agents in a place should be on good terms, and uniform in their rates. We never lose by charging as much as any other company." If Aetna hoped its agents would be models for good practice and demonstrate the importance of gathering and using information, altruism alone did not motivate the firm. Indeed, Aetna worried that imprudent companies could place its own economic interests at risk, and the firm advised agents to show care when insuring property nearby structures insured by an irresponsible agent. Aetna, then, expected its agents to function as models of character *and* practice for the men with whom they transacted business, as well as for the representatives of other insurance companies.[26]

The moral culture of sobriety and discipline that Aetna sought in its customers, as well as in its employees and representatives, was intimately tied to the firm's methods of gathering information and its system of determining premiums. Not only did Aetna want men of virtue as representatives, but it also sought out men who appreciated the company's carefully constructed rates and its extensive intelligence procedures. The company imparted these procedures and rate guidelines beginning in 1825, just a few years after incorporation, when it published a manual for its agents titled *Instructions for the Direction of Agents of the Aetna Insurance Company*. Each new agent received a letter of instruction plus the company manual. Updated repeatedly over the century, the *Instructions* demarcated the responsibilities of field representatives and explicitly advised them "not to transcend" company guidelines. Agents could write individual policies valued at less than $10,000 without special authorization from the home office, though in 1830 this amount was reduced to $8,000. From the instructions, agents learned the rules for writing policies, inspecting property, and corresponding with the home office. More broadly, the manual taught Aetna's culture, especially emphasizing the importance of careful surveillance of the landscape and reinforcing the daily practices that were the heart of Aetna's long-term business strategy. Moral character alone was not sufficient; rather, careful management of information through administrative routines facilitated good business and the firm's expansion.[27]

Constant written communication between Aetna's secretary and company agents further stressed the importance of the instruction manuals and the lessons about proper behavior that they taught. Although initially the board of directors had taken an active interest in daily practice, over time the secretary became the lead figure in writing policies and managing the exchange of information with the field. The secretary acted on behalf of the board and served as the primary conduit through which information and knowledge passed. If the personalities of the men who occupied the position of secretary shaped the tenor of the correspondence, they did not affect the basic procedures that revolved around those business transactions. Regardless of the size of a risk, Aetna instructed agents to return original policies (or renewals) and surveys (revised, if necessary). The returned materials served a basic record-keeping function. If policies were the unit of exchange between Aetna and its customers, correspondence rounded out the file that accompanied the insurance contract. These exchanges revealed the company's deliberations about the risk, the nature of the danger, and the expectations of the parties involved. Perhaps more importantly, though, returned surveys and correspondence advanced Aetna's information-gathering purposes. This conversation contained important intelligence about the fire risks that became part of the company's expanding record of experience.[28]

Correspondence educated agents about the company's rules of practice and gave them performance feedback. In its frequent letters, Aetna evaluated its agents' policy-writing ability and reviewed their work. Was the rate too low or too high? Had the agent forgotten some important detail about the risk? Because company approval was required, though not automatic, Aetna's secretaries quickly offered specific advice and recommendations regarding policies. The company's responses varied widely. Aetna commended or reprimanded agents, and sometimes even specified a section of the *Instructions* to which the agent should pay more attention. For instance, in a letter to a rural New York agent written in 1829, secretary James Goodwin expressed "full" satisfaction with the agent's returns, with a "few" exceptions. Goodwin then commented on four policies and surveys in great detail. About two policies, Goodwin informed the agent that he had rated a property in a more hazardous class than it deserved; he criticized him for charging too little in another instance, when a risk was more dangerous than rated; and regarding another policy he complained that the agent had not enumerated the value and risks of particular items in a specific enough fashion. Aetna did not view this correspondence as a mere formality, but as integral to its business. Indeed, in 1830 Goodwin attributed this agent's higher rate of losses to the "defective mode" in which he prepared his policies and surveys.[29]

The bureaucratic procedures used by Aetna contributed to the development of a programmatic approach to classifying risks—categorizations that became the basic technology around which the firm organized its business. As the home office, field representatives, and customers constantly exchanged information, the company organized its knowledge about fire risk into categories of danger. Each data-gathering activity and exchange of printed matter between Aetna, agents, and clients—surveys, policies, and correspondence—reflected the dynamic interplay between Aetna's home office and the field. Such vigorous interchanges helped company leaders and representatives to produce knowledge about fire danger, and they also helped Aetna as it began to bring the built environment under a program of systematic observation. This surveillance contributed to the evolution of the company's system of classifying risks. With these bureaucratic technologies driving the company's growth, Aetna's leaders continuously refined their ability to organize and to analyze information about fire hazard, and repeatedly altered their program of categorizing it.[30]

Creating categories of risk—or locating a prospective danger within that scheme—was a dynamic activity. Indeed, placing a policy within Aetna's classification system required that agents be familiar with multitudes of manufacturing procedures and commodities. Company manuals provided guidance about rating and were modified according to the existence of particular hazards. However, not all dangers were written into manuals. For instance, representatives were expected to develop "rules-of-thumb" for various hazardous industrial properties, such as tanneries. In other instances, that guidance came through correspondence. In one situation, Aetna's general secretary, Simeon Loomis, imparted the company's "experience" to an agent by reminding him, "We take no risk where the pipe is run up thro the roof or the side of wooden buildings—mind that." In some cases, details as minor as the location of materials within a building affected the agent's evaluation of the property. Such elaborate, and often unwritten, rules indicate that underwriters did more than simply collect information for rating purposes. Through the lens of its leaders' knowledge of fire danger, the company analyzed its wealth of data and applied it to the world. Lists of rates—including conditional statements applied through correspondence—were an expression of only a small portion of Aetna's collective experience with fire—the proverbial tip of the iceberg. Critically these guidelines represented a systematic approach to analyzing risk. By implementing common work routines among its representatives, Aetna had created the basis for a system of classificatory technology—a disciplined program of comprehending and applying knowledge about the dangers of fire in the built landscape.[31]

Aetna's categorizations colored all of its basic business procedures, beginning with the first and most crucial step in the process of taking a risk—making a survey of the danger. Aetna instructed agents on the company's classifications, and it expected them to analyze risk according to those divisions. By the 1820s, surveys had become more than simple written descriptions; they became evaluative documents that reflected the firm's accumulated knowledge. For instance, in its 1825 *Instructions*, Aetna divided risks into seven classifications, which the company wanted agents to use when describing danger. "Buildings" the manual advised agents, "should be described by the class to which they belong, or if they do not exactly correspond with either class, that to which they approach nearest should be referred to, and the variation noted." To Aetna, surveys represented the technical accomplishments of its information-gathering bureaucracy. They were the product of the interaction between carefully selected, prudent men and the company's "knowledge" of fire risk, as represented in instructional manuals and correspondence. The survey provided a venue through which agents and the company communicated, and the classification schemes gave them a common language.[32]

Practically speaking, however, Aetna recognized the limitations of its classification program and did not demand blind fealty to those schemes. Quite to the contrary, the company demanded that agents show excellent judgment, common sense, and above all, a thorough attention to detail that extended beyond the information in its categorizations. Aetna demanded that its agents take stock of everything associated with a structure and its contents, especially its utilization. For instance, the company encouraged agents to describe the materials used in manufacturing, the size of the property (depth, width, and height) and the lot on which it was situated, and how the building was heated. In one case, an agent was asked if a mill under consideration was heated by a stove, what was under it, and how far it was from wood: "Does the pipe enter a chimney, does it have a partition of floor, and how near does it come to wood"? The interrogation continued with a spate of other questions about the manufacturing processes and equipment on hand at the mill. Agents were expected to be experts on industrial processes as well as the built environment. In this particular letter, the agent was asked if there was a drying stove or drying machine. Was there a steam boiler? Was bleaching done in the mills? And, was sizing done in the mill or in a building nearby? The letter continued with still further questions about waste and fuel: "Are the rags assorted and stored in the mill? Is the fuel used wood or coal?"[33]

As Aetna expanded and intensified its surveillance, its methods of surveying a risk became increasingly less satisfactory to the firm's directors. Beginning in the 1830s, Aetna wanted agents to draw a graphic plan of the insured property. To this end, the company's 1830 *Instructions* advised that "sometimes a better description can be given by a plan of the property drawn on the back of the copy, with the distances in feet marked in figures." Like earlier maps, such as those used by the Philadelphia Contributionship in the eighteenth century, these diagrams reinforced the physical descriptions that agents made of property and goods; they were not meant to stand on their own as documentary evidence of risk. By the 1840s, however, the company had begun to imagine an expanded role for the plans that accompanied surveys. Rather than simply being illustrative, diagrams played an important role in the rate-setting and risk-taking process. Increasingly, maps became spatial representations of fire risk, not just added description. In an 1838 letter the company advised the agent to survey "the building requiring the most particular description and give a diagram of the others—marking out the relative situation of other buildings within 600 feet of the one insured." Although Aetna and other insurers had long shown concern with the structures adjacent to property that it insured, they had not been in the habit of precisely specifying this information. Suddenly, Aetna began to associate the landscape surrounding a structure and its particular spatial arrangements with fire risk.[34]

Coincident with its endorsement of graphically representing and interpreting fire risk, Aetna embraced other measures that underscored its more intensive exploration of fire danger. In 1837, the company increased its level of supervision and management of risk by sending its director, Joseph Morgan, on a tour of company agencies and on a scouting mission to cities in which Aetna had no representative. Morgan sought to establish agencies, investigate risks, and adjust losses, but most importantly he wanted to help the firm to develop a better understanding of fire danger. In his first inspection tour, Morgan traveled over a thousand miles, crossing northern New England to New Brunswick in three weeks. In 1840, he voyaged to Norfolk and Richmond (in Virginia) and Wilmington (North Carolina), among other places. Morgan was already an experienced cross-country business traveler in 1842 when he began a ten-week, six-thousand-mile junket that covered New York City, Trenton, Philadelphia, Pittsburgh, Zanesville (Ohio), Cincinnati, Louisville, New Orleans, Vicksburg, St. Louis, Chicago, Cleveland, Detroit, and Buffalo.[35]

Morgan's travels heightened Aetna's administrative surveillance of its agents and risks. The extensive diaries kept by Morgan document his travels and underscored how Aetna's knowledge about fire risk developed within the framework of specific spatial and urban contexts. In the record of his 1843 trip to Montreal, Morgan reported on the company's agency, Aetna's financial exposure to the risk of fire, and the city's fire defenses. As one might expect, he emphasized that agents "know" the landscape and its risk. Morgan's inspection began with a tour of the city with the company's agent, Mr. Jones, as well as his "out of doors" man, Mr. Wood. Morgan commented on their knowledge of the city and underwriting practices: "This morning . . . Mr. Jones has with him in business a Mr. Wood with whom I am much pleased. He in fact does the out of doors business for the office. I took a stroll with him. He says they never take a risk without visiting and examining the premises. He appears like a smart business man." At the end of his visit, before leaving the city, Morgan reiterated the importance of "knowing the business." He especially praised the "out of doors man," Mr. Wood, whose intimate appreciation of the city and its dangers made clear that Wood "knows more about our business than Mr. Jones." Morgan also affirmed the value of his visit: "I now understand much better the nature of our agency here than could be learned from a correspondence." Of course, even though he thought Aetna well represented, Morgan did not leave without giving what "I considered necessary instructions."[36]

If Morgan hoped to acquire intimate knowledge of each city on his route and its diverse fire dangers, he also wanted to learn about each agencies' business and to view the properties insured by the company. During the tour of Montreal Morgan was pleased with his new and *personal* understanding of the particular risks that Aetna took in the city—a knowledge he acquired only after his thorough inspection of the agency and a review of its accounts. Morgan reported, "August 21st, Rainy morning. Called on Mr. Jones, found he had not made out his list of risks. . . . [S]pent an hour with Mr. Jones in looking over his account." After examining the policies and surveys completed by Jones, Morgan accompanied him on a tour of Aetna's many policies in Montreal. He reported, "I then went with him the whole length of St. Paul Street. He has a large number of risks in this street. It is nearly a mile in length, I cannot designate any particular risk on which it is necessary to make remarks. The average rate on first class of building [and] on goods in them is about 50 cents." Morgan evaluated a structure according to its use, construction, and exposure to other buildings—all within the framework of the company's classification scheme.[37]

Morgan's inspections allowed him to instruct the company's representatives about the work and technologies of good underwriting practice and to confirm Aetna's growing fund of knowledge about the problem of fire. In regard to Montreal, he noted that "after taking a lunch with Mr. Jones, I viewed the site of the late fires, Grahams might have been considered an excellent risk. A large stone building, it burned only the building in which it originated. The other fire . . . was a very large one in the best part of the City, it consumed nearly a whole square mostly of brick buildings which all took from a joiners shop in rear, not from one another." Once again, Morgan's thinking about the danger of fire mirrored company instructions, but also reveals his growing awareness that each risk existed in a spatial context. Not only did Morgan emphasize the building's construction, but he also identified its "exposures." In the case of the fire "in the best part of the City" Aetna's loss had occurred because the company had insured a risk that was connected without an interceding wall to a hazardous manufacturing activity that caught fire.[38]

Morgan's overall assessment of the built environment of Montreal provided particular insight into the dilemmas facing nineteenth-century insurers. Buildings in Montreal, he reported, "are safer than ours, each building is in fact a separate risk. No fire, I believe, has ever extended any distance in this [St. Paul] street. His [Jones's] loss in this street since he has been an agent has been very trifling, and he has a very large amount at risk." Clearly Morgan understood the importance of substantial construction and/or nonattachment of buildings in preventing the spread of fire. In addition, by contrasting Montreal's construction style and its lack of fires with "ours"—certainly Hartford but probably America more broadly—Morgan recognized the fragility of the built environment in the United States. Despite demonstrating the extent of his and Aetna's knowledge, Morgan did not take the next step and suggest that underwriters could prevent incidence or, at the very least, spread of fire. Perhaps his diary was not the place to record such sentiments, but Morgan joined the incipient fire insurance industry more broadly in failing to use company's storehouse of knowledge about the hazardous of fire to prevent its occurrence or spread. Like other insurance firms, Aetna focused its attention on indemnifying clients against fire loss without depleting financial reserves.[39]

Of course Aetna did not oppose prevention efforts, and sometimes it told agents that systematic practices and rational manhood could benefit industry and society by providing common-sense fire prevention. For instance, in 1843 Simeon Loomis mused to the company's agent about Milwaukee, "Now it would be the part of wisdom, when these new cities are going up, to plan for the best possible security of property when the population should become large. . . . We trust you will exert a good influence on such matters when on the ground." However, for the most part such thoughts were fanciful. Usually the company explicitly argued the opposite position. Rather than telling agents to seek legislative remedies to the problem of fire, Aetna instructed its agents: "Indemnity is the only legitimate object of insurance." It would be more than fifty years until active programs of prevention explicitly became standard fare in the industry.[40]

Perhaps the greatest dilemma facing insurers was the utility of their classification programs, especially as they related to setting rates. Although underwriters continuously updated and revised classification schemes, such categorizations provided little security against loss. Indeed, the complexity and intricacy of such schedules belie the capriciousness of their origins. Not only is there no evidence that Aetna's—or any other company's—rating tables were statistically based prior to 1850, but everyday rate-setting activities appear to have been only modestly connected to the prescriptions embedded in rating and classification schedules. Quite often rates fluctuated because of a company's financial position or market competition. For instance, firms often raised rates across the board after they suffered excessive losses. In an 1834 letter, Aetna's secretary Goodwin bemoaned the failures of the firm's business in New York State, noting that "our business in New York swallows up our earnings everywhere else. Our whole scale of premiums must be advanced from a third to a half. If that cannot be done we had better withdraw." Goodwin's letter demonstrates the arbitrariness of rates and suggests the precarious state of underwriters' knowledge. Likewise, the marketplace also frequently stymied efforts by Aetna and other companies to have their agents follow regularized practices. In response to market conditions, field representatives sometimes ignored the rates set by their companies, or companies themselves lowered rates. Such conflicts occurred periodically. For instance, in 1819 Aetna's historians recollected that "there was keen competition and reckless slashing of rates that made foreign companies [those not headquartered in a particular city] steer clear of the city [New York], as a prudent man avoids becoming mixed up in a street fight."[41]

Just as often Aetna's nascent bureaucratic systems failed. For instance, after a series of disastrous fires, Secretary James Goodwin chastised the company agent for the "defective" manor in which he returned his surveys and policies. If done correctly, Goodwin argued, the company would not have taken so many bad risks on wooden housing arranged in the same block. Almost as frequently, Goodwin chastised agents for underwriting risks that they knew were bad. In 1830 Goodwin told an agent who had just reported a steam mill fire that "all [steam mills] seem to burn sooner or later. I, however, made my calculation that the fire would proceed from another cause, and it probably would before many years had leaped. I regret that we had not adhered to our first impression and declined it." If Goodwin did not explicitly question the manhood of his agents, the disdain often expressed in his return correspondence tacitly underscored their failure as agents and prudent men. Perhaps more to the point, such letters underscore the difficulties that Aetna had in disciplining its employees and its portfolio of risk.[42]

At best, Aetna's efforts to confront the danger of fire had mixed success in the first half of the nineteenth century. Unable and unwilling to advocate fire prevention as a company goal, the company's leaders frequently expressed fatalism about fire, revealing an underlying belief that fire was beyond their control or comprehension. On one occasion, the secretary wrote an agent, "There appears to be only one course left for us to pursue in relation to your agency and that is to acquiesce in everything that takes place and come to the conclusion that our misfortunes are unremediable." In another, he waxed philosophical: "There is nothing like being used to misfortune. . . . Fires have followed in such rapid succession that I have got hardened, and I can stand it like a philosopher." Almost heroic in his stoicism in the face of ubiquitous danger, the company secretary reveals the pervasiveness of the problem of fire as well as Aetna's—and the industry's—inability to comprehend the danger fully. Moreover, cities throughout the nation experienced dramatic conflagrations more and more often, and dozens, if not hundreds, of insurance firms went bankrupt following such blazes. For instance, after New York's disastrous fire of 1835, twenty-three of twenty-six local New York companies went broke. Similar statistics followed other large fires of the nineteenth century.[43]

For all its difficulties, Aetna subscribed to a long-term strategy based upon rational management practices and a vision of disciplined, careful manhood. Its leaders endorsed a number of relatively straightforward principles: surveying prospective risks thoroughly, developing routine procedures, providing guidelines to a network of field representatives, hiring and transacting business with men of character, and dispersing risks widely and sensibly across the landscape. The firm believed that these practices and its business culture would guarantee profitability, would provide a model to the nascent industry, and would create a market in insurance that eventually would winnow bad from good companies. In the process, Aetna accumulated and assessed information with greater and greater intensity. Agents visited, described, surveyed, diagrammed, and came to know the built environment. As they acquired knowledge about the physical, agents investigated the moral character of their risks. The company monitored agents and visited agencies with equal care and scrutiny. Aetna had taken the first steps in developing an unprecedented surveillance of the physical and cultural landscape that both augured the future direction of insurance practice and transformed the fire insurance industry and the problem of fire in urban America.

Insurance and Public Fire Safety

Nineteenth-century American fire underwriters conceptualized their role within the system of fire protection as secondary to the public work performed by firefighters. The connections between firefighters and insurers dated back to the formation of the first insurance company in North America, and underwriters provided moral and economic support to firefighters as they helped to protect the nascent industry's vested interest in protecting property. In addition, underwriters imprinted their assessment of the importance of community-based fire extinction efforts on company policies. Early policies of the Insurance Company of North America, for instance, depicted a scene of a house engulfed by flames. Above the door of the burning house hangs INA's firemark, the crest of an eagle. The rest of the scene depicts what might be typical behavior at a Philadelphia fire in the first decades of the nineteenth century. In the foreground a line of men pass water from a water spigot into a hand-pumped engine operated by a dozen men. The engine, and another like it, pump water onto the fire as firefighters rescue property and work with axes to thwart the spread of the fire. Similar scenes or metaphoric imagery appeared on other companies' policies, such as the Protection Insurance Company of Hartford and the Fire Association of Philadelphia.[44]

Of course, not all firms imagined their work in relation to firefighting nor did all company symbols—even those that depicted fire scenes—indicate that fire underwriters viewed their own work as unimportant. Aetna's symbol, which was adopted at its formation in 1819, used classical imagery and allusions to articulate the significant role that fire insurance played in American society. At the image's center was a shield with the company's name written across it. Surrounded by two rings of rope, a rising sun and volcano frame the top of the shield, and along the sides and bottom of the shield are the Mediterranean Sea and an unsheathed sword. The *Insurance Index* interpreted the insignia with flowery, metaphoric allusions. The sun represented the strength of the "conserver," signaling the company's role as an "underwriter" but not as a "creator." Mount Aetna symbolized the company's role as a "safety valve," removing dangerous gases and energies, and the Mediterranean signified consistency because the sea "knows no tide." The shield was that of Agamemnon, denoting "singleness of purpose, dignity and daring-do." Perhaps more importantly, the shield was emblematic of defense, "and of protection, vital and instant when needed." Although this interpretation was made a century after the symbol was adopted, the company's founders had clearly intended the seal to reassure its clients of its seriousness and capability through classical allusions to manhood and risk, certainty and steadiness. Balancing the daring nature of its insurance men against their steadfastness, the symbol, like the company, offered protection against the vagaries of nature in a manner akin to firefighters' noble service, if not as a substitute for it.[45]

The early insurance industry's unconditional and often unsolicited pecuniary support for firefighters most clearly indicates their recognition that the system of volunteer fire protection was of crucial importance. Moreover, their advocacy encouraged property owners, municipal governments, and other commercial interests to support volunteer fire companies in a similar manner. During the early nineteenth century, the support of insurance capitalists was generally given without hesitation or qualification. At times, fire companies asked underwriters directly for financial help, but whether requested or not, insurers provided regular and needed assistance to fire companies. It would appear that both parties understood this financial support in terms of a communal effort to check the dangers of fire.[46]

A letter written by the Vigilant Fire Company to INA in 1813 asking for support identifies the boundaries of underwriters' and firefighters' shared obligations. Written by the Vigilant Company's president, the letter documented the company's long history and purpose. Established in 1764, Vigilant claimed to have been "influenced by principles of benevolence and humanity [and] associated themselves for the purpose of protecting the property of their fellow citizens from the ravages of fire." After describing the company's moral commitment, the Vigilant representative documented the "large sums of money" that company members had expended on the public good. In conclusion the writer appealed to INA's economic interest by identifying its geographical location in Philadelphia. Located in the built-up eastern section of the city along the waterfront, the company reported that it was about to lose its lease, which would force the company to relocate. Even more disturbing to Vigilant, several other companies had recently been asked to leave their engine houses in the same vicinity. Unlike those companies, Vigilant wanted to remain in the eastern section of the city, which was "of peculiar importance from the number of valuable buildings there, and the vast quantity of goods stored therein." Vigilant knew of INA's commitment to commercial and industrial properties and emphasized its own efforts to protect just such properties. If its appeal was economic, it also drew upon a shared knowledge of Philadelphia's spatial arrangement. The letter reported that Vigilant was the only company "remaining on Front Street" and that there were "comparatively few [fire companies] eastward of Third." Lest INA officials not comprehend the severity of the danger and the relatively poor fire-protection infrastructure in the city's commercial districts, Vigilant made its point distinctly: "If some prompt and active measures are not resorted to, this part of Philadelphia will be seriously unprotected." Although INA's response to this particular query is not extant, on at least one other occasion it provided a company financial aid for the purchase of a new building.[47]

To a large degree, INA and the Vigilant Fire Company were also bound by a sense of a common battle against a terrible foe. This shared struggle—especially the common desire to avoid conflagration—may well have been as powerful an incentive for insurance company support as were the obvious financial benefits. When insurance companies wrote policies primarily in their local markets, their shared connections may have been especially intense. Indeed, this mutual sense of duty may have colored INA's decisions to offer support to local groups such as the Fire Hose Association.

In its request for support, the Fire Hose Association underscored volunteer firemen's and underwriters' mutual interest in order, duty, and community organization as a means to check the ravages of fire. Nine of Philadelphia's ten hose companies formed the Fire Hose Association in 1807 "for the purpose of securing advantages to the public," which would accrue from better intercompany communication, reciprocal arrangements to share hose, and increased company unity.[48] The association informed INA's directors that it had formed to preserve order and "harmony" and to facilitate communication between companies for enhanced firefighting performance. The letter explicitly denied that the association was acting in its own "self-interest." Finally, the letter informed INA that its contributions would make the association permanent and induce nonmembers to join. Volunteer firefighters expressed a vision of their role in preserving public safety that appeared to resonate with fire insurers. In their discussion of the request, Philadelphia's underwriters recognized the importance of volunteer firefighters' "laudable exertions for the general preservation," even as they recognized that firefighters' work was "especially advantageous" to insurance firms.[49]

Almost immediately following the receipt of the letter from the Fire Hose Association, INA also began to regularly contribute to each of the association's member companies. Between 1807 and 1824, INA subsidized the city's companies with over $300 in donations each year (collectively), in addition to occasionally supporting the odd request from the city's fire companies. Underwriters were not the only contributors to the urban fire companies; merchants, private citizens, other property holders, and even the municipal government made regular donations to the well-being of volunteer fire companies. Further, this relationship between volunteer fire companies and local underwriters, property owners, and municipal governments was not peculiar to Philadelphia. In St. Louis, for instance, one observer estimated that slightly less than half of the volunteer fire department's budget prior to 1850 came from the municipal government. Interested citizens, including insurers, within the community provided the other half.[50]

The close relationship between underwriters and fire companies began to deteriorate during the 1830s and 1840s, with the expansion of the nascent industrial economy. By the 1840s, the flow of solicitations from fire companies had increased to such an extent that INA considered formulating a policy on the matter. Other insurance companies, especially those that did not focus their business in a single geographical region, also began to hesitate in their support of fire companies. Under the pressures of economic expansion, increased competition, and dramatic urbanization, insurance companies increasingly declined to make donations to fire companies. Aetna, for instance, received many solicitations for contributions. Although by the 1840s, the firm had established a policy against making such donations, it nonetheless continued to give money in special situations. For instance, prior to leaving to inspect Aetna's agencies, Joseph Morgan received a request from a Montreal fire company for a $100 donation. After observing Aetna's risks in the city, and the nature of fire danger in Montreal, Morgan decided to offer the city's Union Fire Company Aetna's support. Morgan recorded that Union was Montreal's most efficient volunteer association and would use the donation to purchase of a new engine. Moreover, he noted that Montreal had a relatively weak fire-protection infrastructure, and he provided a finely detailed analysis of why the engine being acquired was a significant advance for the city. But, perhaps most importantly, Morgan's donation came on the heels of evaluating Aetna's financial position in the city, which was substantial. Clearly, Aetna's financial interests benefited from such a small investment.[51]

However, as the insurance industry developed a more systematic surveillance of fire risk, many firms began to view such support as an unnecessary expense. For instance, Aetna began to take this view, which made it unwilling to support local, community-based efforts at fire extinction. This view is perhaps most evident in a letter Aetna sent to one of its agents in 1829 regarding a request that the company support the purchase of a fire engine. The company informed its agent, "It has been found necessary to decline all such applications" because "if we answer them the tax would be more than we could sustain." More to the point, Aetna complained that it would be "taxed" indirectly anyway, because better fire protection caused a "reduction in premiums." Aetna's reply was striking because it adopted a cost/benefit perspective that would not appear again in its company files, manuals, or correspondence for close to twenty years. By the 1850s, such reasoning had become customary. Of course, this did not mean that Aetna had become disinterested in fire defenses. Not only had the company given an agent an additional commission for his "service in securing a fire engine," but by the 1850s it took fire defenses into account when assessing fire risks. Rather, the company did not want to finance improvements in public fire defense because they hurt the firm's bottom line. As firms scrutinized their account books, they began to see financial support for fire companies as a liability rather than an asset.[52]

That Aetna objected to paying for public fire defenses for economic reasons suggests the development of a new way of envisioning fire risk. When companies expanded their business to multiple cities to gain a competitive advantage, they acquired a host of new community obligations. In contrast to firefighters, who had specialized their labor but had not yet removed themselves from their urban neighborhoods, fire insurers became less interested in local community as they developed new bureaucratic technologies to manage information and practice. Insurers became more interested in the community implied in its portfolio of risks than in geographically bound communities. For instance, in Philadelphia, INA officials received requests from their neighbors to support the city's fire defenses but more and more often declined such requests, especially as the scope and intensity of their business activities expanded. Likewise, Aetna's agents, who engaged in face-to-face business transactions and other aspects of community life with their neighbors, funneled requests for fire company donations back to the home office in Hartford. However, faced with the growing number of inquiries, increasing expenses, and an uncertain business environment, Aetna's leadership began to decline all requests for assistance. In some sense, Aetna's decision abrogated a long-held sense of reciprocity between fire companies and their neighbors, including underwriters. Yet, to Aetna, the face-to-face social relationships on which those obligations had been based were becoming anachronistic. Indeed, company officials had purely market relationships with their customers—whom they had never met nor would ever meet. Aetna's obligations to those communities where it transacted business increasingly were not those of a fellow resident, but of a transient. Aetna's interest in the fire safety of distant cities became entirely economic in nature. Thus it reconsidered its responsibility for helping to provide firefighting services, refusing to pay a "tax" for a service that could lower insurance rates, thereby impeding profit taking. The company even began to consider ways to include fire department efficiency in its classification systems. As more insurance companies expanded into the national market, such perspectives gained increased currency, raising questions about the insurance industry's responsibilities for the provision of fire protection.

Conclusion: Building a New Risk Community

In the first fifty years of the nineteenth century, the American fire insurance industry expanded dramatically, but to many insurers in the 1840s, the industry's future seemed far from assured. Conflagrations swept New York, Pittsburgh, and St. Louis, and social crises beset cities throughout the nation for most of the decade. Although incendiary cities certainly made the industry apprehensive, underwriters also began to question the efficacy of their business practices. At Aetna, firm leaders feared that the company's carefully orchestrated strategy had failed. In fact, during the 1840s, the company suspended the symbol of its disciplined culture of manliness and rational administrative procedures: *The Book of Instructions and Tariff of Rates.* Even so, by 1850 the fire insurance industry was rapidly becoming an important factor in the nation's industrial growth—both as a protector of property value and as an engine of finance—but it stood on shaky footing. Prudent manhood, elaborate business practices, and a commitment to public fire safety had not yet proven to be successful strategies for controlling the problem of fire.[53]

Aetna's suspension of its manuals suggests the depth of the confusion that faced the young industry. Aetna's program of surveillance seemed to be foundering, despite efforts to tie it to a culture of responsible manhood among its agents and continued attempts to develop an elaborate technology of classification. Ironically, as its strategy seemed to be failing, Aetna's secretary, Simeon Loomis, complained that agents attempted to follow the printed rules too literally. He fumed that they "just turn to the Book, find something which they concluded to be tolerably well adapted to the case, [and] they would name the rate and make the policy, without exercising any judgment in the case." Loomis's critique was an expression of frustration at the complexity of fire danger. After nearly three decades of correspondence with field representatives, after having evaluated many hundreds, even thousands, of policies and claims, Aetna remained unable to classify risks and to set rates at an appropriate level.[54]

The industry's failures had deep roots and were reflected in bankruptcies and fluctuating rates, which sometimes covered severe financial losses. Variability in rates resulted from a combination of factors. Most simply, firms could not precisely determine how severely the occurrence of fires might affect them negatively, and most did not keep systematic records of losses. As a result, companies often did not know the nature or scope of the dangers that they faced, either in terms of their entire portfolio, between different classes of property, or across geographic neighborhoods, cities, or regions. Additionally, insurers rarely shared information regarding their experiences of paying losses and calculating rates. Thus, despite paying thousands of claims and frequently discussing danger with agents, companies' knowledge remained limited—a condition only exacerbated by undisciplined record keeping. Moreover, without rational information management procedures the emphasis on prudent manhood could not overcome the capitalist zeal of opportunistic insurers. Lacking a credible method of determining rates, market competition frequently drove prices down—often with little regard for fire danger. In addition, the prevalence of both large and small fires appears to have increased throughout the nineteenth century, and the tendency of most companies to underwrite insurance in closely proximate neighborhoods made many firms especially susceptible to huge losses.[55]

Despite the industry's limitations, insurers did not abandon their attempts to organize and manage the problem of fire. Firms aggressively demanded more precise and detailed observations about insured property, its spatial arrangement, and those individuals involved in the business transaction. Among the most pressing problems that insurance companies continued to confront was how to differentiate between risks in very different geographic environments and landscapes. Aetna's Simeon Loomis explained with horror that agents in "New England, Canada, and Florida were working by the same rules . . . it is not easy to imagine a greater inconsistency." Though Loomis did not, and probably could not, identify the exact differences between places, he recognized that fire risks varied across regions. In addition, he recognized that the business of underwriting also differed by locality (within region.) Aetna often urged its agents not to "expose" a great deal of business to loss from a single fire; Loomis's advice generally prescribed that agents "take no risks on wooden blocks having more than three tenements and not to exceed $5000 on merchandise in the same." However, Loomis's recommendations also took account of diverse local conditions. For instance, Loomis explained to one agent that "in a village such as yours" he should not put himself in danger of losing $10,000 from any single fire. Although Aetna had abandoned its instruction book, the company had not given up on making sense of the problem of fire and appreciating its complexity. Though classifying risks proved elusive, Aetna and other insurers gradually developed an orderly approach that emphasized procedure, character, and communication.[56]

Furthermore the insurance industry began to sketch the outlines of a model of security that increasingly differentiated it from the standard definition of safety—the volunteer fireman and his fire company. Rhetorically the insurance industry employed much the same language as firefighters, emphasizing a vision of manliness that balanced prudence, discipline, and aggressive action. However, insurers' nascent culture of risk became increasingly concerned with symbolic order, represented by bureaucratic technologies, such as classification schemes and economic contracts. Whereas firefighters' work concerned the public's safety in the built environment of an urbanizing and industrializing nation, underwriters became less focused on the material landscape and more on their portfolios of risk. Fire, to insurers at least, was becoming an economic matter, abstracted from the physical world. As earning profits and protecting the corporate bottom line became part of the battle to control fire, insurers worked to control risk portfolios by disciplining company practices and agents through bureaucratic technologies and gender norms. More broadly, these strategies offered an entirely new vision for protecting the urban social order from the problem of fire.

During the early nineteenth century, insurers believed that they should work alongside firefighters in helping to eradicate the danger of fire. Even though insurers' belief about their responsibility to the public would wane by midcentury, early on the two groups shared a common attitude toward their risk-taking activities, believing that their efforts against a common and ubiquitous danger set them apart as men and citizens. Although the connection between insurers and firemen developed for a variety of reasons, it most particularly stemmed from their common enemy and commitment to preserving order. Moreover, these men believed that they shared certain attributes. Men fighting fire—either physically or economically—had to be sober and disciplined. They also needed to demonstrate technological competence. When the American landscape became more physically and socially complex, as industrialization intensified, each experienced similar tensions and faced similar difficulties as they reorganized to contain the fire hazard.

Part II / Fire

CHAPTER THREE

Statistics, Maps, and Morals

Making Fire Risk Objective, 1850–1875

In April 1852, prominent Philadelphia lawyer Horace Binney excoriated the fire insurance industry at an odd moment. In the keynote address of a gala event celebrating the one hundredth anniversary of the Philadelphia Contributionship for Loss from Fire, he censured fire underwriters for what he characterized as a haphazard approach to their business. Binney contrasted fire underwriters' uncertain rating methods with the sounder practices of life underwriters, which were based in actuarial method. He ridiculed the industry for not having created a "mortality table" of fire loss, and described the industry's business as unscientific. Binney urged insurers to intensify their program of observing the landscape and to form organizations that advanced the industry's common interests, arguing that such activities and associations would help fire insurers to organize their industry rationally and to understand the problem of fire. He also suggested a wholesale reconsideration of public fire defenses. Amazingly, it is unlikely that his remarks offended the audience. By 1850 many fire insurers realized the limitations of their approach. Even leading firms, such as Aetna, recognized that they could predict neither the frequency and extent of fires nor their financial costs with an acceptable degree of certainty.[1]

Binney's remarks heralded a transition in how underwriters confronted the problem of fire. Shortly afterward, underwriters introduced fire insurance mapping into the lexicon of their tools, continued to make business transactions more standard, formal, and regular, and established trade associations to facilitate intercompany cooperation. Most importantly, perhaps, they began to develop and to apply statistical reasoning to their business, especially to their efforts to categorize and to evaluate danger. As Binney had expected, using statistics transformed the way that fire underwriters produced knowledge in regard to fire risk. Developing actuarial tables marked an important qualitative and quantitative departure for the industry, especially because it occurred in conjunction with the expansion of other information-management technologies. Insurers represented urban fire risk in statistical ratios, objectified fire danger in city maps, and forged new institutional relationships. Within two decades of Binney's critique, fire insurers had altered their routines substantially.

As the industry remade itself, it also remade its approach to the problem of fire and helped to shape the urbanization of North America. In particular, the introduction of two representational technologies almost simultaneously in the 1850s—maps and statistics—reworked how underwriters perceived fire risk, and hence the problem of fire. As insurers studied figures and atlases, they objectified the risk of fire; they turned an incalculable societal threat into something to be scientifically studied, controlled, and managed. Moreover, these tools gave insurers the ability to attend to the minutest details of the built landscape. They could study a wider and more diverse geographic area, and divide and subdivide their classifications into more analytical categories. With its new management technologies, the industry re-envisioned the built landscape as an accumulation of fire hazards, and visualized risk using representational technologies made it conceivable to manipulate the built landscape. As a result, changes in insurance practice were felt beyond the boardrooms or balance sheets of insurance company. As the industry made its administrative procedures more rational, expanded its surveillance, and struggled to create an objective record of fire loss, it helped to reshape American cities. The fire insurance industry's effort to control the problem of fire helped to reorganize all aspects of urbanizing America—from the provision of municipal services to the structure of city landscapes.

At the same time that they quantified the dangers of fire, underwriters offered a scathing critique of the system of fire protection as it had been organized in the first half of the century. They argued that urban communities should not direct fire protection; rather, it should be managed according to the precepts of industrial capitalism, including a new division of firefighting labor that involved paying wages. Strikingly, insurers' new business practices and sudden critique of firefighting coincided with similar efforts by middle-class reformers. In Philadelphia, for instance, a morality novelist offered a stinging evaluation of community-sponsored firefighting. Such moral commentaries became inseparable from underwriters' attempt to rationalize their industry, fire risk, and urban space. Insurers were not alone in their advocacy for paying specialists to extinguish fires, and their suggestions matched broader changes in the division of labor in the industrializing nation. Underwriters offered a new perspective on the problem of fire—one that emphasized individual protection acquired through the marketplace rather than community-based protection.[2]

As insurers reconceived fire protection, they directly and indirectly contributed to the reorganization of life in American cities, as well as the construction of those cityscapes themselves, in much the same way that industrial capitalism was altering America's urban and rural landscapes. But, unlike other, more dramatic, industries—the railroad, textiles, or automobile manufacturing—the insurance industry has remained almost invisible in the historical record, like the commodity it trades. The information technologies, strategies, and organizational arrangements developed by insurers transformed America. When fire underwriters represented safety and danger on statistical loss tables and on fire insurance maps, they created a palimpsest for safety in future American cities. Although the early use of maps and statistics should not be mistaken for organized city plans, these technologies nonetheless gave structure to the urbanization process. Cities—from their physical structures to municipal fire services—began to be structured according to a nascent program for safety. If the outlines remained faint, the underpinnings of a new fire safety discipline were taking root.[3]

Statistics

Horace Binney called on underwriters to be rational, scientific men. He endorsed contemporary procedures—especially those of surveillance and inspection—and recommended a more systematic way of collecting and organizing information. He urged insurers to compile a record of their losses and then to use these figures to re-create categories and classifications of risk. Binney specifically advocated creating a "mortality table" of fire danger, divided into "the case of houses and merchandise, and its variations in the case of particular trades, and in the different conditions of the agents and apparatus for arresting and extinguishing fires." Fortunately, the process of building this database would not require a significant new bureaucratic infrastructure. Already existing practices of observing the landscape and managing risks would help underwriters to develop an actuarial record, which could help them to develop a "rule" around which they could organize their financial portfolios. Thus underwriters eventually would be able to reasonably estimate (and predict) future losses. For the first time, the means of assuring profitability and solvency seemed within grasp.[4]

Binney even provided his audience with an example of the methods he championed. To exemplify his claims, he developed a quantitative profile of the Philadelphia Contributionship's financial losses over a ten-year period. He divided the Contributionship's insured dwellings and warehouses (constructed of either brick or stone) into two categories: occupied at night and unoccupied at night. Next he calculated the average amount of loss on each category during a given year and organized that information into a table. Over ten years of observation he calculated that the company lost an average of 3.6 percent (of the total insured value) on occupied properties against a loss of 9.95 percent on unoccupied properties. Binney argued that the table "awakens [us] to the consideration of the nearly constant relation which the inhabitance or non-inhabitance of a building at night has to the extent of partial loss by fire and it is perhaps explained by the necessary fact, that a fire must in general gain more headway or intensity before discovery in buildings not inhabited than in buildings that are so inhabited." If the finding seemed mundane—namely that property occupied at night burned less frequently—the insights on business method were profoundly informative.[5]

Binney's illustration offered a new perspective on risk taking in the American fire insurance industry. He evaluated the effect of "inhabitance verses non-inhabitance" as an object lesson regarding the benefits of collecting and then applying basic statistics about fire loss. Not only would simple calculations help insurers to place their evaluation of fire danger in space and time, but actuarial data liberated insurers from thinking about risks in the short term because such tables would be cumulative. This approach would become more precise over time, moving from approximations based in "ten years' observation" to those compiled for a "century," which would yield a yet broader perspective on what he termed "plague and cholera losses." Binney pointed out that a statistical analysis of a small sample, such as his example, was not scientifically valid; he noted that "the inference from the table, without further elements, will not strike the mind as conclusive." By observing this shortcoming he underscored his admonition to the industry: underwriters should create a full accounting of fire risk for several decades. He encouraged insurers to keep a continual record of observations and to use them to search for an underlying "law" of fire insurance.[6]

Although it is not clear if the Contributionship ever used Binney's research or how quickly it adopted his perspective, other firms had begun to experiment with such methods even as Binney delivered his speech. In particular, Aetna Fire Insurance Company began to analyze its portfolio using such quantitative reasoning and developed its first crude mortality table of fire loss in 1852. The company began to collect and to tabulate statistical averages of loss within different segments of its portfolio, primarily according to their "use" or "occupation" of a risk—i.e., dwellings, breweries, saloons, etc. In that Aetna did not organize its rudimentary actuarial table in a fashion that reflected all the possible permutations of the many categories within both a property's "use" and its method of construction, the quantitative record of losses was rather limited. Aetna compiled an account of its losses only for categories of "occupation"; in other words, similar "uses" of property were lumped together regardless of the material or method used to construct them. For instance, all textile mills (or dwellings, etc.) were tabulated as a group, whether built of wood, stone, or brick. As a result, the statistical analyses did not reflect minute changes in construction method or building materials. Nor did the rudimentary actuarial table reflect the process through which rates were set. To be sure, Aetna utilized the "use" category to determine a "base rate." However, the firm further modified the rate, sometimes substantially, according to construction materials or the presence of other hazards.[7]

Despite their limitations, Aetna's efforts at collecting and categorizing actuarial data provided an innovative and critical new perspective on its portfolio of risk. To start, the company compiled loss information for the thirty-five property classifications that it normally utilized as a starting point for delineating risks. In addition, the company tabulated the total value of the insured property, the amount of premiums it collected, and financial losses on that type (or use) of property. Significantly, the company also calculated the average premium it charged for a six-month period. For example, between October 1852 and April 1853, Aetna insured coffeehouses and saloons with a total value of $134,599, on which it collected nearly $2,000 in premiums. In the margins, the company converted these numbers into an average rate of premium for saloons and coffeehouses over a given six-month period (by dividing the total value of the insured property by the amount collected in premiums). In the case of saloons and coffeehouses, the company charged a rate of 1.45, which was read either as a percentage or as dollar value per $100 of property insured. Additionally, the company recorded whether it realized a profit or suffered a loss in each category. In the case of saloons and coffeehouses, for instance, it paid over $4,550 in claims, thus experiencing a net debit of $2,602.55 over the aforementioned six-month period. Formally documenting and evaluating such data—both the average rate and the amount of profit or loss—represented a significant step forward in the firm's business practices.[8]

In practice, implementing and coordinating the new mode of assessing danger with everyday business methods proved difficult, and often occurred haphazardly. In fact, although Aetna developed a quantitative portrait of its risks, the company only partially incorporated the new analytic strategies into its daily practices, especially its rate-making activities. Aetna continued to charge different fees based upon the method and materials used in constructing a building and for its spatial relationship to other structures. For example, the company offered modest rates on "detached" dwellings. Similarly, dwellings built of brick or stone were assessed at a lower premium than frame houses. However, as mentioned above, the company compiled loss statistics only for use categories but did *not* cull them for construction methods or spatial layout. Indeed, the company did not collect, organize, or categorize loss data in a manner that could allow it to determine precisely how much less money to charge clients who built with more substantial materials. Even more perplexing, Aetna does not appear to have altered its instructions to company representatives. It continued to collect loss data on property use, and it persisted in encouraging agents to observe types of construction and to diagram a building's location in space, even though it did not use quantitative methods to assess the impact of construction or spatial features on its portfolio of risk and loss.[9]

Despite compiling loss information in such a limited fashion, Aetna nonetheless constructed a broad financial prospectus of its risks. In 1857, Aetna calculated summary statistics of its financial gains and losses, by property use, for a five-year period running from October 1852 to October 1857, and continued keeping such folios through at least the 1880s. The five-year summary statistics resembled those compiled every six months: amount of insurance written by property use, total premiums received, and losses paid. These summary statistics also noted the average rate paid for the first and last six-month period as well as the average rate. Also, for the entire five-year period, the company calculated an "an actual cost rate—five year average." The "actual cost" represented the amount of the company's financial loss (or gain) as a premium, pointing only to the cost of insurance on that category. It did not include an accounting of the administrative costs of writing policies, such as agent's commissions, salaries, and supplies. However, the company leaders seemed to know these costs well; in fact, in its *Instructions for the Use of Agents*, they were estimated to be 20 percent of the premiums received. At the very least, then, Aetna's leaders implicitly included such administrative costs in their assessment of profit and loss.[10]

By developing an impartial record of financial loss using quantitative data, Aetna took the first step toward objectifying fire risk, and it gained renewed confidence in its ability to set rates. With its long-term commitment to gathering loss data set into place, the company boasted that its classification now took into account both the company's record over a period of years and its experience with over "one hundred million" dollars in property, insured annually. The firm believed that this actuarial data provided an accurate method of forecasting loss and a long-term edge in the marketplace. Perhaps most important, establishing a method to predict future losses gave Aetna new confidence to represent the material world as a set of insurance rates. In 1857, the company reissued *Directions for the Use of Agents* for the first time in over a decade. Building on its new quantitative method of gathering loss statistics and classifying danger, Aetna argued that it could now "depend with almost entire certainty" on setting rates that "guarded" the company and its customers from the vagaries of the problem of fire. The historical record of Aetna's losses, depicted numerically, had become the literal embodiment of fire risk. Company officials now delineated between bad and good risks by representing them quantitatively. In addition, these methods provided underwriters their strongest argument against setting rates according to market conditions—either within a local region or within an industry. Further, they served as an argument against setting rates too low and provided firm representatives an explanation as to why they should not lower rates—because it would expose Aetna to long-term risk.[11]

More importantly, Aetna began to use statistical analysis to justify decisions about which types of property to insure. For instance, in its 1857 *Directions to Agents* the company listed seventeen types of property (by use) on which the firm had suffered large financial losses over a four-year period. Based on that quantitative record, Aetna all but prohibited agents from insuring those categories of risk. It warned that "these hazards (i.e. those 'of questionable profit') are most inveigling in their temptations to the inexperienced, and we would sound the note of warning to let them alone, without all can certainly be paid for the dangers incurred." If Aetna's statistical record served as a rationale for avoiding bad risks, that same data warned agents not to succumb to the temptations of market-driven price competition. The company expected that its tabulations would make pricing self-evident, and it believed that when rates fluctuated they were out of accord with observable reality. Aetna not only implored its representatives to keep rates at profitable levels, but it also suggested that in cases in which a "hazard is of questionable profit to insurers" the "rates and rules" should be presented as an example of good practice. By highlighting its incomparable program of analysis, Aetna helped agents to negotiate price and gave them a marketing tool, allowing them to distinguish Aetna from other fire insurance companies.[12]

Such themes emerged more strongly in 1867 when Aetna published a new and much more comprehensive manual, which would serve as the basis of its system of bureaucratic technology for decades. The manual defined good underwriting as a combination of moral character, disciplined daily routines, and rational actuarial science. Compiled by the company's general agent in Cincinnati, J. B. Bennett, the *Aetna Guide to Fire Insurance* appears to have been the first thorough dissertation on insurance practice following the industry's intellectual innovations in the 1850s. Bennett argued that, although not yet fully developed, the industry dealt in averages. According to Bennett, Aetna's "experience with a *great number of risks* in a sufficient area of space, insured for a *long period of time*, is the only safe ground from which a reliable cost of insurance" could be determined. If observed carefully, this record yielded guidance in the everyday practices of setting rates and taking risks. Insurance, according to Bennett, "is a *science* to be worked out." Companies that did not follow sound practices would eventually disappear, victims of not adhering to the law of averages.[13]

Aetna's pricing strategies and underwriting activities reveal the degree to which rudimentary experience tables structured the company's daily practices and philosophy. In fact, comparing the statistical record against the recommended premiums shows that Aetna's rates did not deviate significantly from its quantitative record. In addition, the company appears to have incorporated its administrative costs into the calculus of determining rates. In categories in which the company's rates remained higher than the cost of insurance (losses plus costs) those rates remained in force. On the other hand, if rates in a category slipped below those costs, the company adjusted its prices accordingly. In other words, rates on types of structures that consistently burned and cost Aetna money were adjusted upward to reflect that likelihood. In some cases, such as with "India Rubber" factories, the company prohibited agents from writing insurance altogether—a restriction to which the firm appears to have adhered. For instance, between 1852 and 1857 the company wrote nearly $300,000 of insurance on India Rubber factories and suffered substantial financial loss on those properties. In contrast, between 1857 and 1862 the company took only an additional $20,000 of insurance risk on India Rubber factories and did not pay any losses on those during the five years. Interestingly, rates on property that showed little statistical likelihood of financial loss, such as dwellings, remained high. This suggests that the company used profitable categories as a hedge to balance its exposure to losses in other more risky categories of property. The company's rudimentary actuarial tables, then, became a signal that helped it to insure profitability and solvency.[14]

Indeed, certain categories of risk, especially "dwelling houses and stores," remained steady profit centers for Aetna through the 1880s. In the case of houses, for example, aggregate statistics showed that the company charged, on average, about \$0.83 (per \$100) between 1852 and 1857, against a loss of \$0.23. This category of risk proved to be Aetna's most important and steady source of revenue. Insurance on dwellings accounted for nearly 30 percent of its total risk portfolio (although only 20 percent of its premiums). After keeping statistical records of loss for five years, in 1857 the company separated dwellings into a separate category of risk for the first time, perhaps recognizing this category as a consistent profit center. Within this lucrative class, Aetna varied its rates according to the dwelling's mode of construction, although no statistical record within the general "use" category appears to have been kept. Even for those dwellings most substantially built the company charged about \$.35 (per \$100). This rate was more than sufficient to cover the cost of insurance in this category, underscoring the importance of houses as a significant sector of profit for the firm.[15]

The category of "dwelling houses" played an important role in Aetna's portfolio not just because it generated high profits, but also because it provided income in a steady, unvarying stream. When compared to other classifications within Aetna's risk portfolio, dwelling houses were a model of consistency. The company's aggregate statistics for five-year periods, ending in 1857, 1862, and 1867, reveal that Aetna's average rate remained about constant at \$.85. Loss and "actual cost" also varied little, at about \$.25. As losses in this category gradually increased after the Civil War, so too did premiums (though not as quickly). However, even with the increased losses, this category remained a profit center. Indeed, after removing expenses—estimated at about 20 percent of premium income—the category of dwellings generated regular and substantial profits. Moreover, the regularity of losses within this class of risk, from year to year, may well have been more important than artificially high rates. The lack of variance helped Aetna to keep a regular stream of cash flowing into company coffers that mitigated unexpected losses in other sectors of its portfolio and served as a hedge against economic downturn and/or catastrophic events. Indeed, this broader perspective underscores Aetna's long-held strategy of conceiving the risk of fire in terms of its entire portfolio. Statistical evidence buttressed this perspective, as Aetna used experience tables to verify that rates should be higher in some categories, and to help it identify those areas in which the company was most profitable. Careful examination of the company's actuarial data from 1852 through 1888 suggests that Aetna systematically changed practices in those areas in which losses exceeded premiums—either by raising rates or by more carefully selecting risks; in those areas that were regular centers of profit, practices were kept consistent.[16]

Having established a rigorous system of classifying risks and analyzing loss, Aetna fostered its use by cultivating a work culture that connected traditional notions of middle-class manhood to disciplined risk taking. No longer was it sufficient for insurers to be men of good judgment; Aetna believed that good fire insurance men first had to buy into the company's corporate strategy—its scientific approach to danger. Echoing the tone and content of Horace Binney's 1852 speech, successive manuals published by Aetna argued that insurance was based in rationality: "Many regard the business of insurance as one of *luck* or mere chance . . . but no future contingency can be more confidently predicted, than the reliability of our success and stability, if we will but act in the future on the facts of our past experience." As Aetna utilized its growing actuarial record as a sales tool, the firm also would argue that fire insurance companies—and the men who ran them—succeeded because of their accumulated, quantitative wisdom *as well as* their excellent reputations cultivated over years of careful practice. An excellent character helped firms to offer security from fire, mendacious customers, and sloppy agents. Although reputation had long greased the wheels of commerce, Aetna began to describe the process of investigating men's characters as minimizing "moral hazard." Presented unceremoniously in a brief mention by J. B. Bennett in his huge guide to practice, this expression would come to embody the fire insurance industry's prescriptions about manhood, which extended beyond the parlors of well-built middle-class homes, beyond sobriety and prudence. Insurers had begun to emphasize that economic rationality was the base for business and manhood.[17]

As insurers began to emphasize moral hazard explicitly, they made manhood a more central issue in the insurance transaction, thus structuring how the industry solicited business, disciplined employees, and cultivated consumers. Aetna emphasized that its agents should be prudent and disciplined to attract the right kind of customer and to serve as a model of the ideal customer. Aetna admonished its representatives to seek out middle-class men, because they were better "risks." According to the company manual, such good risks—or good men—demonstrated business discipline and took responsibility for their own safety and that of their property. For instance, such men would not object to the element of Aetna's policies that prohibited insuring more than two-thirds or three-fourths of a property's value. Good risks would recognize that such clauses provided a disincentive for arson, and made insurance companies more secure. Indeed, good risks and prudent men sought insurance from reputable firms, whose stability ensured their security. Likewise, an implicit message resided within Aetna's pitch to its representatives and customers: property insurance was a necessity. It protected financial assets, including men's homes, the haven for a man's most valuable possession—his family. By emphasizing the need for insurance, Aetna indoctrinated its agents and employees, transforming them from salesmen into men that protected the middle-class social and cultural order. Likewise, the company sold consumers an ideal of the secure middle-class home with a man of character at its head. Furthermore, Aetna emphasized that purchasing insurance did not completely fulfill a man's obligations. No indeed; it mattered from whom a man purchased insurance. Men of character—the type of men who procured safety—would want nothing but the best product to protect their assets. They sought to transact business with other men who possessed similar qualities, who worked with companies that practiced business according to the certainty of statistics, science, and system, rather than chance or fortune.[18]

Insurers' emphasis on character also depended upon connecting attributes of manliness—rationality and prudence—with actuarial record-keeping. Aetna transformed its belief in rules and numbers into a moral dictum when it argued that price competition in insurance was unsound both rationally and economically. Indeed, it urged agents not to sell to men who wanted only a low rate of insurance, because anyone could buy insurance inexpensively from a less reputable and less stable insurance company that was willing to cut prices. By contrast, Aetna, with its prices based in the statistical record, did not cut rates simply to please a customer; Aetna chose to emphasize the safety of its assets over the cost of its policies. If it was irrational to purchase insurance from a firm without sound scientific practices, it was also immoral to begrudge a reputable businessman fair compensation. As the firm noted, "All reasonable men are willing to pay premiums for the insurance of their property against loss by fire that will be sufficient to meet the losses and yield the underwriter a fair return for the labor devoted and the capital employed and exposed."[19]

Aetna's logic had an ingenious circularity. According to Aetna, prudent men only purchased insurance from agents disciplined enough to follow the actuarial tables created by Aetna's underwriters. Conversely, any man who bought from a firm that ignored its statistical record was imprudent, just as that company and its representatives were undisciplined. Aetna would not want a man who exhibited such behavior as a customer, though it hoped to educate him to behave more appropriately. Moreover, these precepts applied to Aetna's agents as well. When Aetna's representatives sold insurance using this logic, they sold Aetna's policies *and* its philosophy that manhood was connected to the rationality of numbers. In so doing, Aetna urged insurers and consumers to embrace an era of managerial capitalism that was rapidly transforming the American landscape. Moreover, this new definition of business manhood emerged after midcentury, when fears about confidence men, disorder, and a risky economy, as well as increasingly impersonal social relations, were reaching a zenith.[20]

Near a Fire: Say! Just hold this while I fetch another section, will you (Likely?) (n.d.), from the series of four satirical prints published by Henry G. Harrison and William N. Weightman as *The Fireman.* To promote firefighting reform, Harrison and Weightman drew a series of prints that ridiculed the effectiveness of volunteers, here illustrated by the limp stream of water and the elaborate dress as well as the plea for assistance from a passing gentleman. Courtesy, Library Company of Philadelphia

Aetna's message was both a clever sales pitch and an adroit strategy to avoid price competition, and it also made safety into a consumer product that connected security to the expansion of middle-class cultural ideals. Aetna's customers did not just purchase protection against fire, they bought the knowledge that they were behaving as prudent middle-class men. More importantly, perhaps, consumers began to buy into a new approach to fire risk—one based in economic activity rather than physical labor, one based on individualized protection rather than their communal effort. For a fee, insurance companies collected risks and distributed them throughout their portfolios. By managing those risks effectively, underwriters confronted fire's danger with intellectual labor, information technologies, and an elaborate bureaucracy. Implicit in this approach was an ideal about manliness that was connected to middle-class manhood, but which was far more rationalized than the standard gender norms of the era. Of course, purchasing an insurance policy did not obligate Americans to buy into to such beliefs, just as appealing to the immutable laws of science did not assure the expansion or use of quantification within the insurance industry. Actually the creation and use of a classification system was but the first step in making safety and the business of fire insurance more rational and economically viable. This process intensified as the expansion of the industry fostered a new and complex matrix of business associations, and relationships—not to mention new ways of viewing the built landscape.[21]

Maps

When Binney spoke to the Contributionship, he argued for bringing the built environment under an expansive program of surveillance that attended to the most minute details and that never ceased. He urged an audience that already performed systematic surveys of risks to observe the landscape yet more closely: "New circumstances are constantly occurring to increase or diminish the risk of fires, and all of the phenomena should be constantly and regularly observed." According to Binney, underwriters should inspect the built environment's most minute details and should "keep company with all the changes in the place, its extension, the heights and materials of its buildings, the merchandise contained in them, . . . the nature and management of dangerous trades . . . in fine, daily, and regular observations should be applied to everything that can be supposed to affect either the occurrence of fire or its intensity." To a large degree, Binney's recommendations vis-à-vis creating a program of systematic observation already were contained in basic fire insurance when he spoke in 1852, and Binney offered no concrete suggestions on the best methods through which underwriters could intensify their surveillance. Even so, he performed the important task of connecting surveys of risk and programmatic observation to practices that were more actuarially sound.[22]

Although Binney had not made specific recommendations on how best to observe the landscape when he spoke in 1852, underwriters already had been experimenting with the information technology that would reshape their ability to bring the urban landscape under systematic surveillance—maps. As we have seen, prior to the 1850s, agents sometimes included drawings in insurance surveys for simple illustrative purposes. Later, many firms asked field representatives to return policy applications with surveys, which often included written physical descriptions and sometimes even hand-drawn maps outlining property boundaries. On occasion, agents even depicted structures immediately adjacent to a property and commented on them. By the 1840s, Aetna had begun to ask agents to draw maps that referred to physical details about construction on the back of policy forms. Even so, such diagrams served little analytical purpose; rather they offered visual reference about the position of a risk vis-à-vis its immediate surroundings. Thus, as underwriters refined their categories of analysis to better quantify fire danger, they also began to reimagine the spatial component of fire risk. By placing the problem of fire into spatial context, complete with construction details and other minutiae of the infrastructure, insurers represented potential hazards more comprehensively.[23]

By midcentury this kind of information appeared on a new type of map, produced by specialists, not field agents. Surveyors adopted the insurance industry's understanding and interpretation of fire risk and graphically represented this risk in collaboration with fire underwriters. These maps were not value-free observations about the danger of fire; instead, they embodied insurer's notions of safety, profit, and loss. Atlases reflected underwriters' growing obsession with space and helped reshape fire insurance practice. Rather than chaotic accumulations of danger, cities appeared as an aggregation of many *interrelated* fire risks. The maps further sensitized underwriters to the classifications widely used throughout the industry. Insurers vividly saw the built landscape as color-coded individual hazards, and as districts of greater and lesser risk.

In the early 1850s, a few New York fire insurance companies began to compile information for a map of urban fire risk. The secretary of New York City's Jefferson Insurance Company, George Hope, commissioned a map of the business district for company use in 1852. Shortly before this was completed, William Perris, an architect and civil engineer, approached the company about preparing a map of the entire city. Hope agreed to the proposal and established a committee, comprised of four underwriters, which assisted Perris in constructing the map's format, symbols, and evaluative criteria. By November 1852, Perris surveyed and published the map, which was sold to interested insurers in New York and elsewhere, including to INA in Philadelphia. The practice quickly spread to cities across the nation, and several mapmaking companies sprang up to conduct the new business. In 1857, Ernest Hexamer, who had worked with Perris in New York City, created a map of Philadelphia; in 1859, Western Bascome surveyed St. Louis; and in 1867 Daniel Sanborn represented Boston graphically.[24]

A complex set of social relationships structured the production and use of maps. Commissioned by or drawn in collaboration with underwriters, insurance surveys generally reflected the criteria with which underwriters set rates. Fire insurance maps identified a structure's use—as a dwelling, manufacturing facility, store, or warehouse. Simultaneously, surveyors also depicted construction material. In his 1852 map of New York City, for instance, Perris represented the city in terms of the four classes of construction typically used by underwriters: brick, stone, mixed brick and stone, and frame. Further, over time, map makers employed an elaborate variety of symbols and colors that gave insurance maps the appearance of patched quilts, showing regions of greater and lesser hazard.[25]

Urban landscapes were not the only spaces being reconceived by surveyors and underwriters; industrial landscapes—whether they were isolated or part of an densely populated area—came under the scrutiny of the insurance industry. As with all underwriting practices, companies encouraged agents to include drawings of these sites as part of the survey. Many companies recommended that their agents hire a surveyor to draw such a map, at the expense of the firm seeking insurance. In fact, the Philadelphia Board of Fire Underwriters issued instructions in 1857 that listed drawing a map as one of the many requirements of good underwriting. Not surprisingly, Hexamer and other surveyors began to publish maps of industrial risks as a service to their clients. In 1866, Ernest Hexamer published a map of "special risks" in Philadelphia for the first time; the Whipple Insurance Protective Agency generated similar maps of St. Louis in 1872. At the same time, mutual insurance companies, formed by manufacturers, which had a difficult time obtaining insurance from stock insurers, produced maps that outlined the structural details and fire risks of the properties they insured. Typically, maps of special risks included detailed two-and three-dimensional drawings, as well as detailed verbal descriptions of the property and its fire hazards. Just as insurance maps demarcated "special hazards," thereby drawing particular attention to them, by the 1870s surveyors provided the industry with another more specialized and detailed surveillance tool—atlases comprised exclusively of special risks.[26]

Early maps differed widely for reasons other than subject matter and despite the fact that mapping firms represented fire risk using insurance industry categories. For instance, Bascome's map of St. Louis used much the same coloring patterns and map key as Perris's survey of New York City. However, Hexamer, who had worked for Perris in New York, represented Philadelphia using a different color scheme and classification system. In part the variation in Hexamer's maps may have reflected differences in how Philadelphia and New York insurers perceived the risk of fire. For instance, Hexamer's map represented Philadelphia explicitly in the terms and categories of a standard schedule for evaluating property being developed by a consortium of companies in Philadelphia, the Philadelphia Board of Fire Underwriters. It divided "regular" risks such as stores and dwellings into four categories color-coded by both use and construction: "Brick or Stone Stores," "Brick or Stone Dwellings," "Brick or Stone Dwellings with Stores under," "Frame Dwellings," and "Frame Dwellings with Stores under." These risks were represented using an array of broken, dashed, and whole lines alongside brilliant colors. In addition, Hexamer included a category of "specially hazardous" risks, which he classified according to the danger assessed to them by the Philadelphia Board of Fire Underwriters. For instance, "Bookbinders" and "Brass Founders" were both in the "second class" of specially hazardous risks, while "Bakers" and "Tobacco Manufacturers" were among the "first class." Later versions published by Hexamer would drop the insurance industry's technical nomenclature ("first class hazards") in favor of a simpler descriptive language when describing construction methods (i.e., brick, stone, etc.).[27]

Surveyors' categories for evaluating fire risk expanded during the last half of the nineteenth century. Map keys grew from one page to two pages and demonstrated how underwriters increasingly engaged in detailed surveillance. Such representations became increasingly specific as underwriters' categories for evaluating risk grew more abundant. Early maps listed the width, depth, and height of buildings as well as skylights, boilers, and roof construction. Later maps (those published after 1870) came to include information on street length and width, wall construction, fireproofing, breaks between buildings, shutters, thickness of walls, and other architectural details. Such simple descriptive language provided information that the insurance industry used to categorize risk, or subtly adjust rates. By the 1880s, most maps also diagrammed aspects of the fire protection infrastructure, such as fire plugs, water mains (and their size), and alarm boxes.[28]

Occasionally surveyors experimented with novel ways of representing fire risk. In St. Louis, C. T. Aubin, for instance, expressed the increasingly varied structural character of the city by creating a map that expressed its three-dimensional reality. In 1874, when the St. Louis Board of Fire Underwriters commissioned a map from Aubin, he produced a map similar to the industrial surveys that he completed for the board. The map contained all the information normally identified. However, Aubin's map attempted to capture the increasing variation in building size and structure by using shading to create depth. Where other surveyors simply identified a building's height and number of stories, Aubin represented it graphically. His map's key included a discussion of the map's format and symbols within the context of a three-dimensional representation of a city block. Next to each three-dimensional representation of a structure stood a two-dimensional guide to its portrayal on the map.[29]

The industry incorporated fire insurance surveys into everyday underwriting practice gradually, over twenty years. A communication tool between underwriters and agents located in distant cities, atlases helped representatives at a home office to evaluate premiums. They also encouraged the standardization of information-gathering procedures and risk evaluation. Lastly, as a record-keeping device, insurance maps helped underwriters to manage employees as well as their risk portfolios. By pulling out its map of St. Louis, a company could determine its exposure in a particular neighborhood or block and gain quick access to a record of all its policies, their size, and effective dates on a single form. Indeed, examining original fire insurance maps (as opposed to microfilmed copies) reveals that underwriters and surveyors used the large folios intensely, and on a daily basis.

Historical inquiry about the origins of these maps has been dulled by decades of using them for other purposes—for illustrations, occasional reference tools, or to get students interested in history. Using maps for these other purposes has obscured the context of their creation. Fire insurance maps were produced as a vital tool in the context of the development of the fire insurance industry and its battle to comprehend fire danger. Most obviously, insurance maps provided the basic information about structural features of the built environment that underwriters used to set rates. They identified how a structure was used, how it was constructed, and how it was related in space to nearby buildings. Moreover, mapmakers obliged clients' demands for constant updates. Agencies continuously inspected and regularly resurveyed the built landscape, producing a constant stream of new information. As a result, maps changed frequently, evidenced by the pasteovers covering them. Insurers kept records of the pasteovers, frequently noting them on front and/or back covers or in the frontispiece of the volume. In a sense, the maps were so frequently updated that they stopped being mere representations; they literally embodied the dynamic urban environment.[30]

Both underwriters and agents used the diagrams of the spatial arrangement and material construction of various blocks and neighborhoods to manage their business. By the 1870s, insurance manuals advised field representatives and officials to mark the policy numbers onto the maps, and many display penciled-in numbers consistent with the industry's policy-numbering schemes. In addition to listing policy numbers, underwriters also wrote the dollar value of policies, as well as their effective dates, onto the maps. If a policy was not renewed, then underwriters typically erased the information, sometimes leaving the smudge of a number. Even as these atlases offered insurers a way of keeping track of their risks, they also provided a common connection between agents and insurance company officials—a point of reference for discussing risks and business activities. If field representatives and underwriters sometimes had different interests, insurance maps reinforced the industry's shared routines. Insurance atlases, like other regular business tools, helped to forge the beginnings of a common community of fire insurers.[31]

Insurance maps did more than help to transform administrative routines; they remade underwriters' cognitive understandings of fire risk and space, as the writings of Aetna's special agent suggests. In fact, examining A. A. Williams's diaries offers insight into the how underwriters approached the spatial component of fire risk, how they "read" cities, and how they interpreted insurance maps. Cities became complex assemblages of risk; construction details, property uses, architectural features, etc., had new implications when viewed in relationship to other structures and in the context of the problem of fire. Safety and danger were not just numerical representations of isolated construction techniques but inhered in the very design of cities.[32]

Between 1855 and 1857, when A. A. Williams resumed the company's practice of visiting its agencies, he traveled from New England to Quebec inspecting Aetna's risks and built landscape. Like his predecessor Joseph Morgan twenty years earlier, Williams visited the location of each of the risks underwritten by Aetna, and he commented on the risks taken and the rates charged by each local agency. For instance, at 803 Point-Leve in Quebec City, Williams noted that a policy worth $4,000 was in force. He further remarked that the "part of the risk on Dwelling & Paint Shop—very good—Office & Paint in basement. Storm basement, upper part frame; good new building not much exposure. I say 1 %. The outside risks from 1½ upwards." Yet, Williams's tour differed from Morgan's in important ways. Whereas Morgan had provided only general commentary, Williams offered both a global assessment of the landscape and meticulously documented each of the company's policies in a particular place, such as Quebec City. Second, when Williams passed through space, he evaluated city streets, block by block, and studied manufacturing facilities. Moreover, he connected them within a holistic perspective on the safety and danger of a particular city's landscape. Williams read urban space as if it were represented on a two-dimensional insurance map. Indeed, his descriptions of the built landscape could be mistaken for a graphic depiction of Quebec, much like those available in New York or Philadelphia.[33]

Williams examined changing urban environments according to the invisible logic that organized early insurance atlases. As evidenced by his critical evaluation of Quebec's streetscapes, he adopted the language and physical representations that were becoming standard on fire insurance atlases. His written physical descriptions of the use and construction of buildings varied little from the categories and evaluations encouraged by Aetna's manuals: "Fabrique [*sic*] Street. 17 risks . . . $94,600 . . . Building stone, part wood & part tin covered from 2, 3, & 4 stories high; occupied principally for first class stores; one or two old buildings in this street; avoid them. The risks have been generally fine. Open south upper part of the street; market and open ground in front—French Cathedral stands at the east on upper end of the street." In other instances, Williams recorded the specific hazards that threatened manufacturing facilities. He especially included detailed spatial drawings, not to mention written descriptions, of industrial property. In his notebooks, Williams outlined structures, noted building sizes—both the number of stories and the physical dimension—and diagrammed the distance between structures. He often included drawings, descriptions, and conclusions about rates. These maps depicted the arrangement of the buildings associated with or nearby the factory, and identified building use and construction details. Though Williams's drawings did not use the key or colored ink recommended by Aetna's manuals, his maps' physical characteristics came directly from the sample diagram that the company provided in the 1857 manual.[34]

What is perhaps most striking about Williams's diaries of his trip through New England and to Quebec was the degree to which he brought all the contingencies facing mid-nineteenth-century underwriters together. He did not just describe spatial arrangement and physical construction of risks; Williams included commentary about the company's agents and agencies. Especially illuminating were his discussions of those agencies whose premiums were not high enough. In at least one case, in Wilmington, Delaware, Williams recommended closing the agency, and he "took possession of the books." In another instance, besides describing the structures and the boundaries of the block, he expounded on whether the local agency should pursue further business at that location: "Arsenal Street: one risk in all, $1100, old dilapidated stone buildings, shingled with wood. No more wanted in this street, and the risk now pending may be discontinued unless the rate can be advanced. . . . Smith Street: One risk, $1000—do not renew at any rate and avoid the street by all means—old one story frames in this street entirely. How could you write in this street?" In his discussion of Arsenal and Smith Streets Williams explicitly tied the growing sophistication of Aetna's everyday business practices—including presumably the actuarial tables the company was developing—to the landscape. When Aetna incorporated insurance maps into its repertoire of technologies, it added further depth and intensity to its growing surveillance of its field representatives, its portfolio of risk, and society more broadly.[35]

Reading Williams's diaries is virtually the same as reading a fire insurance map. Williams conceived the city as a collection of discrete, definable dangers, all interconnected. His goal was to distinguish between them, and to write policies on the safest risks. His reading of the physical landscape gives us a hint of how underwriters perused the colors, lines, and symbols of the atlases authored by Perris or Hexamer. From these atlases, underwriters could determine use, construction materials and details, and the spatial arrangement of a block or neighborhood—all the information needed to assess safety and danger, and to quantify a rate. Similarly, Williams's written description takes his readers through the alleys, principal streets, and residential neighborhoods of cities alerting them to the arrangement of space, construction, and use of structures in each block and district. Of course reading his diary achieves an even greater effect when viewed in conjunction with any map of the area.[36]

With the advent and expansion of the insurance mapping business between 1850 and 1880, underwriters added another layer to their everyday practices and acquired another instrument of technology with which to understand the problem of fire. Like statistical record keeping, maps supported the further development and codification of classification systems. Moreover, they helped to embed the industry's classification technologies into the minds of underwriters, thereby reshaping everyday practice. Maps also established a standard point of reference within disparate sectors of the industry. Perhaps most significantly, however, fire insurance atlases began to reenvision the built environment within a coherent intellectual framework—a framework that divided cities into categories of greater and lesser danger. Implicitly, such atlases became city plans that reordered American cities according to the problem of fire.

Trade Associations

As underwriters retooled their business practices, they formed new cooperative relationships that reflected the development of a common interest and shared practices within the industry. The formation of trade associations developed especially from the increasing standardization of everyday practices, and as a way of sharing resources. Once again, Horace Binney had predicted the direction taken by the fire insurance industry when he recommended that firms work together in order to execute the comprehensive observations so important to developing statistical data on fire loss. Moreover, Binney claimed that industry unification would help to create a rigorous actuarial record and lower the costs of gathering information, all the while producing the benefit of more robust data. Even though most insurance trade associations ultimately did not facilitate the development of such grand actuarial tables, they attempted to use their leverage to improve business conditions and especially to improve industry surveillance—statistical and cartographical—of the landscape. Most associations remained local or regional prior to the Civil War, yet even these relatively limited cooperative arrangements often had startling results. Insurance companies met with many difficulties and few successes as they attempted to organize themselves in the decades surrounding the war, but their institutional alliances represented important first steps toward making the assessment of risk more rational and developing a broader common interest within the industry. Indeed, despite the limited power of early organizations, such associations laid the groundwork for the industry's eventual success in implementing a common agenda later in the nineteenth century.[37]

Fire underwriting associations developed as early as New York's Salamander Society in 1819, but no permanent industrial associations appeared until the 1850s. In 1846, following New York's Broad Street Fire, the first National Congress of Fire Underwriters met to examine past experience, to set rates, and to develop measures to better regulate the industry. Like the Salamander Society, though, the National Congress proved ephemeral, and disappeared in 1850. It would not be until the National Board of Fire Underwriters formed after the Civil War that a national fire insurance trade association became permanent. In the intervening years, though, several permanent local associations were formed and intercompany cooperation increased over a spate of issues, including setting rates, mapping cities, and collaborating to implement changes in fire extinction.[38]

As they formed during the 1850s, local fire insurance associations emphasized categorizing risks, assessing them quantitatively, and mapping them, which underscored the industry's growing awareness that fire danger could not be managed without a thorough accounting of fire danger in space and time. Such associations promised to create a common, collective experience that would make business activities, especially the process of setting rates, more certain. In the 1850s, at least, the promise of increased profits through such cooperative methods bolstered support for the new collectives. The first permanent fire insurance trade associations were established between 1852 and 1854 when underwriters in Philadelphia,

Louisville, and Cincinnati met to establish common rates. By fixing rates, insurers in each city hoped to circumscribe market competition. By keeping rates artificially high (i.e., not allowing the market to set the price of insurance), companies also expected to guarantee reliable profits and to protect against insolvency, which frequently accompanied large fires.[39]

Local underwriting associations attempted to forge a common interest among insurers by sharing the information that had become so central to their daily business transactions. A commitment to tabulating a record of their members' combined experience and to using maps to represent urban fire risk became the cornerstones of such organizations. Just days after Horace Binney challenged conventional wisdom at the Philadelphia Contributionship's anniversary, several Philadelphia underwriters met and formed the Philadelphia Board of Fire Underwriters. The PBFU consisted of one representative from each company, which included all but a few of the city's major firms. The association organized itself into four committees, which directly supported the industry's agenda: a committee on statistics, a committee on classifying risks and establishing uniform premiums, a committee on surveyors, inspectors, and appraisers, and a committee to create a "fire police."[40]

Toward the goal of ceasing competition on rates, the Philadelphia Board of Fire Underwriters sought to unify the city's different companies under a common system of rating fire risks. In order to develop a common program of rates, the PBFU first ascertained the prices charged by different companies and then reconciled them into a single set of categories, with attached premiums. Within months of its founding, the PBFU established a set of categories compiled as "Classes of Hazards and Rates of Premium for Insurance against Loss or Damage by Fire in the City and County of Philadelphia." Through combining the experiences of all its members, the board hoped to accumulate a more objective record of fire risk in the city than was available to its members individually. In addition, it would provide a more complete analytical profile of fire risk and supervise surveillance of the city. Toward these ends, the PBFU provided a comprehensive list of categories and classes of hazard. In addition, the organization kept tabs on hundreds of the city's so-called special risks—its manufacturing facilities and warehouses. The ninth article of the organization's constitution codified this system by demanding that member enterprises police one another. When members encountered other firms violating the board's rating agreement, the organization obligated them to report that company. The board would then discipline the noncompliant firm.[41]

Although this differed in content from Aetna's mode of tabulating losses, in practice it accomplished a similar result. In theory at least, the PBFU's rates represented the collective record of fire loss of its member companies, especially if those firms had begun to link their rate-setting strategies to an assessment of their administrative costs, as well as quantitative records of loss. Although certainly distorted, such data nonetheless promised to offer PBFU members an invaluable wealth of knowledge about fire risk, and their competitors. For instance, by perusing this data, firms could acquire a better sense of how their own loss histories on particular types of properties compared to those of other companies. Plus, an organization could determine how its administrative costs differed from other firms by comparing rates in relatively stable loss categories, such as dwellings. Most importantly, the PBFU's rates represented—at least roughly—the shared experience of many different companies, rather than the perhaps atypical experience or arbitrary pricing strategies of a single firm.[42]

Within a year of its formation, the PBFU intensified its surveillance of Philadelphia. In 1855, the trade association made the secretary of the board its surveyor and statistician. The secretary was charged with collecting statistics on fires, which included information on property loss, insurance loss, insurance paid, causes, character of the buildings involved, and how the building was occupied. He classified, surveyed, and mapped various buildings at his own discretion and at the request of individual companies; all of these surveys were to be recorded along with his notes in a ledger available to all association members. The PBFU also appears to have established a relationship with Charles Hexamer, whose firm began mapping fire risk in Philadelphia during the 1870s. Through regular inspections, the PBFU revised rates and reissued guidelines that demonstrated its sensitivity to the city's changing built environment. In addition, the board's surveillance of the city extended to members' activities. Published rates, even if not always followed, subtly encouraged companies to follow local industry practice, and the organization also established common rules for issuing policies.[43]

Over the next decade, the membership of the PBFU waxed and waned—in direct relation to the severity of fires within the city. Even though its ability to enforce common rates often failed, the PBFU provided an invaluable service to the city's underwriters. Through the 1850s, and into the 1860s, the board provided companies with a steady stream of information regarding the incidence of fire. In 1856, the PBFU led city underwriters to sponsor the fire-detective police—a forerunner of the modern fire marshal—in order to curb arson. Led by Alexander Blackburn, the fire detective police compiled lists of the causes and costs of fires similar to those statistics kept by the fire department and the PBFU. In the next year, at the behest of the board, Charles Hexamer drew an insurance map of Philadelphia in which he adopted the PBFU's categories of analysis. As the PBFU became a conduit of information, it helped Philadelphia's insurance companies both to manage their business more effectively and to gain a greater understanding of how to evaluate the dangers posed by fire. Eventually, following the Civil War, local insurance organizations, like the PBFU, would become the model for regional and national underwriting associations as well as the glue that bound the fire insurance community together.[44]

Morality, Technology, and the Organization of Firefighting

When Horace Binney spoke to the Philadelphia Contributionship he did not criticize firefighters, but he suggested that insurers question every facet of fire protection, including the work of those actually fighting the blazes. Binney's recommendation that insurers keep up-to-date regarding the "character of firemen and their apparatus" suggests, too, his awareness of the brewing controversy over the use of steam technology in firefighting and questions about volunteers' ability to keep the peace. If Binney did not offer a direct critique of firefighting, his long discourse on how to change insurance practice provided an alternative vision of fire protection. If both firefighters and underwriters were specialized work communities staffed by male risk takers, the similarities ended there. In fact, the new work routines of underwriters differed dramatically from the physical labor of firefighters. As insurers quantified company loss histories, mapped space, and standardized bureaucratic routines, they objectified the danger of fire as an economic abstraction. No longer a purely physical danger, the problem of fire could also be managed through organizational procedures and information technologies. Moreover, the insurance industry advocated an approach that held distinctively middle-class notions of propriety, which did not mesh with the complex and contradictory culture of volunteer firemen. In the end, however, perhaps the point at which underwriters' approach to fire danger most differed from firefighters was where it located control over fire protection. The fire insurance industry articulated a vision of safety that prioritized economic and contractual communities over face-to-face social relationships based in geographically or socially determined communities.[45]

Quite suddenly, insurers and property owners who had once supported volunteer firefighters became their most vocal critics during the 1850s. They began to argue that fire extinction was a potential cost liability and demanded that fire protection be organized in a more economically rational fashion, akin to the organization of factories. Reformers agitated politically for change by a creating a powerful economic coalition of property owners that included merchants, manufacturers, and insurers. Of course, discontent with volunteer firefighting had been percolating since the 1840s, when underwriters, property owners, and even some firemen had begun to express reservations about the organization of firefighting labor. Although such dissatisfaction had been dispersed across many cities, it presaged later more focused efforts at reform. Led by the fire insurance industry, lingering concerns about volunteer firefighting exploded into full-blown opposition, and fire underwriters articulated a forceful case for reform that contributed to a shift in the provision of fire protection.[46]

Philadelphia's underwriting community mounted a campaign to reorganize the city's firefighting that would become a model for reformers nationwide. Insurers helped to convene and to lead a committee appointed by "a very large meeting of the citizens of Philadelphia favorable to the introduction of a Paid Fire Department." This group, which referred to itself variously as the "committee of twenty-five" or "committee of citizens" (hereafter referred to as the citizens' committee), orchestrated a methodical campaign that "urged the necessity of abandonment of the present voluntary system." The secretive gathering included a broad spectrum of property owners and the city's economic elite. Interestingly, the first meeting of the cloistered committee coincided with Horace Binney's speech and the formation of the Philadelphia Board of Fire Underwriters. Perhaps more important, this group crafted a critique of volunteer fire companies that united three disparate and seeming discontinuous strands of thinking—cost-benefit analyses, scientific objectivity, and moral reform. The committee's efforts spawned or influenced debates in other cities even as it led to the prototypical showdown between Philadelphia's business community and its firemen.[47]

The citizens' committee argued that the system of volunteer firefighting jeopardized the city's future and that fire protection should be reorganized. Generally, it noted that firefighting should be organized along the lines of industrial society more broadly, in which each person performed a particular task. As the committee declared, "The business of protecting the community from loss by fire should as properly be a special business as that of the police or any other occupation." Besides positing a general position about social organization, the coalition also argued that the volunteer fire department increased the costs associated with living and doing business in Philadelphia and created a climate inhospitable to economic development. In this connection it cited "frequent and disastrous fires" and "the disorder and violence manifested by a portion of the firemen." Such problems, the group complained, dissuaded "strangers from making their home and spending their income among us." If disorder discouraged entrepreneurial investment, so too did the potentially high cost of doing business that resulted from inadequate fire protection. The committee argued that because too much property was destroyed, insurance premiums were raised, and that, because of the police and courts necessary to control riots among firemen, taxes were increased to support municipal government.[48]

The committee produced a remarkable amount of evidence to support its case, most of which was generated by Philadelphia underwriters and commercial interests. It argued that the financial and moral benefit of reorganization far outweighed any costs associated with establishing a municipally controlled, and paid, fire department. The increased expenses of outfitting and maintaining such a department could not compare to the large financial outlays that Philadelphia residents paid in high insurance rates. Impugning the performance of volunteer firefighters, the committee complained that local insurance firms expended an extraordinary sum to cover fire losses at industrial locations; "so much so," the committee reported, "that even the high rates of premium charged for this description of risk has fallen far short of repaying it. One office alone, we are credibly informed, has paid within that time [the last two years], the immense sum of three hundred and fifty of thousand dollars!"[49]

The committee also generated detailed cost-benefit analyses to illustrate the advantages of reorganization. Both Philadelphia's board of trade and its fire underwriters produced a cost-benefit analysis during the early 1850s. The board of trade's analysis revealed that insurance rates more than doubled from 1832 to 1852. This report contrasted the average rate of 1852 premiums, which was about $.80 (per $100 of insurance, depending on the type of risk), with average rates twenty years earlier, which it calculated at $.225. Next, it estimated (conservatively, it said) that about one million dollars in property was insured in Philadelphia. According to this calculation, the increase in premiums between 1832 and 1852 cost local industry $575,000. The representatives of the board of trade further argued that this "tax" had been irretrievably lost. Moreover, it had "not been productive of the slightest good, but on the contrary . . . the whole amount has been irretrievably lost, and . . . the loss is the fruit of a system deplorably pernicious to all the other interests of the community." A pamphlet issued by the insurance industry provided yet more dire quantitative evidence to bolster the case of the citizens' committee. It stated that premiums tripled in a single decade; in one instance, rates paid by a bookseller were said to have increased from $.30 in 1844 to $1.75 in 1852. Thus the reports generated by Philadelphia's commercial interests left little room for doubt—the benefits from reorganization outweighed the costs associated with reform.[50]

The coalition of reformers and its backers in Philadelphia's financial community did not end their attack with the cost-benefit analyses. No indeed. They transformed their economic understanding of the incipient industrial order into a moralistic commentary on volunteer firemen, in rhetoric that reflected the values that the nascent middle class held about manliness, family, and work. In a section of its report titled "The character of fire companies, and the causes of disorders," the committee remembered the noble, heroic, and public-spirited duty of firemen past and commended the sentiments of "a large proportion of Philadelphia firemen." However, in the context of the changing demographic character of Philadelphia's volunteer fire department, the committee's assessment seems less a sincere recollection than an expression of middle-class fears of the growing power of immigrants and workers. Perhaps most importantly, the committee especially mourned the loss of face-to-face social relationships: "We have become so numerous that it is impossible for us to have knowledge of each other that may be, and is had in small towns." The only remedies to the social problems and transformations wrought by industrialization and urbanization lay in the introduction of new social values. The values of industry and economy, the committee argued, should replace traditional social arrangements.[51]

Reformers were especially critical of the manliness demonstrated by firefighters. Though it did not doubt firefighters' seriousness of purpose, the committee complained about how they expressed their zeal. In addition, it argued that the volunteer fire department had passed from the hands of "men . . . into those of minors—half grown boys" who delighted in ruffianism, fighting, and other extralegal behaviors. Although the report did not document instances in which boys performed firefighting work, it did not matter, because to a large degree the point could be had in the metaphor it provided. Once men had protected the city, but firemen were no longer men. They were boys incapable of protecting the public safety in a sober, effective manner. This critique also offered an additional attack on the boisterous culture of firemen—whom many in the middle class perceived as working-class thugs. Firemen, the committee insinuated, were not men enough to govern their passions; if they could not contain themselves effectively, how could they control nature? The committee drew upon values expressed by the middle class, which increasingly, as the Victorian age progressed, believed that men should govern their emotion and self-expression.[52]

As debate about the volunteer fire department grew more heated, two artists sympathetic to the reformers' message graphically depicted the shortcomings of volunteer firemen. In 1858, Henry Harrison and William Weightman published a series of four satirical prints, which they titled *The Fireman.* In these scenes, firemen fight fires wearing full parade regalia and display foppish manners. At a time when the middle class eschewed extravagance, such pomp and costume seemed indecorous and even ridiculous. In addition, the prints displayed evidence that firefighters were unable to govern themselves with decorum and modesty—underscoring their inability to control the powerful forces of nature. The satire acquired further power by depicting firefighters as absolutely incompetent. Whether knocking people over or misdirecting hose streams, the firemen bungled basic tasks, and wrought havoc. One print contained a more sinister subtext; *What Boys May Expect When They Get in Firemen's Way* shows a child endangered by a fireman's buffoonery. The city's merchants and underwriters expressed this same sentiment when they wrote that "just as the influence of the young is injurious to the system, so is the influence of the system injurious to them."[53]

Middle-class critiques of firefighting gained their widest distribution in a novel by H. C. Watson, *Jerry Pratt's Progress; or Adventures in the Hose House.* Originally published in the *North American and United States Gazette* during 1853 and 1854, Watson's story argues that creating a paid department was a moral necessity. Jerry Pratt, an unskilled farm boy from Bucks County, journeys to Philadelphia to find work and a future. There he begins a promising life as an industrious apprentice in a tailor's shop, only to be corrupted by the glamour of shiny firefighting equipment, the dandyish swagger of his newfound friends—the "b'hoys"—and the "rowdyism and coarse fun of the engine and hose house." Impressionable and young, Jerry loses his innocence; he begins to smoke, to drink, and to fight. Even the beautiful vision of pure womanhood, Becky, cannot alter the terrible path that leads to his ruin and death. The argument is that, not only were women unable to domesticate the men and boys who served in fire companies, but also engine houses removed those men from the sphere of women's influence—the home. The rough-and-tumble engine houses turned good men into "drunkards, rioters, incendiaries and even murderers." It is helpful to note that Watson made these arguments at a time when *Harper's* reported that "no civilized man is so helpless and dependent in certain respects as an American gentleman, and the reason is obvious: our wives do our thinking."[54]

For the most part, Watson's moral critique did not expressly identify volunteer firefighters as the problem. Although the story is set against the backdrop of a "hose house," it was not actually firefighters who corrupted young Jerry Pratt. Alcohol, tobacco, a dandyish obsession with costume, and a youth gang ruined him. It was these aspects of the story—often associated with working-class community life and appearing in nearly every critique issued by reformers—that Watson emphasized. Thus the relative absence of volunteer firemen (except for their apparatus) suggests that the morality novelist was hesitant to pin Pratt's debasement on volunteers specifically. Perhaps he was among those who feared alienating the thousands of volunteers charged with protecting their property (which perhaps also helps to explain the secretive nature of the committee of citizens). More likely, though, the dearth of firefighters suggests reformers' broader political and social aims. Like other middle-class critics, who criticized firemen's style, not their efficiency, reformers demonstrated little interest in the particulars of fire protection and more in promoting their own agenda. Claims about disorder provided a rhetorical means to assert the values of industrial economy over a society in which diverse local communities held considerable social, cultural, and political power.[55]

An echo of the political tracts distributed by the city's business interests, Watson's novel linked his "true" stories to economic arguments supporting the creation of a paid fire department. In fact, in the preface, Watson restated reformers' cost-benefit equation, noting that municipal control of firefighting would be economically beneficial. Yet, in the novel's final sequence, as Watson once again presented lurid images to his scandalized readers, the lines between morality tale and political tract blur. Watson laid responsibility at the feet of the city's citizens, when he beseeched them to make the changes he advocated: "Fathers and mothers, if this system is continued, you are responsible, for with you rests the power of its annihilation. What is your reply?" With rhetorical flair, Watson presented his readers with the political problem of the day—should Philadelphia support the creation of a municipally controlled fire department? He argued that by supporting order, discipline, and regular behavior, Philadelphia could avert the moral disaster that would come from prurient impulses of physical, working-class culture. However, financial good sense and middle-class morality could protect the city from fiery disaster.[56]

According to the reform coalition, a solution to the spiraling cycle of immorality, growing fire danger, and fiscal decay could be had by adopting the specialization of function characteristic of industrial labor. The committee reported that the "business of protecting the community from loss by fire should as properly be a special business as that of the police, or any other occupation; as it would be more economical to the community." The pamphlet also captured the individualistic ethos so central to the capitalist economic order, suggesting that people should not expect to be shielded by broad communal efforts. Further, critics argued that there was a "special propriety, and obvious economy in every man adhering to his own." As this statement underscored the importance of workplace specialization, it may have possessed also a subtle encouragement to individuals to protect themselves by purchasing fire insurance. Ultimately, however, the citizens' committee focused on transforming the system of fire protection, making three interrelated suggestions: reorganizing the fire department, disciplining firefighters, and adopting new methods and equipment for putting out fires. Almost simultaneously, the municipal government considered the possible role that two new technologies—telegraphic alarm systems and steam fire engines—might play in fire protection, debates that included at least some input from firefighters.[57]

Adopting industrial strategies, the committee argued, would correct the inefficiency that they found everywhere in the city's fire department. It questioned the costs associated with firemen dragging their engines a long distance, workers—mechanics and clerks alike—missing work to fight fires, and the large contributions made by municipal governments and private interests (estimated in 1852 to have been more than $18,000 and $36,000 respectively). As if systematic inefficiency of the organization and financing of labor were not enough, the committee complained that large crowds and competing fire companies further impeded efficient extinction of fires. It reported, "There is a want of harmony and co-operation; a want of that order, discipline, and subordination, which ought to characterize their efforts, to render them effective, and which wastes the energies that ought to be directed by the skill and authority of a general." The work of fire extinction, according to the committee, had been neither performed nor organized in a rational fashion. Discipline and order could be achieved only by creating a special occupation.[58]

For guidance in disciplining firefighters, reformers looked beyond Philadelphia. In Paris, London, Boston, and Cincinnati they found what they expected—evidence that firefighting could be performed as wage labor under tight municipal control. Referring to those places, the committee argued that authority over firefighting should be lodged in a board of control, which would administer the fire department for the entire Philadelphia region. Elected by the councils of the city's regions, the board would select a chief engineer to operate the fire department on a daily basis. In addition, the group recommended that the board be given the authority to select equipment, locate engines, and generally exercise "direction and supervision" over the entire department. The consolidated city (consolidated around the needs of fire protection according to this plan) would be divided into several fire districts. An assistant engineer, reporting to the department's chief engineer, would be appointed to command fire extinction efforts in each section of the city. Candidates for the position would be nominated by the fire companies of the city's districts but selected by the board of control. By centralizing authority of this newly rationalized urban space into a single body, the committee challenged the autonomy of fire companies, and by extension their local communities.[59]

The committee feared that its program would be abandoned if the municipality consolidated into a single unit, as was being considered in Philadelphia about this time. The committee made it clear that the changes it recommended should happen in addition to municipal consolidation, but it advised that consolidation alone would *not* be sufficient to create the change of climate needed in order to make the fire department acceptably efficient. Although the committee acknowledged that unitary municipal control over firefighting was good, merchants and underwriters wanted nothing less than to rationalize all aspects of fire extinction. The committee reiterated its goal of restructuring the department into a single organized, hierarchical unit that stretched across Philadelphia.[60]

Although paying firefighters a wage for carefully prescribed work was central to the new departmental order, this was not the only recommendation made by reformers. They also argued that a formal and hierarchical bureaucratic structure between officers and firemen would lead to more orderly and superior firefighting. Each man would have an assigned duty and work in harmony with the other firemen at the scene—whether from his engine company or not. Competition between companies would be supplanted by a more universal esprit de corps. In addition, firefighting efforts should be commanded more efficiently. A corps of officers would direct well-trained men about where to use axes or apply water. As a result, firemen would no longer be subject to the disorder of the crowd, nor would those immoral elements among them be included in the new department. Strikingly, the committee hatched a plan for organizing firefighting that recalled the sentiments of the founding members of the Philadelphia Hose Company five decades earlier.[61]

The committee's program also included provisions for new equipment that promised to reorganize firefighters' work. In addition to recommending that fire departments substitute "horse power for 'human labor,'" the group especially emphasized using steam technology to extinguish fires. For a number of years, underwriters had supported using steam technology, and in the 1850s steam engines became a critical element in insurers' arguments for reform. Following the recommendations of insurers, the committee argued that steam-powered apparatus offered three primary improvements: steam engines pumped a more powerful stream of water than men could; they reduced the amount of labor required to fight fires; and they could extinguish fires using *steam* itself. In making these claims, the reform coalition looked to Cincinnati's use of steam technology as a model for Philadelphia. Using articles from two Philadelphia newspapers, the committee discussed demonstrations of steam fire engines in Cincinnati. It argued that those exhibitions had succeeded remarkably and had led to more efficient organization of firefighting. Each argument also reinforced the moral dimension of the committee's program and offered a stinging critique of the masculinity of Philadelphia's firemen. According to one report, during a contest with firemen, the steam fire engine threw "a vast body in a solid stream of two hundred and twenty-four feet." By contrast, firemen, using a hand-pumped engine, "bore down and up in quick succession, and strained every nerve, but gave up exhausted."[62]

Even as reformers criticized firefighters' lack of manliness, they argued that firefighters were considered too vital. For instance, the *Public Ledger* reported that steam alone could be used to extinguish fires, and could replace water. Using a section of rubber hose attached to the engine's boiler, steam was applied to the fire—"a vast volume of steam, sufficient to saturate air and penetrate into every crevice where fire could possibly lodge, completely extinguishing fire." Underwriters argued that the application of this technology would reduce water damage, which was an issue of particular importance. The first pamphlet published by the committee to reform the fire department had included a two-page diatribe on excessive water damage at fires. The committee related an incident in which firemen had caused thousands of dollars of unnecessary damage when they directed their hoses into the upper floors of a building with a fire in its basement. The group further related several examples of wasteful water use on property "not really in danger." It concluded that firemen behaved according to their own "discretion" and "from a boyish love of mischief." Once again, the committee had emphasized firefighters' failings as men—only this time their childish lack of restraint produced economic loss.[63]

Using steam fire engines also would reduce the amount of labor needed to put out fires. This promised to make firefighting more efficient and economical because it reduced the city's dependence on undisciplined firefighters. The committee expected to reduce the number of engines and hose apparatus needed by three-fourths or two-thirds, and operating a steam fire engine required a much smaller work unit. To wit, hand-pumped fire engines required as many as thirty or forty men—sometimes more—to keep a steady stream of water flowing, not including the dozen or more additional men required to draw the engine, to guard a hose connection, to tend the hose, or to play water on the fire. By most accounts, operating steam engines required a much smaller work group. One or two men operated the machine, while between six and eight others handled and directed the hoses. Additionally, because the apparatus was heavy and ungainly, most observers assumed that steam apparatus would be drawn by horses not large groups of public-spirited citizens. And, finally, promoters of the new apparatus expected that steam engines would be used only by a well-disciplined, paid workforce, thus eliminating competition and removing the need for firemen to guard hose connections. The *Public Ledger*, for example, reported that work crews could be reduced to a half dozen men; according to its calculations fewer than four hundred men would be required to operate the department's sixty engines, as compared to the six or eight thousand it estimated served as volunteer firemen.[64]

The committee, looking to London as an example, especially recommended that the new fire department use horses to draw its engines. They argued that this would not only reduce the large number of men needed to pull apparatus but also minimize the labor expended "running long distances, to drag the engine," thereby insuring that the firemen were fresh and energetic when they arrived at the scene. Again reinforcing the interrelationship between the moral and economic dimensions, the committee noted that firemen's contests and races to fires were a source of great pride, competition, and rivalry. Rather than seeing firemen's races to fires as beneficial—i.e., promoting fast extinction—the committee emphasized that such vigorous displays of masculine prowess prevented "efficiency." By eliminating sources of competition and conflict between companies the committee sought to regularize the behavior of the men who fought fires. Indeed, disciplining the expressive behavior of firemen was a crucial part of the committee's program for reform.[65]

Adopting steam fire engines drawn by horses, then, accomplished several of the major goals of the committee seeking to reform the fire department. Primary among them, it promised to reduce the number of firemen and placed them within a hierarchical organizational and management structure. At the same time, the cutback in the number of firemen also accomplished a political goal of those affiliated with the committee (though it was *not* central to their program of reforming the fire department). In theory, reducing the number of firemen minimized their political impact as a voting group. Although it is not clear whether firemen regularly exercised great influence on elections, at times they had mobilized themselves as a voting bloc. There is evidence that reformers and elite Philadelphians feared, at least tacitly, the united political interests of firemen.[66]

Additionally, reformers sought to circumscribe neighborhood autonomy and limit the cultural authority that those communities exercised over their own physical spaces. The committee's plan to reorder the city's fire defenses depended upon two related elements: dispersing engine houses throughout the city and joining them together into a systematic network of alarm wires, which would provide more thorough and coherent information about a fire's location and improve the response time. A major, though implicit complaint of the reform committee involved the perceived concentration of the volunteer fire department in space. It argued that, like London and Paris, Philadelphia should have "stations in different parts of the city." Related to the spatial dispersal of the department was a standard system of signaling fire alarms. The group suggested that by dividing the city into different sections, responses to alarms would be more efficient and direct. And, just as horses would reduce company rivalry, so too would a regularized alarm system and routing to fires.[67]

As critics demanded radical change, the city council and firefighters embraced incremental but significant reforms. A new fire alarm system represented just such an innovation, as well as firefighters' and elected officials' continuing efforts toward making the system of fire protection increasingly rational and disciplined. The common council adopted the fire alarm telegraph after a relatively quick review of available options in Boston and New York. After visiting both cities, an appointed committee recommended a modified version of Boston's system of alarm telegraphy. To facilitate the telegraph's use, it urged dividing Philadelphia into seven districts in conjunction with the ongoing reorganization of the fire department. Signal boxes, strategically and regularly located in each district, would be connected to a central alarm station. Once cranked, the box would transmit a signal to the central station, which, in turn, would inform a district's fire companies with another current of electricity. Through the telegraph, the department could alert firemen to the signal box where the alarm originated, thus helping them find the fire quickly. The committee appointed by the Common Council proposed that signal boxes be located within "two hundred fifty yards of every house in the city." Recommending the immediate placement of 150 alarm boxes in the city's seven districts, it provided a map to accompany the report. Signaling alarms by telegraph promised greater efficiency in firefighting by shortening response times and making those intervals more uniform. For reformers, it also promised to remove "moral hazards," such as false alarms. Mostly, though, the telegraph literally and figuratively bound firefighters and the landscape together into a single system. Firemen no longer worked in communities defined by them, but in neighborhoods organized by the city. As the city's fire protection grew increasingly bureaucratic and rational, only one issue remained for reformers: the division of firefighting labor itself.[68]

By 1855, although the pace of change frustrated reformers, Philadelphia's system of fire protection was nonetheless being recast. Philadelphia was subdivided into seven fire districts. In addition, the municipal government passed an ordinance in 1855 that reorganized the volunteer fire companies under a single chief engineer and seven assistant engineers. Each company was given a $400 annual appropriation, company membership was limited, and each was required to elect a director, who reported to the department's new bureaucracy. Although firefighters elected the chief engineer, they came under the city's direct administrative supervision for the first time. And, by 1859, Philadelphia's councils once again explored the possibility of establishing a paid fire department. Its committee corresponded with departments in other cities, consulted with the Philadelphia Fire Department's chief engineer, and reissued the economic cost-benefit analyses of the city's underwriting community. Many of the report's criticisms and conclusions mirrored those made five years earlier by the citizens' committee. Despite the changes, however, underwriters and other reformers remained dissatisfied; they continued to push for technological and administrative reform.[69]

After years of supporting volunteer firefighters financially and morally—even manning engines with them—fire underwriters abruptly had withdrawn their support. If the repudiation of the division of firefighting labor occurred for basic economic reasons, it contained a moral component as well. Fire department reform developed out of the daily experiences of businessmen seeking to profit by selling safety. As underwriters began to apply actuarial principles, they developed a new perspective on the problem of fire. The insurance industry connected fire department reform to broader programs of economic change and ideals about the importance of rational behavior—expressed in terms of management strategies, manliness, and business procedures. The insurance industry had reorganized its practices around quantitative reasoning and intensive information gathering, and it allied with mercantile interests to demand new institutional arrangements. Designed to preserve the integrity of the built environment and protect capital, these new arrangements also affected cities more broadly. Fire insurance maps and the fire alarm telegraph represented the spatial dimension of this change. Disparate geographic communities were literally and figuratively joined into a single urban network. In addition, through its agenda of moral reform Philadelphia's insurance community sought to implicate the city's firemen, and by extension all citizens, in their categorizations, calculations, and geographic representations. Together with other reformers, underwriters articulated an alternative vision of social organization in which economic rationality characterized community, masculinity, and social relationships.

Conclusion: A New Vision of Fire Risk

In the 1850s fire underwriters took their first tentative steps toward making the insurance business more rational when they and their representatives gathered information more intensively and filtered it through the lens of quantitative reasoning. If sound actuarial practice imbued everyday practices with new significance, insurers also implemented more invasive modes of surveying property and gathering information. By the 1870s, underwriters and their agents still visited clients seeking insurance, and they still inspected property. However, few insurance agents produced handwritten descriptive diagrams as they might have in the 1840s. Instead, specialized mapping agencies, such as Sanborn, precisely mapped cityscapes and factories. These diagrams complemented the extensive classification systems that firms now connected to statistical data. As insurers read these maps, statistics, and categories of risk, they revised their understanding of fire risk. Maps and categories expanded in detail and coverage and urban landscapes became colored by the dynamic interplay of danger and safety. At the same time, maps, categories, and statistics represented quantitative portraits of company loss histories. The problem of fire was becoming an objective, financial risk divorced from the material reality of urban life.

As insurance capitalists used information technologies to reorganize daily practices, they reconfigured the relationship between cities, public safety, and manhood. Insurers reached beyond their boardrooms and offered an alternative vision of environmental and community order in which technology, bureaucratic social relationships, and economic exchange became the basis for fire protection. Fire insurance maps categorized the city according to classifications devised, in part, from observable statistical experience. By objectifying the problem of fire in terms of statistical categories, underwriters created order where seemingly none had previously existed. In the same manner, the urban fire alarm telegraph network and fire districts re-created the city as a single, rationally organized unit. Volunteer firemen—exuberant heroes within their local communities—became obstacles to this vision of a well-ordered society in which functional specialization dictated economic and social relationships. Underwriters especially sought to reform the division of firefighting labor. Like industrialists, fire insurers argued that individualism and function specialization were the best way to organize society, including public services like firefighting. If underwriters' emphasis on making firefighting rational and efficient ran contrary to volunteers' rhetoric, it nonetheless matched the trend already under way among volunteer firefighters toward the same end. In fact, the vision of safety that underwriters developed—in which risk was rationalized, quantified, and objectified—mirrored the broader cultural shifts of the industrializing nation. The informal bonds between individuals and society were replaced with economic and contractual obligations—represented by wages or insurance policies.[70]

Ironically, although insurers sought to circumscribe the role of urban neighborhoods in fire safety, the responsibility for the problem of fire remained a community endeavor, although its nature was changing. Insurers established new types of connections between people, organized around economic risk rather than based on particular geographic locales or social categories. The industry distributed the risk of fire beyond local neighborhoods, individual cities, or sectors of the economy throughout the entirety of American society. To a degree, the purchase of insurance policies—a commodity bought and sold in the marketplace—repudiated a vision of environmental order in which everyone in society, if not the polity, shared equally in the danger of fire. Even so, individuals and businesses that purchased insurance did not reject the idea that environmental risk should be shared. Rather, they supported insurers' methods for organizing and managing the problem of fire by spreading responsibility among others who had purchased safety via insurance contracts. Certainly, the boundaries of this financial community were circumscribed by the relatively limited scope of the industry, which underwrote risks primarily for businesses and well-healed property owners. In fact, the purchase of insurance policies and safety may well have been one of the many consumer choices through which the middle class began to define itself in the 1850s. More significantly, as middle-class Americans forged communities of risk defined by corporations, they contributed to the narrowing of public responsibility for safety.

CHAPTER FOUR

Muscle and Steam:

Establishing Municipal Fire Departments, 1850–1875

In May 1855 over fifty thousand Philadelphians braved a torrential downpour to watch a contest between muscle and steam that symbolized the broader conflict over the provision of fire protection in the United States. Crowding the streets in front of Dr. Wadsworth's Church, throngs gathered in windows and on rooftops along Arch and Tenth Streets. They witnessed a philosophical and technological trial arranged by the Philadelphia City Council, pitting a steam engine manufactured in Cincinnati against Philadelphia's top hand-pumped fire engines. The steam engine was tested first. Viewers waited for over eight minutes, as engineers lit the engine's furnace, and then were treated to fifty-seven minutes of action. The engine threw continuous streams of water up and down Tenth Street, further drenching the crowd. Twice engineers turned the hose upward to the church steeple, obtaining a height of 120 feet. After the engine had been fully tested, the steam was let off and the "grate skwirt" was taken from the grounds. A hush came over the crowd as thousands of eyes anticipated the work of firefighters. Philadelphians especially turned their eyes toward the Diligent engine, the pride of the city, and they were not disappointed. A full complement of volunteers pumped furiously on the engine's brakes, quickly developing pressure and expelling a steady stream of water from its pipes. The men worked with great resolve and twice sent streams of water shooting above the church steeple, attaining a height of 133 feet.[1]

The competition exposed conflicts that lay just beneath the surface of American life in the middle of the nineteenth century. Across political, economic, and cultural domains, Americans reconsidered the best methods for organizing society, including the provision of public fire safety. At the broadest level, the debate over firefighting pitted the interests of business and the incipient middle class against the interests of urban working-class and immigrant communities, but the street-level battle to control firefighting reveals a more complex portrait. Although firefighters often emerged victorious in ongoing contests with insurers and the middle class, they ultimately embraced many of the same values espoused by their opponents. Volunteers, for instance, claimed to be specialists in firefighting, with distinctive claims to the public trust. They worked to increase their efficiency and, as firefighters reorganized their labor, they moved inexorably closer to establishing firefighting as one of many specialized occupations in America's industrializing economy. And, although firefighters sometimes argued with underwriters and merchants about the merits of using new technologies, such as steam fire engines, just as frequently they did not debate the usefulness of innovation. Rather, they debated who would control the new instruments. As a result, steam engines did not remake American fire departments; firefighters did. Ultimately, volunteers dictated the pace of departmental reform. They seized opportunities presented by debates about fire protection to lead and direct fire departments that formally compensated them for what had been an avocation. Then, volunteers deployed new machines and management strategies to assert their authority over the revised organizations.

The newly reorganized fire departments did not jettison the principles that had guided volunteer fire companies, however. Rather, firefighters built upon the previous organizations' ideals of manhood, technology, and service. They especially continued a trend that volunteers had begun earlier in the century—the process of specializing and reorganizing firefighting. Volunteer firefighters, then, did not fade away simply as a result of some broad shift in community values. To the contrary, firemen led the charge to make public fire safety more bureaucratic and the purview of specialists. They reorganized firefighting as paid labor directed by skilled experts and organized departments under municipal control. This transformation did not signal an end of a mythical era as much as it indicated a revision in how firefighters approached their service to the community.

More broadly, the creation of municipal fire departments reveals how public safety, the expansion of city governments, and the process of urbanization were each shaped by attitudes about manhood and technology. Movement away from volunteer firefighting occurred at a time when cities began to extend their administrative capacities, to create more specialized programs of public safety, and to employ skilled professionals to provide those services. However, the growth of municipal bureaucracies was not imposed from the top down. Quite the contrary, the demand for and provision of these services, such as firefighting, developed within urban neighborhoods and work communities. Everyday Americans initiated and directed a shift in the process of keeping order in cities. Alongside their neighbors, firefighters played an important role in shaping the functions of city governments in the middle of the nineteenth century.

Firemen, the Middle Class, and Urban Disorder

In the middle of the nineteenth century, cities throughout the United States became seething powder kegs as a result of repeated economic crises (like the depression of 1837), the influx of new migrants, sharpening political divides, and rapid spatial change. In Philadelphia, civic unrest was marked by antiabolitionist rioting in 1838 and anti-immigrant turbulence in 1844. Not surprisingly, firefighters and other agents of order came under scrutiny for their role in these disturbances. Intimately connected to their neighborhoods, firefighters shared the same passions as their neighbors, and their demonstrative and rough culture heightened attention to their part in conflicts. As a result, melees between companies received inordinate coverage in local newspapers, which complained of firemen's riots alongside lurid descriptions of "lower-class" life. Though these battles resulted from the same factors that were destabilizing the society in general, the local judiciary, not surprisingly, singled out firemen for especially stiff punishments. After all, firefighters played a crucial role in keeping order. Firefighters, according to their own claims, did not just preserve physical order, they upheld the moral integrity of cities.[2]

In particular, the emerging middle class linked firefighters to the crisis of order in mid-nineteenth-century cities. In its quest for greater social and political power, the nascent middle class seized upon and amplified reports of disorder, including violence among firefighters. The middle class expressed concern about the rough physicality of firemen—a portrait that contrasted markedly with the social values that they emphasized. The middle class defined manhood as dispassionate and rational, and saw it in terms of white-collar work, all of which separated middle-class men from the rough-and-tumble world of the streets and the darker impulses of competitive industrial culture. The middle class sought to create a society that was not only more mannered but that also was organized around rational economic markets and values of sobriety and efficiency. Additionally, the class began to define itself in terms of wealth and income, to which fire threatened to lay waste. As middle-class families placed a greater premium on economic well-being, fire danger intensified. If they were predisposed to finding disorder everywhere, the middle class nonetheless had an ever greater stake in effective fire protection. For this reason, the middle class—not to mention most nineteenth-century urban residents—held legitimate concerns about firefighters' participation in rioting, especially since it could compromise firemen's ability to restore order. Thus the concerns of the middle class extended beyond questions of style to substantive material matters. Ineffective fire protection potentially exposed their economic capital, often invested in the built landscape, to dangers beyond the capricious market economy. When they expressed doubts about the organization of fire protection, then, middle-class Americans sought to reform the social and economic order of urban America.

The middle class especially found fault with those communal fire companies that expressed themselves in the idioms of working-class or ethnic communities. The general attitude toward firemen is exemplified by a former Philadelphia newspaper reporter, George G. Foster, writing in the *New York Tribune* in 1848. Foster described firemen as comic caricatures hardly worthy of his attention, but he observed them with fascination nonetheless. Though it was clear to critics that firefighters could not control themselves and exhibited excessive aggressiveness, Foster paradoxically questioned firemen's manliness—because he believed that their fisticuffs were staged, not real. Such claims about bogus fracases raised issues about the hardiness of firemen and, along with firefighters' elaborate costume, led Foster to describe firemen as being as "vain as women." He reported, "The minute and detailed notoriety which the reporters (hard run for subjects) conferred upon these riots in the newspapers was exactly the kind of fame they coveted." Rather than identify violence as the problem, as the citizens' committee had done, Foster's curious argument offered an equally, if not more damning critique of firefighters. Whether real or staged, fights raised global questions about firemen's manhood and, because of their significant role in public order, about the well-being of cities.[3]

Following a similar strategy, reformers construed volunteers as immoral and menacing by linking fire companies to street gangs. In mid-nineteenth-century Philadelphia, bands of working-class and immigrant youths seemed especially to threaten state authority and the middle-class sensibilities of leading citizens, such as Eli Price and Horace Binney. Although their precise numbers or influence is not clear, Philadelphia "gangs" such as the "Killers" contributed to the sense of chaos that reigned on frenzied urban streets in antebellum America. However, the political and social criticism stated or implied by these groups of young men vexed citizens more than their audacious attire and expressive style. In particular, the "Killers" came to symbolize the city out of control, no doubt because of the group's radical political message, which was distributed in a popular novel. Both in the novel and in reality, urban youth appear to have been drawn to the exploits of fire companies. For instance, the "Killers" were connected loosely to the Moyamensing Hose Company (as well as to the Democratic Keystone Club and the infamous political boss Squire McMullin). The link between fire companies and gangs, no matter how distant, tainted the ability of firefighters to keep order, at least in the eyes of the middle class. Indeed, the *actual* strength or pervasiveness of such connections mattered little in polite parlors, where apprehensions about social and political power appeared in widely read advice books and sensationalist novels. The everyday life of the street mattered less than appearances of impropriety. H. C. Watson's popular morality novel, *Jerry Pratt's Progress*, had made this point clearly. If firemen did not directly corrupt young Jerry, the connection between fire companies and gangs made it possible for incorrigible gang members to ruin the youngster.[4]

Theaters became symbols of the widening gap between the middle class and workers, and in the case of firefighting this chasm was embodied in a popular character of the 1840s and 1850s. Mose the fireman first appeared on New York stages in the late 1840s in Benjamin Baker and Frank Chanfrau's *A Glance at New York*. Chanfrau played Mose in more than a thousand performances in theaters throughout the nation. The performances delighted working-class audiences, who viewed Mose, protecting the community and rescuing damsels in distress, as a neighborhood character to whom they could relate. However, middle-class audiences saw him as an embodiment of serious social unrest and the growing societal power of workers and immigrants, and perhaps as evidence of a noble institution gone bad. In New York and Philadelphia, "respectable" audiences scorned Mose, and when the play toured in Louisville, reformers used Mose as an example of the behavior they wanted to correct. Perhaps middle-class audiences, with their focus on the expanding industrial economy, feared that firefighters like Mose, who focused on protecting people, would become less effective at protecting property from fire—which they viewed as the proper function of specialist firefighters. Either way, as he fueled debates about reforming fire departments, Mose came to symbolize the widening differences between working-class virility and middle-class rationality.[5]

Just as Mose's reception suggests, critiques of both firefighters and working-class life cannot be dissociated from broader issues of societal power. Quite sim-

A Paid Fire Department As It Is Likely to Be under the Contract System, ca. 1853, J. L. Magee. As debate about fire department reform intensified, volunteers also commissioned prints harshly critical of reformers' plans. In one image, Philadelphia businessmen watch the city burn while wage laborers—depicted according to period racial and ethnic stereotypes—operate steam fire engines ineffectively.
Courtesy, Library Company of Philadelphia

ply, immigrants and workers threatened the social and political capital of economic elites and the middle class. Sidney George Fisher expressed concerns about the growing power of volunteer firemen for just such reasons. Fisher estimated that as many as five thousand made up this "dangerous body of men." If the large numbers of volunteers troubled Fisher, their personal qualities distressed him more. When Fisher characterized firemen as hardy, young, and vulgar and worried about their brawling, he really expressed concerns about the increasing social clout of Philadelphia's working-class and immigrant communities. At a time when the political authority of the social elite in central-city Philadelphia had decreased relative to the influence of immigrants and workers, Fisher worried especially that politicians had "cherished" firefighters into "pestilential growth": "They are incorporated, they have property, they are numerous, are bound together by esprit de corps, they are *armed & disciplined*, and they have *votes*." Fisher used the metaphor of the body—one long used by volunteer firefighters themselves—to underscore the principal issue facing the middle-class and business interests. He asked, perhaps rhetorically, "How is such a body of men to be controlled by a democratic government?"[6]

The underlying importance of broader issues of social and political power to the debate about firefighting is underscored by the fact that few ever questioned firefighters' effectiveness. Even ardent opponents, such as Sidney Fisher, believed firemen defended property with great zeal. Fisher described the scene at a fire in 1841: "In a short time the fire was got under command, six or eight engines playing on it & hoses carried into the building. They managed it very well, the engines throwing the water clear over the house & keeping Coxe's roof wet on the other side. The fire department here is very energetic and efficient." Just as the novel *Jerry Pratt's Progress* did not challenge the work techniques or efficiency of firemen, neither did the citizens' committee and other reformers question the effectiveness of firefighters on the fire ground. Nobody asked whether common firefighting strategies, such as throwing water onto or over buildings, were effective at bringing blazes under control. They only questioned the character of the firemen, reflecting their broad concern with how best to protect the nascent industrial social order.[7]

Whatever the validity of such commentary, firefighters felt compelled to restate their commitment to the urban social order, often adopting rhetoric and images similar to those used by their critics. For instance, future mayor and Philadelphia Hose Company veteran Richard Vaux urged his fellow firemen to remain true to the specialized vision of firefighting and manhood that his father had helped to invent fifty years earlier when he formed the company. Like other firefighters, Vaux surely recognized that the economy and cityscape were unstable, and that fire danger seemed to be increasing everywhere. In 1850, a great fire destroyed several hundred structures along the Delaware River, and the year before a massive conflagration had nearly destroyed St. Louis. Also, firefighters certainly worried that, whether or not battles between companies made the city more dangerous, the inability of firemen to discipline themselves diminished their authority as the protectors of community and the social order. In this context, Vaux demanded "public good first, private interest last; *but the common honor, fame and usefulness always.*" He wanted firemen to bring order to a city in crisis by reaffirming the vision of brotherhood and duty—honed in competition—that had made their specialized service so integral to early-nineteenth-century American culture.[8]

Firefighters attacked the proposals of middle-class merchants and insurers by commissioning prints that depicted the potential consequences of having a paid fire department. One such print, drawn in 1853 as *Jerry Pratt's Progress* was being peddled in the city's bookstalls, showed incompetent firemen whose hats and apparatus were marked with the label "paid." The print's central figures reversed the caricatures made by the committee agitating for a paid fire department and *Jerry Pratt's Progress*, and it used language with special meaning for urban artisans fearing the degradation of their craft. By depicting paid firefighters as Irish and African American hirelings, it connected firefighters' impotency (the engines could not even get up a stream of water) to the debasement of wage labor and slavery. This image drew upon the idiom of nativist politics and appealed to the racism that was often implicit in the free labor movement. The message was that volunteer firemen were not depraved immigrants, unskilled wage laborers, or slaves; they were independent native-born men who commanded their own labor and technology.[9]

Yet, even as firefighters defended themselves, they continued to exhibit great diversity in their approach to their public service. While some urged their peers to sublimate individual interests for the community good, others viewed the expression of local community identity as the basis for active democracy, and fire companies spoke in a cacophony of different voices at parades and in the public space. If extraordinary differences in style and expression gave firefighters' culture a reputation for disorder, firemen remained steadfast in their battle against environmental risk. The middle class attacked their culture, but only rarely challenged firefighters' effectiveness as workers. Indeed, although volunteer firefighters found little common ground on which to unite, their job performance remained exemplary. In the face of escalating fire danger, they continued to innovate and control the pace of technological change in their workplaces. If anything, the premium that they placed on physical power and technological competence grew stronger as firefighting work grew more technically complex and physically demanding. Volunteers initiated an increasingly specialized division of labor, embraced organizational complexity, and sought more efficient ways to manage and execute their labor.[10]

More broadly, the debate about the merits of volunteer firefighting in American cities transcended the critiques of the new middle class. Indeed, creating and preserving order occupied most everyone's mind, from the industrialists to dockworkers to politicians. Not surprisingly, then, discussions about fire protection occurred at the intersections of broader efforts to quell disorder, especially those concerning municipal administration and political economy. The nature of government—federal, state, and local—was changing in America. At the municipal level, politicians began to form political machines and debated more active administrative control over the provision of daily services, such as police protection, sewers, and public health. Meanwhile, in the nascent industrial economy employers replaced traditional work relations with those mediated by wages and the market. When Philadelphia's citizens' committee urged hiring wage labor to perform firefighting, it hooked into ideas that were gaining currency elsewhere in the nation. Likewise, volunteer firefighters fed into these broader trends when they emphasized technological change, embraced organizational complexity, and sought more efficient ways to manage their labor. Just as factory floors would become organized into particularized functional units, so too firefighters made fire protection increasingly specialized and bureaucratized. As they refined their methods of controlling fire danger and the city, they offered a vision of order that was increasingly appropriate to industrial society.[11]

Bureaucratic Technologies

Firefighters in Philadelphia, as well as elsewhere in the nation, defended their service by creating elaborate and more expansive new management structures. Much as they had done in previous decades, Philadelphia volunteers hoped to restore order by initiating bureaucratic reforms. This time, however, they responded in the context of two potent and new external developments. On one side, firefighters felt squeezed by the secretive committee appointed by the city's economic elites, and on the other side, the region had begun to contemplate "consolidation," which would join the entire Philadelphia region into a single political and municipal administrative unit. In 1853, in response to these conditions, a "Firemen's Convention" met and attempted to establish a single administrative body for fire companies throughout the region. Composed of representatives from thirty-nine companies, the convention established a "Board of Directors of the Philadelphia Fire Department." The convention introduced a new term into firefighters' vocabulary; for the first time, firemen began to refer to themselves as part of a single "volunteer fire department." Likewise, the convention signaled a change in how firefighters wielded their political clout. No longer content to exercise it independently, they expressed a unified voice in the polity. Building upon previous failures at forming permanent management associations, firefighters also provided the new governing body with more bureaucratic authority over the department.[12]

The board of directors sought to reorganize firefighting by placing fire companies under the command of a central governing body, exempt from the interference of local political life. As a first step, in October 1853, the board asked the state legislature to authorize the expansion of its powers and to exempt it from local political control. It further requested that, although funding for fire companies would continue to emanate from local governmental units, those municipal officials would be barred from intervening in the management of the department.

Rather, control over fire protection, including how companies spent their budgets, would be vested in the board. These provisions enhanced the board's power relative to the region's fractured political units. By acquiring financial control over the department, the board effectively prevented local communities from meddling with the association, thus removing what previously had been major obstacles to reform: neighborhood politics, ethnic conflicts, or class disputes. In return for such unprecedented authority, the association formalized the department's management into a clear hierarchy. The region's firemen elected officers, including a chief engineer and assistant engineers, to manage the department. Although electing departmental leadership reproduced the power structure evident at the level of company organization, locating final control over the department in the hands of a single official broke with a tradition that had vested authority in the companies independently.[13]

Of equal significance, the board connected the fire department to the local municipal governments and to the fire insurance industry in an increasingly formal manner. It demanded that the chief engineer (and department more generally) record information about the department's performance at fires. The chief would furnish "an account of all fires as aforesaid, the number of Companies in service at each fire, the amounts of loss and insurance, which reports shall be printed annually." Additionally, the board wanted the statistics submitted to the "President of Councils, the Board of Commissioners of the Districts, the Presidents of the Fire Insurance Companies, and the Presidents of each Fire Company in the City and County of Philadelphia." Whether such rules made the department more efficient is not clear, but they made fire protection more regimented. Public reporting made firefighters accountable to one another, to elected officials, and to economic leaders. Moreover, the association's plan organized the city's fire protection in a manner that supported underwriters' efforts at understanding better the dangers of fire.[14]

The board of directors also established strict work guidelines that superseded the traditions of volunteer firefighters. For instance, new rules forbade racing to fires and disallowed other informal competitions—emphatically forbidding brawling. In addition, the board made the provision of firefighting more rational by dividing the city into three fire districts. It prohibited companies from attending blazes outside of their districts, and specified work routines to make labor at fires more efficient. Of equal significance, the association established itself as the mediator of disputes between companies, supplanting the longstanding tradition of settling disputes informally. In direct appeals to firemen, it requested that companies adhere to a common set of rules and especially to embrace greater discipline.

The organization experienced many difficulties in trying to reform the fire department. Almost from its inception in 1853, for instance, the board condemned members for making disputes public (in the court system), and expressed dismay at the lack of self-control among firemen. The board's efforts exposed a rift between companies that sought central management and those that preferred using community-based solutions to settle disputes. If many firefighters wanted to create a unified departmental interest, others emphasized local community connections, as had long been customary among firefighters. As volunteers debated how best to settle disputes, a new approach began to emerge. Support for centralized authority took precedence over company autonomy.[15]

When merchants challenged the board of directors regarding the utility of its reform proposals, the board made it clear that it had reconceptualized firefighting service. It admitted what many reformers believed—that firemen were out of control—and it emphasized the importance of management to keeping order and quelling disorder. The board portrayed itself as the solution by using the metaphor of the body, emphasizing the centralization of bureaucratic authority within itself as the "head": "The simple question for consideration now is the incorporation of a body for the management of a department composed of over ten thousand members that, according to the remonstrance referred to, is without head or management." While firefighters—including the Philadelphia fire companies that had reinvented volunteer firefighting early in the nineteenth century—had once imagined themselves in terms of the body politic, that ideal gave way before a notion of a body as a system—of bureaucratic management and hierarchy. Of course, the board did not neglect to mention the basic democratic values that firemen embodied nor did it fail to emphasize that firemen protected the public good on behalf of all citizens. Stressing firefighters' performance of public duty, the board claimed a moral high ground, questioning whether the individualistic approach of the business elite was a suitable way to organize society.[16]

As the board of directors challenged customary firefighting practices and encouraged reform, it altered the debate about fire protection. A growing portion of the department supported the more centralized administrative structure. In unanimously endorsing the board's appeal to the state government and its resolutions demanding greater self-control, firefighters acknowledged a more circumspect notion of brotherhood. By regulating themselves, firefighters reasserted the self-discipline that had always been present in firefighters' culture—at least rhetorically. Additionally, they directly countered the criticisms of insurers and merchants. Indeed, just as the coalition of reformers and property owners had asked firemen to control themselves and pump water on fires more strategically, the

An engraving of the steam engine Young America, 1855, from *Report of the Subcommittee to Put the Steam Fire Engine Young America in Service.* In 1855, the Philadelphia City Council sponsored a contest between a steam engine—the Young America—and fire companies operating hand-pumped apparatus. Although the city's hand-operated engines won the duel, Philadelphia business leaders purchased the Young America and offered it to the city, on the condition that it enact radical reform. Ultimately, they refused to let the volunteer firefighters use the machine. Courtesy, Historical Society of Pennsylvania

board of directors demanded that volunteers stop participating in both the rough and tumble of community life and many firefighting rituals. If firefighters did not champion reform because they thought it necessary, they may have acted out of expediency. After all, the board promised greater benefits to individual companies in the form of municipal funding and continued legitimacy in the eyes of the state. They may have sensed the growing power of merchants and middle-class citizens, choosing to preserve their organizations rather than lose them altogether. Whatever the reasons for firefighters' support, the board reoriented the debate about urban firefighting. It knit firemen into a bureaucratic network that diminished company autonomy. It explicitly separated volunteers from the community by replacing informal problem-solving procedures with fixed rules and by quashing expressive competitions. Its scheme also incorporated firemen into underwriters' efforts to contain the problem of fire through bureaucratic processes. No longer just a communal response to the problem of fire, firefighting was becoming an urban service with a rational bureaucracy and written work rules.[17]

Much to the board's surprise, its managerial proposals faced fierce resistance not from firefighters but from so-called reformers. In promoting its legislative program in the state capital, the board utilized the political savvy and clout of firemen. Capitalizing on "their influence," fire companies and board members first canvassed local politicians (in both Philadelphia and its districts). After obtaining significant support, the board wrote a bill that lawmakers presented to the state legislature. In May 1854, the board watched the "Firemen's Bill" sail through "several gradations" and move toward final passage, only to have it stopped as it awaited the signature of the Speaker of the House. Meanwhile, business interests aligned against the bill called in their political chits to have it killed, and on the last day of the legislative session, the bill was reconsidered and postponed indefinitely. Firefighters once again found themselves squaring off against Horace Binney and Stephen Colwell—two of the city's most prominent leaders. Earlier in the spring, Binney and Colwell wrote a scathing letter urging "the abolition of the volunteer system" and encouraging the state legislature to defeat the bill. If the near success of the "Firemen's Bill" validated middle-class fears about firefighters' influence at the ballot box, it also showed that firemen had powerful enemies. Once again the Philadelphia business community questioned whether firefighting should be the responsibility of volunteer workers.[18]

Though Philadelphia's volunteer firemen lost the battle in the state capital, the legislative proposal that firefighters introduced was part of a broader political strategy designed to ensure that volunteers would control firefighting even as the Philadelphia region moved toward consolidation into a single municipality. In fact, as the board shepherded its proposal through the legislature, it kept one eye focused on Philadelphia, awaiting the outcome of a drawn-out effort to consolidate the city into a single municipal unit. Modeled after the law that created the marshal's police in 1850, the board's proposal was designed to govern the fire department only until consolidation passed. Upon consolidation, the issue of a paid fire department would be placed squarely into the hands of voters. Undoubtedly, volunteers believed that their years of dedicated service, their political clout, and their support of reform would sway public debate in their favor. Meanwhile, had consolidation failed and the firemen's bill passed, the result would have been the regional fire department desired by the board. Alternately, if both consolidation and the proposal failed, then the status quo would have been maintained. Firefighters had orchestrated a crafty strategy, which ultimately helped to preserve the system of volunteer firefighting when Philadelphia consolidated in 1854. Not only did the consolidated region not create a paid fire department, but it also generally followed the board of directors' plan for organizing the department. However, firemen had committed to a course of reform that undermined their autonomy and relationship with their urban communities.[19]

As expected, when Philadelphia consolidated, the fire department experienced a wave of reforms. The municipal government chose not to implement a paid fire department, but it did not involve firefighters in planning the new department. This led to months of contentious debate. By choosing not to consult with the board, the newly consolidated government failed to anticipate the concerns of volunteers. Not surprisingly, the resulting ordinance met with resistance from the majority of fire companies. Only twelve of forty-seven companies liked the ordinance well enough to support it, and the board of directors published its disapproval in local newspapers. Firemen opposed three parts of the ordinance: the creation of seven fire districts with corresponding assistant engineers; making the chief engineer a political appointee; and disallowing contributing members from participating in company life (thereby removing the incentive for their large payments to fire companies). Despite mounting criticism, the municipality reorganized the fire department in January 1855. However, rather than fight the city, the board of directors shifted its position, and decided to become part of the nascent bureaucracy. In March 1855, the board disbanded and reconstituted itself under the auspices of the city government. Next, the board invited "companies *accepting*" the ordinance to appoint representatives to its ranks. Nearly 70 percent of the newly consolidated city's fire companies—fifty-two out of about seventy—capitulated to the expansion of municipal power. The majority of firefighters chose to work within the new framework rather than to oppose it.[20]

Despite widespread acceptance of the new ordinance, there were still many companies opposed, and even accepting companies demanded significant revisions to the ordinance. Twenty companies dissented vociferously. Labeled "nonaccepting" companies, they continued to perform service, but they refused to accept both the ordinance and financial contributions from the municipal government. By forming an alternative "Board of Directors," nonaccepting companies challenged the legitimacy of the municipal government's decision, as well as the newly organized fire department. However, the protest was short-lived and largely ineffective. Several of them, such as the Northern Liberty Engine Company, quickly petitioned for admission to the department. As nonaccepting companies challenged the new ordinance, accepting companies hoped to see it revised. In particular, firemen wanted candidates for chief engineer to fulfill minimum service requirements and to be directly elected by firemen. Dissension within the department simmered for months, and appeared to be dissipating when it reignited in

September 1855, after the city council named Benjamin Shoemaker as the chief engineer. Fire companies had elected another candidate (by a vote of thirty-two to twenty-nine). In the ensuing controversy, several accepting companies withdrew from active service because of the decision. The Southwark Engine Company even recommended that firemen strike unless the city council gave in to the department's demands. Responding to the groundswell of protest, the city council met volunteers' demands and altered the ordinance. The chief engineer would be elected directly by fire companies, and contributing members would be allowed to actively participate in company life. These concessions effectively ended protest against the newly reorganized department, and gradually, between 1857 and 1860, most "non-accepting" companies appealed for admission. The board of directors continued to represent the interests of fire companies, taking an especially active role in recommending new technologies and organizational arrangements. Volunteer firefighters endorsed the new arrangement, but had done so on their own terms.[21]

As Philadelphia's fire companies accepted the new order in the 1850s, they became ensnared in the web of municipal bureaucracy, which diminished their autonomy. A well-established and powerful third party now mediated companies' interactions with one another. Not only did firefighters have to follow rules, but their companies would come to depend upon the state for a measure of their funding and legitimacy, which brought them further under the municipality's authority. However, even though becoming part of the expanding bureaucracy circumscribed company independence, firefighters benefited from the new organization. Larger budgets meant better equipment; they also allowed volunteers to purchase new costumes, commission artworks, and design elaborate membership certificates. Firefighters also maintained the right to vote for departmental and company leaders, which buttressed their control over their workplaces and work routines. Perhaps most importantly, the new organization united firemen into a formal collective that explicitly defined itself as a fire department.

Although formed by an act of the municipal government, the creation of the Philadelphia Volunteer Fire Department represented the assertion of a common political and social identity. Firefighters tested the boundaries of their new collectivity by blunting the demands of underwriters and merchants. At the same time, they continued to make firefighting more efficient and specialized, especially in regard to the use of new technologies. Indeed, the way in which Philadelphia's volunteer firemen confronted the advent of steam engines reveals the degree to which firefighters were intimately involved in remaking the work of firefighting and fire departments at midcentury. Moreover, this technological innovation did not cause the reorganization of firefighting; firefighters were already doing that themselves.

During the 1850s, firemen would continue to negotiate the future of firefighting in Philadelphia and other American cities.

Steam Technology and Manhood in Philadelphia

As the municipal government and the city's firemen approached rapprochement, Philadelphia's mercantile interests ratcheted up their calls for a radical reorganization of the fire department. Merchants and underwriters, reflecting sentiments spreading throughout the nation, advocated hiring paid workers to use steam technology to replace volunteer firemen. The idea received a significant push forward when Cincinnati artisans built what may have been the first effective steam fire engine, which the city's volunteers tested. Political officials in Cincinnati then selected a prominent local industrialist and volunteer fireman, Miles Greenwood, to lead the creation of a municipally administered fire department that paid a wage to firefighters. Insurance companies, the Cincinnati mayor, and the manufacturers of the steam engine—Abel Shawk and Alexander Latta—effusively praised the new department. They reported that, in Cincinnati, a disciplined industrialist and a steam fire engine physically bested disorderly volunteer firefighters and saved the city from being ravaged. Insurers, reformers, and industrial leaders throughout the nation adopted this legend, repeating it again and again, to justify their claims that volunteers should be replaced by steam engines operated by wage laborers.[22]

In this context the Philadelphia City Council investigated the "subject of steam fire engines" in the spring of 1854. A specially formed committee announced that its purpose was to study the use of steam technology, but not to consider "the organization of the department, though [the joint special committee] believe that very great change is needed in this particular." Like many cities that considered adopting steam technology, the committee journeyed to Cincinnati to visit with Miles Greenwood. After a practical demonstration, the committee concluded that the Cincinnati Fire Department showed "the most perfect system and arrangement," but it noted that the system used in Cincinnati was "not attributable to the character of the apparatus." Even so, it recommended purchasing two for Philadelphia.[23]

Once again, Philadelphia firemen found themselves struggling against the social and political agenda of Philadelphia's mercantile community. This time the confrontation took physical form, when the city council arranged the contest between their hand-pumped engines and a steam engine manufactured by Cincinnati's Abel Shawk. The council invited three engine companies from the volunteer corps: Assistance, Diligent, and Weccacoe. Those choices were *not* coincidental. On the streets of Philadelphia, folk wisdom held that firemen from the Diligent Engine Company operated the city's most effective engine. Patrick Lyon—Philadelphia's own public-spirited artisan and firemen—had manufactured Diligent's first apparatus, and John Agnew—the city's most renowned engine maker in the 1840s—had improved upon Lyon's handiwork. The company even captured this history of innovation by reproducing John Neagle's print *Pat Lyon at the Forge* on its membership certificates. In so doing, Diligent linked Lyon's technical prowess to its own. Not only had Pat Lyon built the company's first engine, but also his spirit had forged their Fire Company. Philadelphians understood that the company's apparatus, and by extension its men, embodied the traditions of firefighting in Philadelphia—efficiency, citizenship, physical power, and technological innovation. Local newspapers sensed the dimensions of the contest. They reported, "The respective merits of the steam and hand engines will be fairly tested, as the Old Diligent Engine, the masterpiece of Patrick Lyon, will have her powers displayed at the same time." As we have seen, Philadelphia's firemen proved their manhood against the industrialists' steam fire engine. Local papers reported, "The Diligent was fully manned, a fine stream of water shot with lightning speed from her pipe. The men worked with incredible spirit and strength, and after two efforts succeeded in attaining a height of 133 feet" (besting the steam-engine Young America's 120 feet).[24]

Although the results seemed to speak for themselves, interpretations of the day's events varied widely. To many observers, the results demonstrated firemen's manhood and the folly of underwriters' plans to purchase a steam engine. One report compiled by the Phoenix Hose Company, which collected newspaper articles published in the days after the event, billed the contest as "the Grand Test of Muscle Against Steam—the Weccacoe (2nd Class Muscle) ahead of the Great Steam Squirt," in reference to one of the smaller Philadelphia engines. Another account stated that "second-class muscle" had "whipped the great humbug that some thoughtless person would like to see thrown on our citizens at an enormous price." Indeed, the Weccacoe's performance especially revealed the limitations of the steam engine because even this small machine with less power—or second-class engine as it was technically called—beat the steam apparatus. In addition, the steam engine had a much larger bore and cylinder than Diligent's hand-operated apparatus, and in theory, the Young America should have thrown a stream of water "at least double the distance reached by [Diligent's machine]." Thus, not only had the steam engine failed to best the technique and power of Philadelphia's firemen, but also its mechanical quality was suspect. Technical competence and phys-

Engine of the Diligent Fire Company, 1852–55, Wagner and McGuigan (printers), G. G. Heiss (lithographer). Originally manufactured by Patrick Lyon in the 1830s and reengineered by renowned engine maker John Agnew in the 1840s, the Diligent Fire Company's engine was reputed to be the best in the region. This status as well as the engine's history lent great significance to the contest between "muscle and steam" held in 1855. To the delight of thousands of onlookers, the Diligent outperformed the steam engine. Courtesy, Spruance Library of the Bucks County Historical Society; gift of the Volunteer Fireman's Association of Philadelphia, 1919

ical power had triumphed over the engineering ability of Cincinnati manufacturers. Clearly the manufacturing and design abilities of Philadelphia's artisan firemen far exceeded those of their rivals elsewhere in the nation. According to all the criteria by which firefighters judged themselves—standards that connected physical strength, technique, and technical innovation—volunteer firemen had vanquished steam technology.[25]

However, one voice in Philadelphia challenged the dominant view that appeared to be forming about the contest. Although this newspaper too articulated the inevitability that steam would replace hand-operated apparatus, it rejected the methods used to assess firefighting effectiveness. Adopting a derisive headline—"The Grand Aquatic Exhibition," the paper presented a contrary view of the trial: "We do not design giving the number of feet the water was thrown, the size of the nozzles used by different apparatus engaged in the experiment or other facts of similar character." This was a repudiation of the ways volunteer firefighters determined the quality of their manhood, technology, and service. Reversing the argument made by firefighters (and even the citizens' committee in its allusions to the impotency of the fire department), the paper argued that manhood could not be measured by the length of hose streams. Rather, it claimed that true efficiency lay in the implementation of labor-saving devices and reorganization of labor. Therefore it urged the city to purchase the steam fire engine, whether it had been made in Cincinnati or Philadelphia. The newspaper echoed insurers' demands for a fire department manned by steam engines and paid labor rather than large numbers of volunteers.[26]

Curiously, neither the public nor firefighters viewed steam engines negatively, and they embraced the new machines—but on their own terms. In fact, despite the victory, newspapers argued that steam technology inevitably would be used in firefighting. Two of the city's dailies challenged the city's artisan firemen to higher standards of innovation, service, and manhood, urging "Philadelphia mechanics" to "put their inventive genius to work" and to produce an engine worthy of the city's firemen. They further drew a parallel between the manufacture of steam fire engines and steam locomotives: "Philadelphia mechanics who make the best locomotive in the world have not yet tried their hands upon a steam fire engine." The contest had demonstrated nothing more than the limitations of Cincinnati manufacturers and men. Purchasing a steam engine would not be a mistake, though buying the Young America would be folly because of its technical inferiority.[27]

Although Philadelphians debated the merits of his engine, for Cincinnati manufacturer Abel Shawk the trial had its desired effect. Convinced that steam technology would "prove invaluable, and a great auxiliary, to our own efficient Fire Department"—especially for large fires along the city's valuable riverfront—Philadelphia eventually acquired the engine, but it did so through an unusual route. The municipal government could not afford the engine, but it accepted a lease on the machine, courtesy of the largesse of a group of "public-spirited" citizens. A coterie of business leaders raised $10,000 to purchase the Young America. Led by J. Cowperthwait and William Pettit, who also had helped to organize the citizens' committee, this group leased the engine to the city, but they attached strings. They demanded that the city hire paid employees to operate the engine, and they expected that the company running the Young America would form the backbone of a new type of firefighting organization. This placed the business leaders in conflict with municipal officials, who did not support replacing volunteer firefighters.[28]

Fire companies' interest in using the new machine further complicated the situation, in that several vied for the right to operate the engine. The Vigilant Fire

Company offered the most compelling and forceful appeal to use the Young America, and the Committee on Trusts and Fire Department (charged with maintaining the engine) recorded that Vigilant "was very anxious to take charge of the Steam Fire Engine." The only stipulation that Vigilant placed on the arrangement was a request for the city to "furnish horses, one Engineer and a Fireman." The Committee on Trusts agreed to put the engine in service for five years with the company and informed business leaders of the decision. Surprised by the volunteers' request, Joseph Cowperthwait, chairman of the "Trustees of the Steam Fire Engine," expressed reservations about placing the machine with a volunteer fire company. Concerned that it might be "neglected or perverted" from protecting the "safety of property and general good," Cowperthwait refused to loan the engine. He gave two reasons. First, the trustees wanted it to be used for more than five years. Second, Cowperthwait repeated his demands that the engine be used efficiently. To the trustees, efficiency meant that it would be operated by "practical and competent hands, under the direction of an intelligent Engineer all of whom shall be in the employ, under the direction of, and responsible to the public authorities." Echoing the demands made by the citizens' committee, the trustees would not lease the machine unless paid laborers operated it. They contrived to use steam technology to implement their political agenda—to force the formation of a "paid" fire department.[29]

Cowperthwait's opposition stymied the municipality's plan to improve public safety, but the city nonetheless went ahead with its plans. It began to build a house for the steam engine, to purchase horses, and to hire engineers—all of which would be placed under direction of the chief engineer of the fire department. By May 1856, the city council passed an ordinance authorizing payment of workmen to operate the engine. Yet as late as 1860, Fire Marshal Alexander Blackburn reported that "from causes which the Fire Marshal never could comprehend," the engine remained idle. Technical shortcomings accounted in part, as did the city's not having procured horses promptly enough. Also the chief engineer, who was elected by a vote of firefighters, may have contributed to the delay by not pushing harder to use the Cincinnati machine. Ultimately, the Young America never saw regular service. The debate, however, exposed the plans of the mercantile interests. They were less interested in efficiency than in replacing the volunteers.[30]

If many firefighters had been skeptical of steam engines, fire companies quickly warmed to the idea of using the technology. Indeed, a small but significant portion of Philadelphia volunteers saw steam engines as an opportunity to enhance their performance as firemen. Moreover, in a culture that valued a man by his equipment, such apparatus offered companies a new opportunity to exhibit their manhood. If properly constructed, steam machines would allow fire companies to throw more robust streams of water than ever before. The complex apparatus also offered a technological challenge, in that their operation and maintenance demanded skill. Furthermore, because of its expense the new technology tested the esprit de corps of companies, encouraging them to find more creative ways to finance their operations and sharpening their relations to their communities. Driven by a desire to prove their competence as men and firemen, firefighters continued to innovate, and competed to develop the best steam engines.

An increasing number of fire companies that sought to use the new technology did so independently of the municipal government. There were several reasons: the city lacked funds, mercantile interests continued to haggle with the municipality over the conditions under which the Young America might be used, and firefighters questioned the quality of the machine manufactured in Cincinnati. The Philadelphia Hose Company, which had introduced hose technologies to the city, began to experiment with steam engines first. They had even helped test the Young America, seeing its potential value. Following their experience operating that machine, the Philadelphia Hose Company actively pursued acquiring steam technology. Fulfilling its "reputation" and "fame" for innovation, in 1857 the company asked local manufacturers and mechanics to submit plans for a steam engine. It adopted the plans of Joseph L. Parry, a local mechanic and fireman, and a local manufacturer, Reaney & Co., constructed Parry's plan. In January 1858, the engine went into active service; later in the same year, the company won a contest between steam engines that was held on Boston Common.[31]

The Philadelphia Hose Company's acceptance of the new technology signaled a shift in how the city's firemen perceived themselves and worked at fires. When it "manfully bore the brunt of opposition" to its steam engine, Philadelphia Hose instigated a new point of competition among the city's fire companies. In this "spirit of laudable rivalry," according to Fire Marshal Blackburn, the city's most powerful hand engine company, the company that had vanquished the Young America—the Diligent Engine Company—asked to use the engine. Finally the city's mercantile interests relented. After having a local mechanic significantly alter it, Diligent placed the Young America in service in 1859. In short order, Hope Hose, Hibernia Engine, Good Will Engine, Weccacoe Engine, and Delaware Engine all procured steam apparatus. By 1860, nineteen fire companies had bought such machines—some even purchasing a second. Not only did these acquisitions represent a renewed devotion to efficiency, but they also reflected firemen's strong commitment to protecting the city's economic interest. Indeed, companies located nearest to property of especially high commercial value adopted steam technology earlier than those companies located primarily in residential districts. For instance, Fire Marshal Blackburn noted that Hope Hose had "resolved to obtain a steamer for the special protection of the immense amount of valuable property along the Delaware front in the south-eastern portion of the city."[32]

As volunteer firemen began to use steam engines, they continued to denounce the plans of the city's mercantile interests to replace volunteer fire companies with paid laborers. Once again, they commissioned a print to critique middle-class manhood and the division of labor recommended by businessmen. Titled "A Paid Fire Department As It Is Likely to Be under the Contract System," the print expanded on the earlier caricature. At its center, it depicts a steam fire engine surrounded by laconic, disinterested, and underfed men. The steam engine stands idle as Philadelphia burns and as well-dressed and fat middle-class men watch, helpless to stop the destruction. Not only did the image symbolically criticize middle-class men for their lack of action, but it commented on the actions of Cowperthwait and other merchants who refused to allow the Young America to see service. The print made equally clear that no technology—steam or otherwise—could protect the city if it was operated by lazy, inept laborers. In depicting paid firemen as degraded Irish and African American workers, it drew upon two highly charged and popular political and social beliefs of the period—nativist's xenophobia and concerns about the impact of freed slaves on northern labor markets. Speech bubbles underscored the caricature, revealing that the paid workers spoke the patois of the urban immigrant working class and southern migrants/slaves:

"O! I'll sing you a lottle song;
And it shant be very long,
Bout some good people in de town of Boston,
Dey got up a paid Department,
When our councils set dere hearts on't.
Soon day wish'd em to de toder side of Jordan,
Tuck in de trousers, an roll up de sleaves,
Jordan am a hard road to trabbell,
Loaf bout de Injinehouse, do as de please,
Jordan, am a hard road to trabell, I believe."

Another figure, clearly drunk and with a full bottle in his hand, considered:

"To be, or not to be, is not
the Question, wether tis
nobler in the mind to take

several good square drinks
and then put out the fire,
or to put out the fire first,
and take several numerous
drinks afterwards, *that
is the question.*"

The print's imagery, language, and sentiments capitalized on the worst fears of working Philadelphians, who were buffeted by extraordinary societal change. Appealing to their anti-immigrant beliefs and their racism, the print criticized the new division of labor proposed by industrialists. Firefighting service, the image suggested, should not be performed by wage laborers without a stake in community life.[33]

Strikingly, the print did not impugn steam technology. To the contrary, it depicted a beautifully rendered steam engine that bore striking resemblance to the machine that the Philadelphia Hose Company had commissioned from local artisans, and which the company had celebrated in its own laudatory print. The caricature argued that effective new technologies, such as steam engines, were useless unless manned by firefighters with technical prowess, dedication to community and efficiency, and esprit de corps. Ironically, however, as firemen enacted their competence as innovators, they furthered the goals of reformers and the fire insurance industry. Firemen explicitly chose to make their labor more specialized and to reorient it around the power of steam apparatus, continuing a century-long trend of innovating and improving efficiency. However, adopting new technology had unintended consequences. It wove firemen more tightly into Philadelphia's municipal bureaucracy. Gradually, fire companies became dependent upon the specialized and costly expertise of engineers and/or stewards to operate and care for the equipment. Maintaining and purchasing steam engines increased fire company costs dramatically, sending them scurrying for additional income. They found the municipal government willing to exchange financial support for bureaucratic control. In fact, the city refused to support fire companies unless they became part of its administrative umbrella.[34]

Although its impact was great, steam technology did not suddenly change the nature of firefighting work. Rather, it accelerated long-term trends in its organization, especially in the development of specialized levels of service. It particularly reinforced a trend evident as early as the 1840s, in which firefighters established different levels of commitment that devolved responsibility for firefighting labor into the hands of fewer men. Over time, an increasingly smaller number of fire-

"Phoenix Hose" work group, albumen print, ca. 1860s. Over time, fire companies wore increasingly similar costumes, bordering on uniforms, as the differences between fire companies diminished because of common departmental management associations, close bureaucratic and financial ties to municipal governments, and new technologies. Courtesy, Spruance Library of the Bucks County Historical Society; gift of the Volunteer Fireman's Association of Philadelphia, 1919

men performed active service for the broader communal safety. When companies introduced grades of membership and service—active, honorary, and contributing—a new set of responsibilities defined firefighters' relationships to their brothers. Over time, the proportion of active members declined precipitously. In 1842 the number of men active in the department nearly equaled the combined number of honorary and contributing members. By the 1860s, honorary and contributing members outnumbered active company members by a ratio of four to one. This trend was more pronounced in companies that adopted steam fire engines. In the 1860s, the ratio of contributing members to active members was roughly 1:1 in companies operating hand engines, but almost 3:1 in companies with steam fire engines. Companies not only sought additional revenues from the municipal government to operate steam engines, but also accepted a greater number of contributing members in order to support their use of the new equipment. More importantly, perhaps, the actual performance of firefighting labor became the purview of men with a taste for danger and excitement, or those who could not buy their way out of serving.[35]

Volunteer firemen also enmeshed themselves in the web of the municipal bureaucracy by adopting other new technologies in addition to steam engines. The fire alarm telegraph, for instance, helped to tie them more tightly to the authority of the municipal government. By networking the city, the telegraph wove the built environment into a system that connected evenly dispersed alarm boxes with firehouses and an organized system of labor. At the same time, the network regularized the signaling of alarms and improved departmental response time. According to municipal officials, it also removed the moral and economic dangers of false alarms. In other words, municipal leaders hoped that the telegraph would decrease informal expressions of firefighting culture, such as racing. If the telegraph brought greater discipline to firefighters, it nonetheless helped to make them more self-consciously unified. Rather than being an agglomeration of individual companies, volunteers became members of a single fire department, sharing common goals, work rules, and connected workplaces.[36]

As firefighters changed the terms of their service, they also expressed themselves differently in the artwork and costumes that they purchased and in their rituals. As firemen chose to emphasize efficiency and departmental unity rather than company autonomy and neighborhood, the expression of political and social differences became less pronounced and provocative. Between 1840 and 1860—as companies accepted greater numbers of contributing members and as municipal funds became more generous—firemen produced an increasing amount of material culture. At regularly held parades, balls, and other scheduled events, firefighters showed off more detailed and ornate costumes, membership certificates, and engine decorations. The choice of symbols grew less controversial, and these artistic expressions became more standard and stylized as well. Companies more often employed common symbolic language and depicted more neutral subjects, such as tools, apparatus, and fire scenes. Of course, contentious symbols did not disappear altogether, as is evidenced by the municipal government's ban on the use of "provocative" symbols in 1857. However, this prohibition, as well as firefighters' continued attempts to forge a unified identity, fostered both the shift toward greater decorum and the decline in spontaneity. A shared expressive palette replaced the cacophony of difference.[37]

Simultaneously, firemen replaced informal competitions, such as races and fisticuffs, with formally scheduled events, such as footraces and attendance contests. Those rituals that the middle class labeled as inefficient and morally suspect grew into events sanctioned by the department hierarchy. Although rioting diminished, department leaders still punished companies participating in melees, as in January 1869, when department leaders swiftly suspended the Niagara Hose Company and Franklin Steam Engine Company for fighting. The nature of firefighters' parades demonstrated a similar shift, becoming more restrained and more carefully orchestrated to demonstrate departmental unity. Firemen paraded triennially from 1843 to 1852, but did not celebrate their service in 1855, probably because of all the controversy surrounding the department. Parades were held again in 1857 and in 1865. Over time, these processions grew less discordant and kinetic, and later parades acquired a formal protocol. More broadly, firefighters expressed their ethnic, political, and class affiliations less prominently and certainly less provocatively after the 1850s. Although demographic differences between fire companies likely had not diminished, firefighters had begun to view themselves as members of a single department rather than a collection of different associations. Firemen's identities increasingly centered on their labor, equipment, and service.[38]

In the 1850s, volunteer firefighters in Philadelphia had thwarted outside efforts to reorganize firefighting. However, in the process, they created a unified fire department that embraced an increasingly stratified and specialized division of labor resembling the plans laid out by critics. Activists—especially mercantile interests—nudged firefighters forward, but volunteers stood at the vanguard of change, taking those first tentative steps toward creating a new occupation themselves. As firefighters embraced efficiency, technological solutions led them to implement a more complex and formal network of safety. Moreover, as firemen unified into a single brotherhood, they gradually disconnected their service from the metaphor of a body politic and their connections to their urban neighborhoods grew more distanced. Performing public service became an increasingly specialized task, an end in itself as "every man adhered to his own."[39]

Steam Technology and Administrative Reform in St. Louis

The 1850s witnessed similar public conversations about the organization of firefighting across America. In the wake of the horrific conflagration that nearly destroyed St. Louis in 1849, residents reexamined the question of how best to protect their city. Widely heralded for their labor at the blaze, firemen initiated the public debate and created a plan for reorganizing the city's fire defenses, culminating in a new organization. The Fire Association (FA) signaled a shift in how firefighters sought to manage their labor at fires. They abandoned more informal relations, in which they forged authority anew at each blaze, in favor of more centralized control. The new organization unified the city's diverse and independent fire companies into a common "fire department," complete with administrative rules, within a formal management hierarchy. By forming the FA, the city's fire companies created a mechanism for demanding more regular and greater financial support from the municipal government. Firefighters traded the independence of neighborhood organizations for the collective strength of a management association; they did so to improve the quality of fire protection but also to enhance their ability to press their claims for public authority and recognition. As firefighters reformed their service, they articulated a new collective identity.[40]

The FA launched a sweeping critique of the city's fire protection. Led by prominent St. Louis business leader Hiram Shaw, the FA celebrated firefighters and scolded the municipal government and St. Louis's citizens for insufficient support of the city's fire companies. Shaw recalled a record of public service by firefighters, producing a narrative of collective heroism and altruistic service that he contrasted with the behavior of the city and its residents. Yet, even as he looked backward to a mythic past, Shaw began to subtly reconfigure the relation between the city, its citizens, and its firefighters. Shaw raised questions about how best to conquer the problem of fire in a world in which support for firefighters came only reluctantly. As he began to work out alternatives, he produced a plan that reshaped fire protection in St. Louis, and reflected the direction of change in fire protection throughout the United States.[41]

At the heart of Shaw's appeal stood the long-standing belief that firefighters shared a reciprocal relationship with the community. Firefighters protected their fellow citizens, who supplied fire companies with moral and financial support. Shaw argued that firemen "give our services, in winter's cold and summer's heat, at midnight and at mid-day, at a moment's warning." With Thomas Targee's death fresh in the minds of his neighbors, Shaw did not need to restate that firefighting was physically arduous and dangerous work. He did, however, feel compelled to reassure his fellow citizens that firefighters were invested in the community; they were not simply men who loved adventure. Perhaps spurred by concerns about conflict among fire companies, Shaw assured the community of the department's character. He wrote that it was "composed of men who have some stake in the welfare of the city; they have cast their lot here with the intention of making it their permanent home, and having warm hearts and able hands they are disposed to render aid to their fellow-citizens."[42]

Despite firemen's commitment, Shaw told his neighbors, neither the community nor the municipality upheld its end of the social contract with firemen. In particular, he complained that, in its lack of support, the city was failing to fulfill its responsibility not only to firemen but also to its residents. Although the municipal government had paid more than half of the department's expenses between 1838 and 1850, Shaw argued that the city's fire companies were ill equipped. In particular, he reported that the firefighters did not have enough hose. Such laxity, according to Shaw, could have dire consequences and potentially threatened the foundations of the social order. His contentions surely resonated in the wake of the recent disaster. Was the blaze and the disorder in its wake a result of inadequate municipal support? More generally, Shaw questioned the city's long-term commitment to firemen and public safety by criticizing the methods through which fire companies obtained financing from the municipality, which he described as "begging." The unmanliness implied by begging only underscored how the city demeaned firemen, violated the social contract between fire companies and the community, and endangered public safety. Shaw demanded unquestioned, regular pecuniary backing. This support, he told the city council, would "do away with the necessity on our part of *begging*, and of the disagreeable annoyance of having a firemen's *begging* committee calling on you once or twice a week."[43]

Shaw's critique continued, with an interesting twist. He argued that the state's approach to funding shifted the burden of preserving the social order onto the community in an unequal fashion. When he stated that an especially "heavy burden" was placed on a small portion of the community, Shaw did not mean firefighters. Rather, his point was that "the merchants of Water and Main streets pay a large proportion of what is collected." Although Shaw did not document precisely how much St. Louis's merchants offered to fire companies, he and other FA delegates undoubtedly were aware of the amount, because most of the FA delegates were business leaders, engaged in commerce along the city's waterfront. Likewise, Shaw certainly knew that merchants benefited far more than most citizens from firefighters' exertions. After all, in the first half of the nineteenth century, volunteer firefighters focused not on saving lives but on protecting property—and the increasing wealth of the business community certainly exposed it to greater jeopardy. Shaw had begun to shift the terrain of arguments about fire protection. Instead of evaluating fire danger in terms of the entire community's well being, he imagined it as primarily a threat to the middle class and the nascent industrial social order. He assuaged growing middle-class concerns about urban disorder and immorality by underscoring that firemen had "a stake in the community" and "owned property." In so doing, Shaw connected firemen's culture to that of the middle class. Nevertheless, he also expressed apprehension that the department could "fall into the hands of those whose principal delight is in riot and fighting" if his "fellow citizens" were not vigilant at demanding greater support from the city council.[44]

Shaw also introduced to the city a new formulation of the connections between communities and firefighters. He argued that the relationship between firefighters and the community should be governed less by benevolence and more by economic self-interest. He did not appeal to an abstract notion of community or idealized civic virtue; he argued that preserving property was the reason that citizens should support firefighters. In a subtle but important way, Shaw remade the equation between firefighters, communities, and cities. Representing not just himself but St. Louis firemen through the FA, Shaw argued that communal service bounded by a culture of manhood was not sufficient to protect the city's property. Firefighters required additional and regular financial support to provide an increasingly specialized and expensive public service. Through Shaw, volunteer firefighters singled out economic risk as a principal element in the danger that fire posed to community. Shaw's letter provides the first indications that, at least in St. Louis, the language of the market was gaining ascendancy in defining firemen's relationship to the community.[45]

Just as Shaw's letter began to remake the rhetorical connections between firemen and their neighbors in St. Louis, so too the Fire Association restructured the institutional relationship between firefighters and the municipality. This transformation began in the spring of 1850, when the FA petitioned the city council on behalf of the city's fire companies for $1,500 per annum, per company, and up to two thousand feet of new hose per company, as well as free heating gas for each company. The petition further requested that if any company's apparatus was destroyed, the city would agree to pay for its replacement. The association recommended that its members take up a "subscription paper" to improve their financial situations. By May, the FA's plea had an effect. The city council promised a $1,000 appropriation for each fire company (paid quarterly), provided up to one thousand feet of new hose per company, and agreed to pay each company's debt. The delegation reported that this new financial relationship would be acceptable to the municipal government provided that "an inspector appointed by the city" could examine each of the companies quarterly. The Fire Association accepted the city's proposal with one modification. The organization would appoint the inspector, who would be "confirmed by the City Council." This adjustment guaranteed the FA institutional independence from the municipal government.[46]

When the city council had not yet appropriated the funds in July, the FA became militant, precipitating a crisis in the city's fire department. The association petitioned the council for the money needed to "place [the companies] in an effi-

Philadelphia Hose Company steam fire engine, ca. 1858–60, Wagner and McGuigan (printers), G. G. Heiss (lithographer). The Philadelphia Hose Company set the standard for nineteenth-century volunteer firefighters by regularly introducing new firefighting innovations, including the use of hose, grades of membership, and steam engines. In the 1850s, when the company acquired a steam fire engine, it and other like-minded volunteers created the basis for the city's professional fire department. Courtesy, CIGNA Museum and Art Collection

cient situation," and it threatened to strike if these funds were not forthcoming. In "twenty days" the organization would "recommend to each company to close doors until such time as the city council comes to some financial determination." Such militant language emphasized that firemen had defined a new and common interest as a fire department, and it also brought a swift reaction from the municipal government. Mayor Luther Kennett, a member of the Missouri Fire Company and generally sympathetic to firefighters' demands, replied sternly. He expressed confidence that the city would help the department, but he also bristled at the FA's threat. According to Kennett, the city council "would lose sight of the respect due to the position they [firemen] occupy if they are in any way influenced by the threat of closing doors." Kennett asked for the association's assistance in quickly appointing an inspector and requested patience vis-à-vis the appropriation for the department. The controversy ended abruptly with the former volunteer's rebuke, although the organization appears to have retained the right to name the inspector. Shortly thereafter, the city's fire companies elected Edward Brooks—a member of the Central Fire Company and an insurance agent—as inspector. At the same time, the FA bolstered its position at the head of the body of firemen by establishing regulations to "prevent difficulties" and to "restore and preserve . . . efficiency."[47]

The department's status remained unsettled, however, because the municipality wanted greater authority in exchange for its heightened financial role. By February 1851, Kennett had agreed to a yearly appropriation, but the city had not yet settled the companies' debts. When the FA asked the city about the matter—this time in a demure, nonthreatening letter—the mayor replied quickly and with bluntness. Kennett took Shaw's rhetorical arguments for an exchange relationship to their logical conclusion. He wrote that the city would not "pay the debts of the Department and never will, until the companies one and all come under the supreme control of the City Government." Company independence had long been central to firefighters' cultural discourse, and Kennett's edict slapped directly at firemen's control over their service. Edward Brooks, who had assumed the presidency of the Fire Association, announced that the city council had "proposed to adopt a paid department in place of the present volunteer system." Although the FA did not accede to such a dramatic transformation, the association ultimately acquiesced to the city's demand. More tellingly, perhaps, a small but vocal minority of the FA—led by Edward Brooks and several economically prominent members—supported the motion to increase the power of the municipal government. These sympathies became especially evident when he stepped out of the president's chair temporarily in an attempt to table the FA's opposition to the city council's "supreme control." In St. Louis, discussions about the provision of firefighting had changed.[48]

Although the prospect of a "paid" fire department disappeared as quickly as it had materialized, the ensuing compromise transformed the relationship between St. Louis's fire companies and the city forever. In a compromise engineered by Edward Brooks, the Fire Association approved a new ordinance that invested final authority over the department firmly in the hands of the state. The FA retained administrative power over the day-to-day operations in exchange for a generous one-time donation to each company. The ordinance allowed the organization to investigate difficulties between companies, but the city council and mayor retained final administrative authority over fiduciary matters, disciplinary actions, and selection of the department inspector. Claims of reciprocity and autonomy took a different hue under the new arrangement. If men fought fires as part of their obligation as citizens, firemen became implicated in the growing authority of the state and served with its approval. The relationships between fire companies and their communities was now mediated by an intermediary bureaucracy. The change mirrored the broader shift in nineteenth-century social relationships. Legal contracts replaced social contracts.[49]

Of course, not all fire companies adhered to or approved of the new order. The Union Fire Company especially sensed that the new ties prefigured a transformation in firefighting, and decided to disband. Composed of many of St. Louis's most respected citizens, the company operated the city's most effective and revered engine. During the 1840s and 1850s, the "Emperor" and the Union company reigned supreme. Even though Union could not abide the new system, it appears to have supported the shift toward more specialized firefighting. It embraced radical change in 1854, when it initiated the next wave of fire department reform. The company quit the volunteer fire department, sold its assets, and donated a steam engine to the city on the condition that a paid fire department be formed. Just as had happened in Philadelphia, volunteer firemen's approach to technological innovation predicted the reorganization of firefighting, albeit with much different results.[50]

Union's decision reflected the first step in the process that firefighters set in motion as they developed more formal contractual relations with municipal governments and continued to create specialized labor. Following the recommendations of the Union Fire Company and other leading firefighters, St. Louis established a municipally controlled fire department in 1857. The pace of reform sped forward in 1858, as the city's insurance and business community became involved, supporting the change with a concrete assessment of its economic impact. When in 1858 a committee of local insurance companies made their visit to Cincinnati, it returned with glowing accounts of steam engines, noting that the machines helped firefighters to confine blazes to a single structure and that Cincinnati's business community felt more secure because of their use. It recommended that St. Louis acquire the machines because "the people of St. Louis will regard $3000 per annum as a trifling change to secure the efficient aid of six such machines." Even better, insurance rates would lower and "two or three hundred thousand dollars worth of property can be saved from fire each year." Even before they had returned to St. Louis, the insurance committee contracted with Cincinnati manufacturers for two steam engines.[51]

Metaphorically, Union's decision to disband symbolized the transformation of fire departments from 1850 to 1870. At one time, the company, its engine—the Emperor—and its motto IN UNION THERE IS STRENGTH had expressed the possibilities of unified male action and celebrated the political order of the nation. By relinquishing its separate identity, the company repudiated the community organization of fire protection. It also challenged the concept of the collective male body as an appropriate organizing principle for an industrial society. By casting its collective body aside in favor of a new division of labor, Union embraced the increasing role that specialization and individualism would play in the expanding market economy. In many ways, this decision foreshadowed the social and political fragmentation that wrenched the nation during industrialization and the Civil War.

Volunteers, Innovation, and a New Occupation

Too narrow a focus on methods of remunerating firefighters or on steam engines fails to explain adequately the transformation of firefighting during the middle decades of the nineteenth century. It mischaracterizes the change, conceals the dynamic role that volunteers played in the transition, and especially obscures the degree to which firefighters invented their own occupation. Firefighters had been in the process of transforming firefighting from a broad community endeavor to a specialized public service from the earliest days of the nineteenth century. Gradually, they had been reorganizing their work, their relation to one another, and their association to the local municipality. With the creation of municipally administered fire departments, the role of volunteers in firefighting did not diminish; in many cases, volunteer firefighters actually organized, operated, and directed the new departments. Likewise the organization of firefighters' work did not change fundamentally or abruptly. Quite often, city-operated departments employed hand-pumped engines as well as part-time laborers to extinguish fires. For instance, in Cincinnati, which is typically labeled the first "paid" fire department, volunteers initiated the transition to a professional department by publishing a letter in a local newspaper suggesting just such a shift. Afterwards, a former volunteer fireman and prominent industrialist worked with a cadre other former volunteers to structure the new department's organization and work rules.[52]

As firefighters pushed forward with department reform, volunteers faced a critical decision: should they give up their current employment and take up a new occupation—as firefighters? Although relatively few volunteers joined the new organizations, those who did supplied both the labor and leadership so critical to the restructured fire departments. In Philadelphia less than 5 percent of the city's approximately three thousand firemen chose to or were given the opportunity to earn wages as firefighters in 1871. However, that small crew wielded an extraordinary amount of influence in the development of the department. In fact, approximately

40 percent of the members of that new institution—165 of approximately 400 men—had been listed in the city's enumeration of volunteer firemen in 1868. Likewise, in St. Louis former volunteers performed a vital role in the new department. Though their prevalence is difficult to quantify with precision, evidence suggests that about one-third of the firefighters in the municipally administered organization had previously served with the volunteer department. More importantly, though, former volunteers occupied nearly all of the skilled positions in the newly created St. Louis Fire Department.[53]

When St. Louis firefighters created a municipal fire department, they not only assumed key leadership positions but also provided skills essential to the operation of the city's first steam fire engine, which represented the department's future. As the city continued to move toward reforming its fire department, it relied on the expertise of four volunteer firemen who served on the city council and had nominated themselves to the Board of Fire Commissioners. The fire commissioners appointed Henry Clay Sexton, former president of the Mound Fire Company, as the chief engineer. In turn, Sexton appointed Richard Beggs and John Bame, former officers of the Franklin and Phoenix fire companies and leaders of the Fire Association, as assistant engineers. With volunteer firefighters placed in prominent leadership roles, the newly appointed Board of Fire Commissioners acquired equipment, infrastructure, and labor from volunteer fire companies. Almost immediately, at least three of the city's ten volunteer companies—the Franklin, Washington, and Mound fire companies—supported the new organization. They complied with the ordinance establishing the paid department, disbanded, and sold their equipment and property to the municipal government. Equipment from these companies supplemented new purchases (of steam apparatus) and helped to provision the department.[54]

In its final step toward reorganizing firefighting, the St. Louis Board of Fire Commissioners obtained the most valuable commodity—the skilled labor of former volunteer firemen. Although the numbers varied by company, many volunteers joined the organization. The records of the Franklin Fire Company, the volunteer company for which there is the most information on membership, indicate that approximately one-third of its men joined the paid organization, and most signed on with the new department's Franklin Fire Company. More importantly, of the (municipal) Franklin Fire Company's eight full-time employees, all but the engineer were affiliated with the earlier volunteer organization. Twenty-one of Franklin's members—over 75 percent—had served as volunteers. Former volunteers also occupied at least 60 percent of the new department's most responsible jobs, and they commanded all of the department's six engine companies. By contrast, men without previous experience served as part-time laborers—or "call men"—on hand engines.[55]

The organization of firefighting in St. Louis showed continuity with its preceding form, but the former volunteers who led the department remade the division of labor in important ways. Departmental pay stubs demonstrate that operating hand-pumped engines—once a marker of manliness—became less valued as firefighters transformed their beliefs about technology, skill, and physical labor. Indeed, the vast majority of employees in the department occupied part-time, low-wage positions as physical laborers operating hand-pumped apparatus. By the winter of 1857, approximately 80 of the 130 men employed by the department worked as "privates" in the four hand-engine companies—the Jefferson, Washington, Mound, and South St. Louis fire companies—which employed twenty privates each. Privates, or call men, were allowed to hold outside jobs, but they were required to attend fires when called. Privates earned a fixed salary of about $.27 daily, just over $8 monthly. However, if they did not appear at fires, they were fined as much as $1. In January 1858 one firefighter accumulated $2 in fines, more than 25 percent of his salary.[56]

Skilled workers earned significantly higher salaries and won greater privileges than physical laborers. Each fire company, whether operating a hand-pumped or steam-pumped apparatus, usually employed seven "stewards." Disproportionately drawn from the ranks of the volunteers, stewards maintained the apparatus, directed the horse-drawn engine to the blaze, and manned the hoses. Stewards earned five times as much as call men, but were prohibited from holding a position outside the department. They took home $1.33 daily, $40 monthly, and $480 yearly, which did not vary according to the type of equipment their company operated. Captains occupied the space between stewards and call men. Fire company leaders received lower pay than full-time stewards did, but their salary of approximately $250 to $300 yearly does not tell the full story. Like the department's part-time employees, captains could hold jobs outside the department. This arrangement suggests that the department may have had difficulty prying the most skilled, or at least the most respected, volunteer firemen away from their other careers.[57]

As happened in fire departments throughout the country, steam engines became a critical element in the organization of firefighting in St. Louis. Even before the city assumed complete control of the fire department, the municipal government began operating the steam engine that had been donated by the Union Fire Company, and within two years the city had acquired six additional steam apparatus. In 1857, the city hired a volunteer firefighter with the Phoenix Fire Company, Richard

Life of a Fireman. The New Era. Steam and Muscle, 1861, Charles Parsons (draughtsman/artist), Currier & Ives (lithographer/publisher). The physical work of hand-pumped engines contrasted remarkably with the mechanical power of steam engines, as shown in this Currier & Ives print from the 1860s. Steam engines and paid laborers replaced hand-pumped apparatus and volunteers gradually, and sometimes these divergent methods co-existed as cities transitioned toward a new division of firefighting labor. Courtesy, Spruance Library of the Bucks County Historical Society; gift of William B. Thomson

Mawdsley, to serve as "superintendent" of the engine. Mawdsley directed a labor gang composed of eight workmen plus horses; for his expertise he earned $75 monthly. The crew—two drivers, two firemen, three "pipemen," and a watchman—were each paid $40 monthly. All told, the city spent approximately $470 each month to operate its steam fire engine. Similarly, after the new department had been organized, those men operating steam equipment continued to receive higher salaries, underscoring the importance of steam technology to the new organization. For example, by 1865, as the principal operators of the city's expensive and sometimes temperamental steam fire engines, engineers received $2.74 daily, over $80 monthly, and $1,000 yearly. Meanwhile, men who led hand-engine companies earned nearly 33 percent less than the captains of steam engine companies. They received about $17 daily and $200 monthly, while their counterparts earned $25 monthly, and $300 yearly. The pay disparity suggests the relative scarcity of the technical skills held by steam engineers in the 1850s, even as it points to the value of steam engines within the department. The department considered these machines its most important resource and viewed the ability to operate them as a most critical skill. Once again, firefighters and municipal leaders emphasized a man's specialized ability in the use of technology as the organizing principle for firefighting.[58]

Departments compensated engineers so well because they possessed both diagnostic and practical technical skills. Most obviously, engineers operated the steam apparatus at fires, which involved stoking them to full steam and sustaining adequate pressure. In addition, engineers' expertise extended to the nuts and bolts of keeping the engine in running order. Many early steam operators maintained their apparatus, although local machinists performed the most serious repairs; by the twentieth century, departmental machine shops would typically perform most maintenance activities. In 1874 at the annual meeting of fire department leaders, St. Louis Fire Chief Henry Clay Sexton recommended hiring engineers who were "practical machinists." Firefighting handbooks argued that engineers should possess broad knowledge of the science of steam engines, including knowledge of hydraulics. Although wage differences between steam engineers and rank-and-file firefighters eventually diminished, they nonetheless reflected the degree to which steam engines were the backbone of newly reorganized departments.[59]

As firefighters constructed the new occupation, many customs carried over from volunteer to municipal fire departments. For example, firefighters continued to emphasize speed and efficiency in transmitting water from city mains onto a fire. Just as volunteer firemen had measured their manhood and service by arriving quickly at fires and by spewing as much water as possible, so too paid firefighters construed their headlong dashes as markers of virility, and reveled in playful competition. In fact, long after firefighting had become a well-established occupation in large metropolitan areas, firemen competed so ferociously that departments issued orders to prevent racing to fires. So strong was this impulse that, at late as 1915, Philadelphia's training manual for firemen admonished companies to work together. "It is a natural and proper thing for a company to take pride in being the first to play a stream on the fire," the manual stated, "but at times it is necessary to forego such personal glory to help another company."[60]

The drive behind the desire to arrive early and to play first water was manifest even in firemen's battles over departmental reorganization. In fact, this competitive streak structured the contest for power between volunteers and professionals during the 1850s. In St. Louis, the municipal department's ability to control the city's water supply played both a symbolic and a practical role in establishing its authority. In 1857, a number of independent volunteer companies remained active in the city, challenging the hegemony of the new fire department. The volunteers believed that acquiring hose connections first might discredit the municipal organization because custom dictated that the first company to control a hydrant commanded the fire scene. And, the independent companies held a critical advantage; they were located near to the city center and the business district, whereas the engine houses of the paid department were situated further away. With a geographical edge, the independent companies often gained charge of *all* the plugs near to a fire. Rather than seeking legal remedy, the city's municipal firemen engaged the independent volunteers on their own terms. Taunted as "hirelings" by men with whom they had once worked side by side, the wage-earning firefighters competed for control over the water supply. They converted a four-wheel hose tender into an apparatus that could be horse drawn, and two firemen (Dressell and Marquis) patrolled the business portion of the city, especially at night, awaiting alarm of fire. Using the superior speed of the horses, Dressell and Marquis would take possession of as many fire plugs as possible before the independent volunteer fire companies arrived; they had "outwitted" the volunteers.[61]

This contest underscored the ongoing importance of innovation in firefighting culture and outlines the connections between early municipal firefighting organizations and their predecessors. In their continued emphasis on technological skill, paid firefighters defined their occupation by drawing upon the cultural legacy created by the specialized corps of volunteer firefighters. Just as volunteers altered their work and culture by inventing and/or appropriating technologies, wage-earning firemen developed and implemented a variety of new workplace tools. The use of horses to gain an edge in the contest over hose connections was just the first in a series of innovations undertaken by successive generations of firefighters that later resulted in technologies such as the "quick-hitch" harness, chemical fire engines, and aerial ladders.[62]

More broadly, when they reorganized fire protection, firefighters created a new occupation that was grounded in ideals of manly control and service. They took command of fire protection by asserting their control over nature and the city. By literally wresting control of the city's hydrants from their competitors, they established their mastery of water and the infrastructure—tools of signal importance in the battle against an environment run amok. Even more striking, St. Louis's firefighters did not establish their authority through state mandate. Rather, they displayed manly competence—signaled through their use of superior technology and their extraordinary dedication to public service. The same principles guided the new firefighters' activities, and eventually would propel them deeper into burning buildings and into more dangerous situations.

Conclusion: Constructing Technology and Identity

Changes in the provision of firefighting paralleled, if not preceded, the growth of other municipal services at midcentury, such as police departments, public health departments, and sewer and water systems. No single event precipitated the expansion of such services, although their extension certainly derived from a heightening sense of danger and hazard—of human disorder, fires, disease, and accidents. Urban populations and political leaders developed the belief that cities had an obligation to provide for public safety, which became more pronounced as municipalities took more responsibility. The growth of urban firefighting services depended upon a confluence of factors: the development of a labor force of specialists, the activism of a middle class, the support of business and mercantile interests, and the willingness and financial ability of local municipalities to become more bureaucratic and service oriented.[63]

In this context, volunteer firefighters initiated and led the transformation of their service. From early in the century, firefighters frequently remade their labor and adopted new technologies, making their service more efficient and bureaucratic. Volunteers claimed to be noble public servants and to possess a special expertise in fighting fires. Gradually, firefighters' developed a collective interest and identity that undermined the communal origins of their service. The emphasis on innovation and efficiency embodied the spirit of the industrial era, compelled firefighters into a more formal relation with municipal governments, and propelled them toward an organization of labor that resembled the proposals favored by the fire insurance industry. However, the particular trajectory of change and the shape of firefighting reform efforts varied widely, depending upon locale. For instance, in Philadelphia, with its long tradition of voluntarism, firemen incorporated steam technology into their companies, created an administrative bureaucracy, and established paid positions for department leaders and skilled steam engineers. In St. Louis, however, volunteers first changed their relation to the municipal government and later moved to adopt steam engines, which paved the way for, but did not cause, the transition to a municipally administered fire department. In other cities—Boston, Chicago, New York, and Cincinnati—volunteers played equally prominent roles and were deeply engaged in creating paid fire departments.[64]

Although paying firefighters wages and using steam engines mattered to the reorganization of fire departments, these factors reflected a larger process through which firefighters were crafting a new occupation. Becoming a firefighter meant possessing and exhibiting a special set of skills and attitudes, embodied in a new division of labor. In St. Louis a group of specialist firemen, almost all of them former volunteers, directed a fire department composed of volunteers and "call men"—ordinary laborers with other occupations—for several years after the city reorganized the department in 1857. Likewise, Cincinnati's fire department did not employ full-time firefighters other than departmental leaders until the 1870s. For twenty years, the city paid workers in the department a wage, but still expected them to earn their livelihoods from other jobs. Gradually, the department raised wages to discourage firefighters from taking outside employment, but it did not prohibit such employment until 1873. Strikingly, although preindustrial traditions slowed the shift toward the creation of paid firefighting forces in New York and Philadelphia, in most respects those departments resembled St. Louis and Cincinnati more than they differed from them in the 1860s. In Philadelphia volunteer firemen entered into a contractual relationship with the municipal government, initiated the use of steam technology, encouraged a telegraphic alarm system, and adopted strict work rules and a bureaucratic hierarchy to manage the department. A coterie of skilled workers directed the operation of the fire department, and each individual company received a government appropriation to pay its bills, which included hiring specialists to operate steam engines. The new division of labor, with formal bureaucratic routines and wages for departmental leaders, represented a significant departure from the informal culture of neighborhood fire companies. Firefighting in American cities had begun to show striking similarities in organization, which were advanced more quickly by the shift toward using expensive new machine technologies. By the 1880s, steam engine manufacturers—Amoskeag, Ahrens, Silsby, and LaFrance were the most prominent—had sold thousands of the machines to departments nationwide. Widespread use of steam apparatus reflected how firefighters had begun to remake their service from an avocation into an occupation, with common tools, formal work rules, hierarchical leadership, and well-defined skills.[65]

More broadly, as firemen and underwriters debated how best to control fire, both groups contributed to the reorientation of the American society around the values of industrial capitalism. Indeed, even as they argued about the nature of fire protection in newspapers, on broadsides, in art, on the streets, and at fire scenes, insurers and firefighters adopted an approach to the environmental hazard that placed authority over the problem into fewer hands. Fire insurers made containing the economic damages caused by fire their central business concern, providing individuals an opportunity to protect themselves financially. At the same time, firefighters had established themselves as the first and only line of defense against the fire's physical hazards, filling a critical functional niche within the urban industrial world. As they worked to make fire protection more effective in the first half of the nineteenth century, firemen forged a brotherhood built around physical and technical competence, in which the performance of efficient duty gradually eclipsed a man's connections to his local community. Interestingly, even as firefighters and underwriters narrowed the responsibility for fire protection, the risk of fire nonetheless remained a collective problem—represented by firemen's bodies and insurer's economic portfolios—but one increasingly managed by specialists.

Part III / Water

CHAPTER FIVE

Disciplining the City:

Everyday Practice and Mapping Risk, 1875–1900

Infected by the culture of unfettered laissez-faire capitalism and emboldened by its discovery of actuarial method, the fire insurance industry pursued a market-oriented solution to the problem of fire in the last three decades of the nineteenth century. With a few notable exceptions, most companies paid little heed to preventive measures until late in the century. Indeed, as late as 1894, the leading industry trade association argued that it was "not the province of the insurance companies to regulate fires," but rather that of ordinary citizens. The industry claimed that it recommended methods for preventing fire, but ultimately, insurers could "make our profit out of these conflagrations, for we can figure them up, and assess them to the public." Gathering information and managing it properly became increasingly crucial to understanding how fire affected company risk portfolios. Toward this end, underwriters developed standard administrative procedures and forms, created classification systems to order risk, and used representational technologies, such as maps and statistics, to organize everyday practices. Through common routines the industry sought to develop a unified self-discipline that would produce economic security for firms and policyholders.[1]

The industry argued that security from fire would increase only if it developed standard practices of inspecting and rating properties and then applied them in an unvarying fashion. Such consistency would encourage the long-term process through which safety would be "burnt in" to the landscape, to borrow industry leader J. B. Bennett's phrasing. As Bennett phrased it shortly after the Civil War, "experience prompts improvements everywhere, and substantial edifices of stone, brick, and metal, gradually replace the hastily and cheaply erected buildings of pine." Bennett further argued that property owners would build safely and draw lessons from the "science of insurance" in order to reduce their expenses and exposure to danger. However, he noted emphatically that this would happen only if the insurance industry was united in its method of apprehending and rating fire risk—if, for instance, firms within the industry followed the unilateral policy of leveling large premiums on badly constructed property. According to industry leaders, then, setting high rates and maintaining them would diminish losses, raise profitability, and even encourage the spread of safe building practices. In addition, many firms advocated insuring less than the full value of property and goods, usually three-fourths of the total. Throughout the last decades of the nineteenth century, C. C. Hine, J. B. Bennett, and other experts warned against "overinsurance." Although not new in the 1870s, this emphasis took on new significance because of the industry's attitudes about safety. In short, underwriters believed that laying part of the financial responsibility for loss directly on the insured would provide property owners with additional incentives to prevent fires.[2]

Even as the industry eschewed direct responsibility for fire public safety, it remade itself by extending the business strategies developed by industry leaders in the 1850s. Between 1870 and 1900, the industry forged a common set of practices and management procedures by increasing the intensity, extensiveness, and depth of its surveillance of the danger of fire. Underwriters implemented more detailed administrative routines, used statistical principles to calculate insurance costs more accurately, commissioned ever more expansive and detailed maps of the built environment, and continued to create elaborate systems of classification to bind each of these technologies together. These dynamic activities—managerial, actuarial, representational, and classificatory—cannot be dissociated from the broader system of administrative technology being worked out within the fire insurance industry. Indeed, as underwriters took a risk, they used each of these technologies to objectify and analyze danger. Embedded in these knowledge-gathering technologies lay a set of interlocking safety standards, which thoroughly demystified fire risk. Information technologies had helped to make fire danger routine, and the fire insurance industry had begun to reimagine a safe city, at least from the vagaries of this one environmental hazard.

Equally important to fostering unity of purpose and procedure was the establishment of industrywide administrative associations. Formed by the industry's largest and most influential companies in 1866, the National Board of Fire Underwriters (NBFU) became the most important (though by no means only) such organization formed within the industry. Building on the development of standard business procedures, the NBFU specifically advocated setting rates derived from the statistical record of losses and hoped to standardize rates of premiums and expenses across the industry—both integral elements to improving the industry's financial position. Although the NBFU never achieved its goals, and suffered through nearly two decades of retrenchment from the mid-1870s through early 1890s, it spawned a network of local and regional underwriting associations that would have a lasting impact on the industry. These organizations fostered a groundswell of cooperative activity among insurance firms, including collaborations on managing information. They filled the void left by the NBFU, and they generated intimate connections between the insurance companies and local environments. Eventually, the interrelationships fostered by the local societies would reinvigorate the National Board of Fire Underwriters. Building on the extraordinary familiarity that the industry had developed with fire danger and its relationship to the built environment, the NBFU would even begin to experiment with fire preventive measures in the 1890s, even as it stated publicly that fire insurers had no responsibility for public fire safety.[3]

The National Board of Fire Underwriters

When in 1866 over one hundred fire insurance companies organized the NBFU in order to pursue a "common interest," the new organization did not explicitly identify a single overarching organizing principle, though its program nonetheless contained a coherent logic that defined the mainstream fire insurance industry's goals for over thirty years. The NBFU most pointedly expressed concern that insurers did not manage information very well, whether it was in reference to business costs, losses, or income, and its statement of purpose focused on developing standard daily practices that would allow it to improve profitability. At its first annual meeting, the NBFU described its mission explicitly in terms of establishing and maintaining "uniform rates," establishing "uniform compensation" for agents, repressing arson, and devising a means to increase cooperation between individual firms.

Each element of the NBFU's agenda supported a common principle revolving around understanding how fire affected the loss portfolios and financial ledgers of companies in the industry. For instance, creating standard commissions lowered administrative expenses and made them predictable across the industry, helping companies to manage their costs more effectively. Likewise, setting common rates would increase premium income, or at the least make it more predictable. Finally, more effectively prosecuting arson promised to reduce losses across the industry. Taken together, then, the NBFU's advocacy of these three points was about setting in motion a strategy that was primarily about reducing and/or controlling expenses, increasing and/or regularizing income, and reducing losses.[4]

When the NBFU advocated creating "uniform rates" it hoped to establish an actuarial record of fire loss that would serve as a guide to the industry and that would generate a set of common practices across the industry. The industry did not—at least initially—seek to create uniform rates simply to improve profitability, although the benefits of standard rates were obvious. The NBFU wanted to create common rates in a systematic fashion that offered an objective portrait of fire danger; it wanted to establish and oversee an industrywide "mortality table," which it believed could become the basis for the development of rational and scientific principles to guide the industry, eradicating intercompany differences and local particularities.[5]

Creating a master actuarial table demanded extensive industrywide cooperation in gathering the requisite information. Toward this end, the NBFU developed a partnership with local boards of fire underwriters (and encouraged the formation of such boards). Local insurance boards served the industry in seemingly mundane ways; for instance, they assigned committees to keep statistics, to classify risks, to conduct surveys, and to adjust losses. At the same time, the NBFU hoped to utilize the actuarial data gathered by large insurers, which had begun to use statistical methods into their daily operations with greater intensity. Using standard administrative routines, following a national fire-mortality table, and heeding information received from local underwriting boards increased the likelihood that insurance companies would arrive at similar conclusions about a fire risk. That is, these practices would provide underwriters with a unified front in the battle against fire. The NBFU and its members imagined an insurance business predicated on statistically self-evident rules. Not only would differences within the industry disappear, but also business practices predicated in the certainty of statistical science would make the problems of fire insurance vanish.[6]

The strategy of setting common rates especially benefited the large nationally oriented companies that dominated the NBFU. Indeed, as much as rates of insurance reflected company losses, they also reflected a firm's expenses, and the nation's largest firms had a network of agents and a (relatively) large bureaucracy, which increased those costs. For instance, companies such as Aetna needed to set premiums at a level high enough to both pay losses and cover the expenses of their exceptionally large network of agents. As a result, the initiative to establish standard rates worked to the advantage of large insurers, especially if the rates were fixed at a relatively high level, because it eliminated price competition, which worked to the benefit of small, usually local companies with low overhead. Likewise the program of setting standard commissions for agents was designed to regularize and reduce administrative costs, thereby improving the market position of large national insurers. Generally speaking, two major tenets of the NBFU's agenda—setting uniform rates and commissions—worked against the interests of small local and mutual insurance companies, which operated without the large staff overhead of national insurers.

Through the last years of the nineteenth century vast industrywide differences thwarted the NBFU's program to create uniform rates. Some companies organized themselves along the principle of mutuality; others raised capital by offering stock. Some competed for business primarily in local or regional markets; others hired vast numbers of agents to prosecute their interests in faraway locations. Though there are many exceptions, for the purposes of this study company organization can be divided into two basic types: mutual and stock. Mutual insurance companies typically served particular geographic communities or regions but sometimes attended to communities of similar economic interests, such as textile mills. For the most part, these companies employed few agents, and company officers solicited business and inspected risks themselves. Less bureaucratically complex than many stock companies, mutual companies also functioned on a different principle. Unlike stock companies, mutuals returned dividends to the insured (less administrative costs and losses) rather than to stockholders. Over time, mutual insurers gradually expanded beyond local boundaries and adopted practices more characteristic of stock companies, but this did not occur until well into the twentieth century.[7]

Another distinction typically made within the industry was the organization of a company's business—local, regional, or national. When the NBFU spoke of a common interest it referred to a relatively select group of large companies "doing an agency business." The term *agency business* designates companies that required the services of business agents located in distant places to solicit new business, inspect risks, write policies, and adjust losses. Companies that did business across a broad geographic area typically hired agents to serve as intermediaries between the underwriting company and the insured. As a result, the NBFU represented especially the financially secure, well-capitalized firms doing business nationwide, often headquartered in places like Hartford, New York City, and Philadelphia. By contrast, local companies had a more direct relationship with the insured and often dealt directly with their clients. Among other things, these different relationships produced different cost structures. Agency companies relied on an extensive, expensive bureaucracy to conduct business; they compensated brokers or agents from 10 to 20 percent of the premiums they collected. Local firms avoided this expense by doing inspections and policy writing themselves.[8]

Of course, agency companies and local firms disagreed vehemently on which mode of organizing was superior. For example, many smaller insurers criticized larger, so-called agency companies for gouging consumers. In addition, these regionally focused firms often accused big, national concerns with removing capital from their local economies—usually represented in terms of money moving from southern and western towns to eastern metropolises. In response, executives of those large organizations, such as Aetna's J. B. Bennett, argued that nearly all the capital spent on premiums within any given city was returned to that community through payment of losses, taxes, and commissions to agents. Bennett further argued that the extensive portfolios of larger firms distributed fire risk across the national economy, thus affording greater protection to customers. J. B. Bennett described this benefit by noting that "no experience can be cited to warrant the expectation that five hundred [separate risks] will burn at once." Bennett further claimed that well-capitalized companies were more cautious because "capital, by its very nature, is conservative, and holds to the more sure and the safer modes of business." Lastly, he argued that if managing large volumes of business "required a higher order of experience," it also meant that large firms suffered many losses. According to Bennett, the resulting glut of data made insurers with major firms more knowledgeable about the problem of fire. Thus substantial portfolios of geographically dispersed risks actually buttressed good practice, which, according to Bennett, "was governed according to certain laws of average, only obtainable from a broad and varied experience."[9]

Despite the clear advantages of such an approach, the fire insurance industry remained a risky investment in the final decades of the nineteenth century, which reveals the industry's deep divisions and speaks to the difficulties of enacting what was sometimes termed "scientific underwriting" in a competitive market economy. The NBFU's ongoing struggle exemplifies the waves of success and failure that the industry faced following the Civil War. Immediately after its formation, the board experienced explosive growth and modest success in its attempts to enforce a rating discipline. Over 135 underwriters drawn from 99 companies and 32 agencies attended the NBFU's 1867 meeting, and the executive committee reported that it had organized "over two hundred local boards with rates more or less advanced, and uniform in character." Within three years, the NBFU had established an extensive administrative apparatus for the express purpose of advocating its "basis rates" and enforcing them among agents and companies. By 1869, it had formed a Rating Bureau, divided the nation into six departments staffed by paid employees, and enlisted the support of 475 new boards. In Chicago, thirty-seven companies formed the Chicago Compact and pledged to remove local agents if they violated the NBFU's rates a second time.[10]

The organization's mandate, though, was short lived, and it declined in the 1870s as rapidly as it had grown. In particular, the board's authority to raise and standardize rates lasted for a very short time, and in 1869 it disappeared abruptly in a flurry of rate cutting. Even those companies that had signed the Chicago Compact evaded its provisions. Forming a "Special Committee on Reorganization" in an attempt to maintain its authority, the NBFU reported "bad faith among those who have been understood to consent to the legislation of the board," and proposed a solemn oath be signed by all members—"The Articles of Association and Obligation." Although a few companies signed the articles, most did not. By 1870, the NBFU acknowledged defeat when it allowed members and local boards to "modify, suspend, or declare advisory any or all rates fixed by them."[11]

Just as quickly as the NBFU had lost rate-making authority, it regained it in the wake of the disastrous Chicago and Boston conflagrations of 1871 and 1872. The story of the massive destruction wrought by those fires is well known. After the Chicago fire, more than one hundred fire insurance companies went belly-up; similar consequences followed the Boston blaze, and even the largest insurance companies did not escape unscathed. The Insurance Company of North America paid nearly $1 million in losses to Boston policyholders. The losses shook the fire insurance industry. One company official worried that "one more such conflagration will strip this country of every fire-insurance company." The NBFU, whose membership was galvanized by the conflagrations, raised rates in small towns (population less than 50,000) by 30 percent and those in large cities by as much as 50 percent. Its plan worked. In 1873, according to the memories of prominent St. Louis underwriters James Waterworth, rates throughout the nation had increased by 25 to 50 percent.[12]

For several years following the disastrous fires, the NBFU held broad cultural authority, as firms that survived the conflagrations joined the organization in droves. By 1874, the board represented 90 percent of the nation's insurance premiums and 95 percent of the nation's total insurance capital. Representing such an enormous amount of the industry's capital provided the organization with enormous political and social leverage, which was heightened by popular concern about the potential devastation of future conflagrations. The NBFU enacted its extraordinary power through the office of a newly appointed general agent, Thomas Montgomery, who enforced the board's edicts with great vigor. During 1875, for instance, Montgomery and his supervising agents held 29 trials, convicted 120 local agents for violating NBFU rates, and collected nearly $6,000 in fines. Shortly after the St. Louis Board of Fire Underwriters (STLBFU) formed, Montgomery demanded that that organization adhere to the national organization's rating schedule. The St. Louis board complied immediately. The NBFU's history reported that in the 1870s "the National Board constituted a virtual business monopoly" and was perhaps "the first great American trust."[13]

The NBFU used its renewed power to demand better methods of construction and seemed on the verge of becoming a powerful advocate of public safety. In fact at the organization's 1873 annual meeting Henry Oakley, the board's president, spoke of the need to "shape legislation which would benefit our own interests . . . by securing such wise and salutary laws as might prevent the recurrence of other destructive conflagrations." In conjunction with local underwriting boards the NBFU showed intense interest in how factors such as fire defense, water systems, and urban design affected the problem of fire. In 1872, for instance, the organization published a rate sheet that recommended lower rates for towns with water supply systems, and by the 1880s, making such distinctions had become commonplace. In 1888, the American Water Works Association, which monitored rate differentials in fire insurance, reported that towns with waterworks experienced rate reductions of between 20 and 50 percent. Nonetheless, the NBFU and the insurance industry applied little direct pressure on cities to develop or modernize their water systems. As much and probably more pressure came from a variety of other sources. For instance, the growing American public health movement and engineering professionals were among the most vocal proponents of improving water systems. The American Public Health Association, formed in 1872, and the American Water Works Association were especially forceful proponents of the expansion of city water systems. Between 1870 and 1890 the number of waterworks more than tripled, and by 1890 over 70 percent of American cities had waterworks. Similarly, insurers showed brief but intense interest in improving construction practices in the wake of the Chicago and Boston fires. In fact, the NBFU argued that Chicago and Boston were being rebuilt improperly and demanded that those cities rebuild according to stricter rules. Even though both municipal governments succumbed to NBFU pressure by mandating stricter construction standards, those codes were rarely enforced.[14]

Once again, however, the NBFU's authority was short-lived. Lack of industry consensus and consumer displeasure with high rates challenged the hegemony of the national organization. One local board, the New England Provisional Committee, recorded this dissatisfaction when it told the NBFU that it had received over four thousand applications for revisions in rates and new ratings. Competing on rates once again became a familiar strategy of agents and companies seeking business. According to one observer, by 1875 over one hundred new "rate-cutting" companies formed. NBFU members were not exempt from the changing business climate; twenty of its members resigned, eighteen were dropped, and two were expelled. According to the NBFU's historian, the "fire insurance autocracy" had met its death. In 1877, the Phoenix Insurance Company's D. W. C. Skilton identified the cause of the decline: "It is plainly evident that the future success of the National Board of Fire Underwriters depends wholly on a full and firm confidence among the members in each other, . . . it being granted that it is now practically lost." In the wake of this loss of legitimacy, the NBFU appointed a committee on retrenchment to examine the future. The committee advised abandoning rate-setting functions—a measure approved by a vote of thirty-one to nine. The committee further recommended reducing the budget from $113,000 to $15,000. In the wake of the budget cuts, the NBFU declined precipitously. By 1888, only twenty-five members remained; eight companies attended the annual meeting.[15]

Ironically, the NBFU's sagging fortunes came in part because its goal of creating uniform rates was out of touch with underwriters' and their agent's growing knowledge of fire danger. Insurers' daily encounters with fire suggested to them that the problem might be more complex. Those underwriters who had field experience or worked daily in assessing risks especially became sensitive to the ways in which their particular locales may have been exceptional. According to many local underwriting officials, uniform rates—or "flat rates" as critics derided them—treated the landscape with a simplicity that was being outpaced by the problem of fire, as well as the industry's increasing understanding of it. Almost two decades after firms had begun to use statistics in evaluating risk and had expanded their programs of methodical observation, insurers continued primarily to emphasize internal dangers when setting rates, although they realized that fire risk contained an external, environmental component. That is, underwriters realized that fires and losses did not occur, or mount, simply because of how a property was used or because of its poor construction. However, the NBFU's program of uniform rating relied almost exclusively on such a determination of this intrinsic risk. In addition, this program implied that hazards did not vary, even in different places, and were *not* modified by environmental conditions. Although the NBFU's proposals were not so inflexible as critics argued, they nonetheless were out of step with an industry that viewed fire danger through an increasingly complex lens.[16]

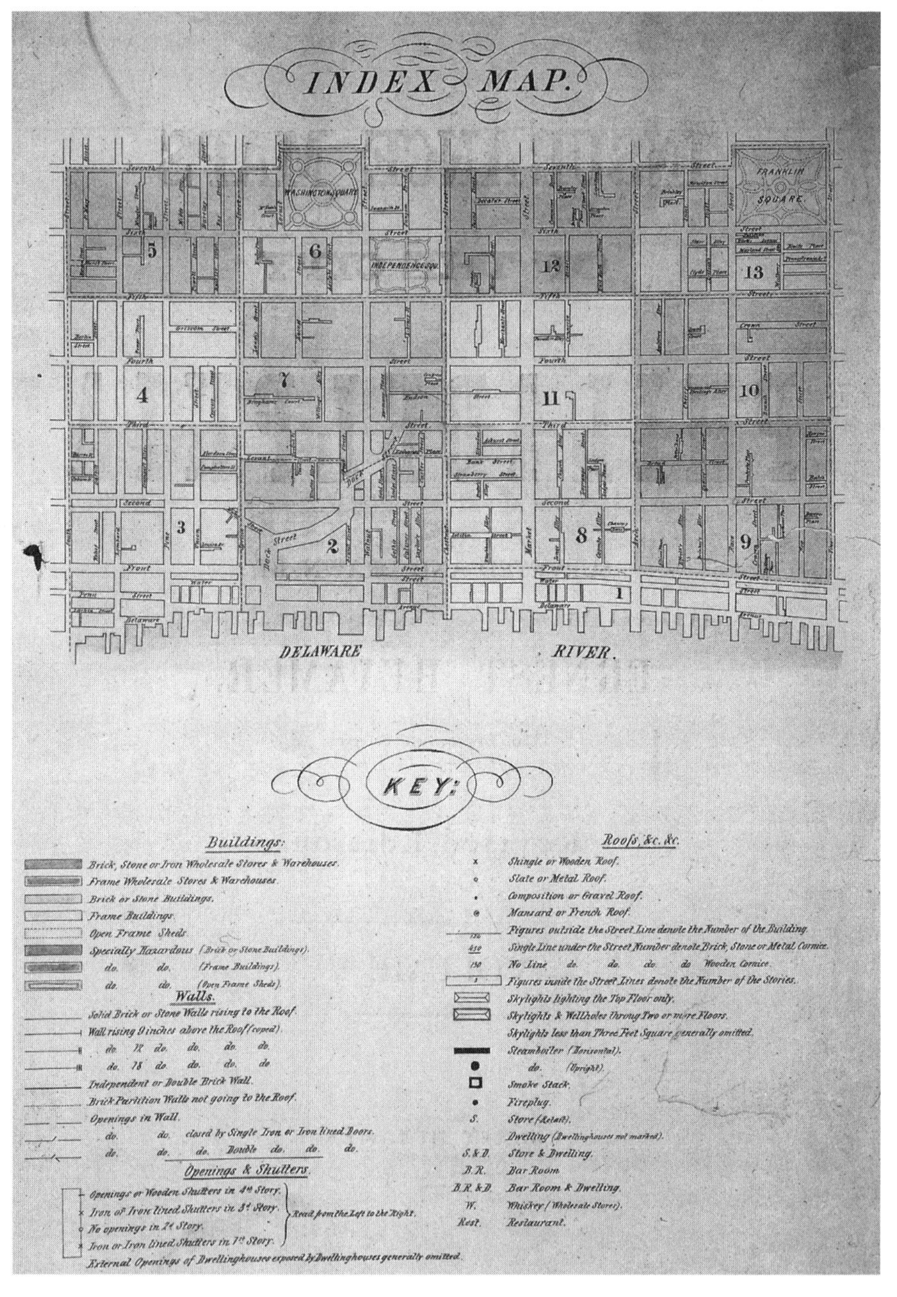

Charles Hexamer, *Fire Insurance Map of Philadelphia*, volume 1, key, 1872. Fire insurance surveys helped underwriters to better comprehend the hazards of fire in distant cities; they represented danger using the categories that underwriters commonly used to distinguish between risks. Maps especially emphasized how a property was used and its basic method of construction, as suggested by the key from this 1870s Hexamer atlas of Philadelphia. Courtesy, Geography and Map Division, Library of Congress

The shortcomings in the NBFU's approach were especially evident in its disagreements with local underwriting associations about the program of uniform rates. Flat rates, according to detractors, were out of step with the everyday practices of local agents and community leaders. In particular, insurance consumers and underwriters with extensive experience in particular locales (especially representatives of small geographically centered firms) chafed at the lack of discernment shown by the NBFU. For instance, according to the St. Louis Board of Fire Underwriter's president, James Waterworth, customers and underwriters in his city had never been "satisfied with the flat rate system, knowing it to be discriminatory and indefensible." The historian of Philadelphia's insurance industry, J. A. Fowler, echoed this sentiment. He argued that the system of flat rates ran contrary to scientific rationality. That is, it "was in subordination to the rule that each locality should be rated primarily in accordance with its own fire experience." Although one might be able to predict how often stone warehouses burned throughout the nation, underwriters like Waterworth and Fowler argued that such tabulations could not be translated into a rate which made sense in different urban environments. In St. Louis such warehouses might burn infrequently, whereas in Spokane they often caught fire. As a result, the rates on stone warehouses in St. Louis and Spokane did not reflect adequately the differences in their likelihood to burn, which depended on local peculiarities.[17]

Local underwriting boards' familiarity with their cities provided the basis for their disagreement with flat rates. For example, the St. Louis Board of Fire Underwriters actually sought to develop accurate local rates and actuarial tables, even during the periods in which it supported the higher, flat rates advocated by the NBFU. Shortly after it was established in 1872, the St. Louis board requested that one of its members, Western Bascome, prepare a plan for rating the city's fire risks. Bascome, who had surveyed, printed, and distributed the city's first fire insurance map in 1857, was, according to St. Louis insurance historian James Waterworth, "full of the idea of rating by schedule[;] he had in fact been laboring on it privately for some years." Within four months, Bascome and a rating committee presented the STLBFU with its schedule of rates, together with principles of operation, illustrated with examples. The board adopted its own "St. Louis Schedule" and requested the NBFU's financial support to do additional surveys to make the city's rating schedule more accurate. Although the NBFU provided the resources necessary to fund a surveyor, it advised the STLBFU to adopt the NBFU's "basis schedule." The St. Louis board officially abandoned its own rating schedule, but it "silently and continuously" continued the project by appointing C. T. Aubin, a prominent local insurance surveyor and civil engineer, to develop an accurate schedule. Somewhat paradoxically, the influence of the NBFU waned in direct relation to underwriters' growing knowledge about the complexity of the problem of fire. As underwriters learned more about the particular components of the fire problem, the uniformity of approach sought by the NBFU began to seem misguided, if not impossible. Ironically, the NBFU's recommendations for standard procedures in inspections, mapping, keeping statistics, and common forms had empowered local underwriting associations and diminished the relevance of the national organization.[18]

Daily Routines, Bureaucratic Standards, and Manliness

As insurers sought to remake their industry by developing standard business practices, prescriptions for behavior also spread throughout the industry. In the last decades of the century, the prescription for achievement showed striking similarity to notions of manliness developing among workers in life insurance and other financial industries. In these areas, according to historian Angel Kwolek-Folland, manliness "could be expressed through rational application of mechanical business laws of profit and loss." This ideal of what it meant to be a man combined with common practices and business activities to bind company executives, agents, and other industry specialists into a coherent whole, and to foster the creation of a new economic calculus.[19]

The "visible hands" that shaped the new division of labor within the fire insurance industry drew predominantly from Aetna. One the industry's pioneering firms, Aetna employed hundreds of representatives throughout the nation and led the effort to establish standard procedures within the industry. J. B. Bennett, who served as general agent for Aetna in the western United States from 1853 through 1870, rewrote the company's booklet that instructed agents on company procedures. In 1857, his *Guide to Fire Insurance* replaced preceding instruction books. Eventually known as the "Aetna Bible," the guide was a landmark accomplishment in advocating for the company's interests and training its agents. Like Bennett, influential publisher C. C. Hine worked for Aetna during the 1850s, serving as special agent between 1858 and 1865. His experiences provided him a wide opportunity to survey and define Aetna's practices across a wide region. During the same period, Daniel Sanborn, a pioneer in the insurance mapping industry, began his career surveying manuscript maps for Aetna. Well before such maps became standard, Sanborn helped Aetna to develop graphical representations of risk that would be useful in daily practice.[20]

The fire insurance industry's largest companies advocated the development of common administrative and underwriting routines. Internal procedural manuals like the "Aetna Bible" became less common as industry leaders published manuals of practice that were disseminated to a wider audience. Developed in the 1850s, national business periodicals such as the *Insurance Monitor* and the *Weekly Underwriter*, expanded their readership dramatically following the Civil War. By the 1870s, these publications became important proponents of standard procedures and began to propagate the ideal of good practice based in information gathering and in prudent and statistical evaluation of fire danger—in short, according to the principles of "science." These publications especially encouraged companies—no matter their type or size—to adopt standard routines that involved a number of explicit procedural steps and which relied upon a bevy of administrative documents. In 1865, C. C. Hine collected eighty-six example forms in a book, *Fire Insurance, Policy Forms, and Policy Writing*. By 1870, Hine had collected 139 forms, and his 1882 edition contained 328 forms. The decades following the Civil War saw the number of policy forms increase tremendously, mirroring a similar explosion in management information and the use of bureaucratic technologies in the broader economy.[21]

Hine played an especially significant role in formalizing daily routines throughout the industry. Publisher of the influential periodical the *Insurance Monitor*, he spoke definitively on the subject of practice. Continuously republished for twenty years, Hine's procedural guide—*Fire Insurance: A Book of Instructions for the Use of the Agents of the United States*—was adopted by many firms, including the Insurance Company of North America. The degree to which these firms followed Hine's suggestions is unclear, but the guidelines he established 1870 remained essentially unchanged in subsequent nineteenth-century industry literature. Another guide, Francis Cruger Moore's *Fires: Their Causes, Prevention, and Extinction; Combining Also a Guide to Agents Respecting Insurance against Loss by Fire*, originally published in 1877, varied little until 1903, when Moore changed the title to *Fire Insurance and How to Build: Combining Also a Guide to Insurance Agents Respecting Fire Prevention and Extinction, Special Features of Manufacturing Risks, Writing of Policies, Adjustment of Losses*. The title notwithstanding, Moore recommended only slightly different administrative procedures than had Hine. At the very least, industry experts and the insurance press had formulated and widely promulgated a clear and unvarying set of principles for daily practice. The continuity between Hine's and Moore's volumes and their wide publication also suggest that bureaucratic routines gained increasing use and regularity during the last decades of the century.[22]

In his "working handbook," Hine divided standard practice on insuring "ordinary hazards" into six steps, which are almost self-explanatory: Ask whether the risk was desirable or not by taking a "survey"; take an application and write the policy in the agency "record book"; write the policy; return the "daily report" that same day to the agency; collect the premium and deliver the policy to the insured. These administrative tools—daily reports, record books, and monthly returns—possessed all the particulars about the business transaction, including the amount and specifics of insurance polices. They were especially critical, because in the taking of "ordinary risks," the policy forms were not returned to the insurance firm's headquarters. Rather, daily reports provided the home office with all the information that it would ever know about a risk. Not surprisingly, then, instruction manuals emphasized careful record keeping on the part of representatives, instructing them to keep a record for themselves in the agency record book, to return daily and monthly reports promptly, and to fill out the proper forms with careful attention to detail. Underwriting "special hazards" involved a few separate steps that required a higher level of surveillance. Rather than simply taking an application and surveying, Hine urged agents to be especially thorough; they should, he wrote, "TAKE AN APPLICATION ON THE PROPER BLANK, full and complete, Diagram included, and send it to head-quarters, with your view of the risk." He reasoned that those companies willing to underwrite such properties—which usually involved significant financial exposure—were willing to lose the business rather than take an imprudent risk. If the "home office" accepted the risk, then Hine allowed the agent to write the policy.[23]

The proliferation of standard routines and bureaucratic forms corresponded to the expansion of the industry, with its increasingly specialized division of labor. Indeed, following the Civil War, and certainly by the 1880s, the job title *underwriter* or *insurer* could no longer accurately reflect the increasingly complex organization of labor in the fire insurance industry. As early as the 1850s, Aetna had organized its management activities in a manner that prefigured the organization of the fire insurance industry later in the century. Just prior to the Civil War, Aetna employed over five hundred agents, many of whom also may have represented other firms. Few agents worked full time, but all labored under the direction of a management team that numbered fewer than ten men. However, by 1900, Aetna's practices no longer were unique, as firms in both the fire insurance and the life insurance industries commonly employed field representatives to sell insurance. Although it is difficult to determine precisely how many firms used agents, they played a key role in the expansion of both industries. This shift is evidenced by the flourishing number of publications that explicitly spoke to an audience of agents and company managers and by the advent of professional associations for insurance agents beginning in the 1890s. In addition, the United States Census first identified "insurance agents and brokers" as an occupational category in 1890, when it showed 8,554 people employed primarily as insurance agents; in a separate category, there were 1,467 "insurance company officials." By 1900, the number of people who listed their occupation as insurance agents and brokers had grown to nearly 120,000, and the number of insurance company officials had increased to over 5,000. In 1910 there were over 88,000 fire insurance agents licensed to do business in the United States, and by the 1930s, there were over 250,000 property insurance agents throughout the country. As the number of full-time agents grew, part-time agents continued to play an important role.[24]

The competing agendas of field representatives and company managers threatened to throw the industry into chaos, but those different interests were contained by a common work culture. Although the two groups debated commissions and other issues, both came to depend upon one another in a team effort; as manuals observed, "The interests of the agent and the company are identical." Their interests converged because of their economic ties and common ideals of proper male behavior in the business world. If this shared financial fate helped insurers of all stripes to negotiate the tricky world of Gilded Age capitalism, instruction from industry publications provided workers with plenty of specific guidance about how to behave properly.[25]

As the industry prescribed good practice, the expression *moral hazard* gained every wider currency, and evaluating the character of risk—specifically the men involved in the business transaction—became both more explicit and more central to industry practice. At its most basic, moral hazard remained a catch-all for the process of evaluating a relatively ambiguous component of fire danger—the reputations of customers and agents. Nonetheless, the term became more precisely defined and specifically linked to quantifiable aspects of the underwriting activity. Most noticeably, perhaps, insurers began to minimize this cultural risk through a simple economic calculation. According to insurance expert Francis Cruger Moore, moral hazards occurred when the amount of insurance on a property exceeded its economic value. This financial disjunction—easily measured and determined—provided consumers an incentive to burn their own property or to avoid practicing fire safety. To remedy this problem, the industry pushed aggressively a financial, quantitative solution, the "three-fourths clause," which stated that only three quarters of the value of the property was covered. By deliberately underinsuring customers, the industry removed temptation for a quick profit and encouraged them to be responsible and disciplined in their home and business affairs, especially as it regarded fire safety. Of course, the danger of "overinsurance" had been discussed for many years, but it took on renewed vigor as the industry attached it to moral hazard and its broader programmatic initiatives. More significantly, by defining moral hazard using language of quantifiable economic rationality, the industry was able to link its efforts to minimize both the cultural and physical component of risk, thereby making its strategies for dealing with the problem of fire more systematic.[26]

Accompanying the objectification of cultural aspects of risk—the moral hazard—was an increased emphasis on gathering information and using rational, economic decision-making processes. The industry's emphasis and daily use of a systematic method for gathering and evaluating knowledge was the foundation for calculating the manhood of underwriters, agents, and their customers. Managing information was vital. As Francis Cruger Moore urged, "That he may properly distinguish between safe and unsafe risks . . . it must be evident that the information of an underwriter should be varied and extensive. . . . There is probably no calling requiring so intimate a knowledge of every other as this." In a world in which soliciting new business was seen as the only way for the industry to expand, "knowing" one's business—the built environment, customers, and the company's policies—proved indispensable. Urged to cultivate new business aggressively and with persistence, agents were instructed to persuade potential clients with substantive knowledge about fire risk and the economic benefits of insurance. To the degree that intelligence, effort, and assertiveness were key to an agent's success, it was "the duty of the agent to qualify himself, by close observation and study." Without the ability to make "clear and logical argument" and use "apt examples," agents could not sell insurance policies effectively. Command of information allowed underwriters to balance aggression with risk aversion, individual interest with corporate needs, and intellectual dexterity with rule-based behavior.[27]

As underwriters systematized daily work routines, they contributed to a broader redefinition of middle-class manhood, which became focused increasingly upon economic matters—financial success, market rationality, and business acumen. A number of factors contributed to this broader reorientation: the intense development of managerial capitalism, the rise of scientific management theory, the increasing use of bureaucratic technologies in American workplaces, and the dramatic expansion of white-collar work more broadly. At the same time, however, the unrestrained drive for profit that characterized laissez-faire capitalism complicated the lives of middle-class men everywhere, including the men who labored in the fire insurance industry. Try as they might to follow their industry's standard practices or to adhere to rhetoric about unity and rationality, fire underwriters found themselves competing to earn a living, to support their families, and to achieve wealth. This countervailing tendency found expression in frequent bankruptcies among firms in the industry. In the three decades before 1900, over eight hundred companies failed. Frequently, fire underwriters—agents and company leaders alike—were not the paragons of virtuous rationality that industry manuals demanded.[28]

Nonetheless, the industry's emphasis on economic rationality and risk-averse behavior indicates a change in how middle-class insurers perceived their work. Standard practices and cultural ideals harnessed the manhood of middle-class, white-collar underwriters to an increasingly rationalized and bureaucratic society. Moreover, standardizing work routines heralded the development of new attitudes about safety. Fire underwriters created a vast bureaucracy designed to observe themselves, their customers, and the landscape. These observations tied underwriters' battle against the problem of fire to behaviors driven by an economic calculus. This surveillance offered the insurance industry the possibility of using quantitative financial factors to discipline itself, its representatives, customers, and even the built landscape. Acting according to a cold economic rationale would increasingly become the basis for dealing with danger and safety.[29]

Mapping the Landscape

Of the many technologies used by insurers in their everyday practice, fire insurance diagrams may have had the most profound impact on the industry. During the nineteenth century, fire insurance maps became an important way that the fire insurance industry disciplined itself and the urban landscape, and they also represented one of the many ways that industry practices became more standard. More significantly, the expansion of specialized mapping agencies, the growing level of detail in the maps, and their increasing uniformity helped to objectify danger. That is, they transformed danger from something physical into an abstraction that could be quantified and then sold. Maps also fostered the development of unity within the industry by providing a common visual language of risk, serving as an important and common point of communication between insurance companies and their field representatives.

Such atlases further underscored and increased the effectiveness of each of the industry's other strategies to standardize underwriting practices. They placed the insurance policy and daily report within a clearly delineated and carefully drawn spatial context. In fact, insurance atlases were so commonly used in the industry by the late nineteenth century that industry experts, such as C. C. Hine and Francis Cruger Moore, instructed their readers on how to "read" maps and even taught them how to draw their own diagrams of risk. As these specialized firms actively enumerated, accumulated, and quantified the incidence and causes of fire, they fed the insurance industry's insatiable appetite for information. And, over time, much of the underwriting community's capacity to evaluate danger grew from the dogged thoroughness of mapping agencies. More importantly, they encouraged dynamic interplay between the built environment, statistical records of loss, and classification schemes. As much as the industry continued to struggle with questions about how statistical knowledge and categories of risk should affect rates, fire insurance maps represented safety and danger in a manner that was intertwined with actuarial charts.[30]

Mapmakers' categories for assessing fire risk expanded during the last five decades of the nineteenth century as the fire insurance industry developed a better understanding of fire risk. Map keys grew from one page to two pages as underwriters became more involved in specifying this increasingly minute surveillance of the urban landscape. Such representations became more detailed according to the special conditions that underwriters believed shaped the incidence of fire. Maps listed the width, depth, and height of buildings as well as skylights, boilers, and roof construction. Later maps, those published after 1870, would include information on street length and width, wall construction, fireproof construction, breaks between buildings, shutters, thickness of walls, and other architectural details. Finally, most maps diagrammed the relevant fire extinction infrastructure of the city, such as fire hydrants, water mains (and their size), and alarm boxes. The growing sophistication of insurance maps occurred as a result of a dynamic process between surveyors and underwriters—an ongoing give-and-take about the problem of fire.[31]

The business records of a fire insurance surveying company—St. Louis's Whipple Insurance Protective Agency—reveal that fire insurance maps served a purpose beyond mere representation. Insurance mapping companies placed cities under constant and comprehensive observation. Through a program of inspections, mapmakers helped to police compliance with the insurance industry's standard practices and made insurance contracts strictly enforceable. Perhaps unique among mapping companies, Whipple published a widely distributed newsletter that takes us inside the extensive surveillance provided by insurance mapping companies. Begun in 1875, *Whipple's Daily Fire Reporter* offers an astounding comment on the intensity and scope of the insurance industry's surveillance of fire risk in the urban landscape.

Whipple sold subscriptions of its *Daily Fire Reporter* to customers that included local firms, national insurance companies, agents, owners of commercial and industrial property within the city, and the local municipal government. Through the newsletter, Whipple performed many services for its subscribers. First, the company inspected and reinspected property (typically in the business district or along the city's wharves). When property owners failed to adhere to practices commonly associated with fire safety, their names, locations, and variances were published. Second, Whipple also collected, and daily disseminated, information about the origins, causes, and responses to all fires and false alarms in the city. At the end of each year, the *Daily Fire Reporter* listed the aggregate statistics and often compared them to events in previous years. Finally, subscribers were also apprised of the opportunity to purchase Whipple's other services. For instance, Whipple could also be commissioned to survey individual commercial or industrial properties and produce an extensive report of fire risk at that location.

The single most important aspect of Whipple's business was its frequent inspections of property in the city. The program of inspection served the practical purpose of keeping maps updated but had more far-reaching implications as well. Inspections constituted a surveillance of property owners, which coerced them into behaving according to the insurance industry's standard of conduct regarding the risk of fire. In 1882, for instance, Whipple performed 20,632 "inspections and re-inspections" of 3,267 buildings, almost exclusively located in the business district of the city. Indeed, at every location on the block, inspectors carefully examined "every story, cellar and attic of every building," surveying the buildings for any and all dangers of fire, "be it rubbish, dangerous gas burners, defective flues, defective furnaces, careless disposition of ashes, careless use of open lights, etc., etc." Each day company representatives examined various "blocks" in the city corresponding to the "block" organization of Whipple maps. When Whipple announced it had inspected "Block 123" on September 1, 1882, it directed readers' attention to an area along the 600 block of Franklin Avenue:

> On the second floor of No. 610 Franklin Avenue (R. B. Tunstall & Co.), a stovepipe is not safely adjusted in flue.
>
> On the attic floor of No. 612 Franklin Avenue (A. Mohl), there is an open stovepipe hole, the flue is used on first floor in cold weather. In the rear of first floor willow baskets are suspended over a gas jet."

Naturally Whipple's reports employed the language used by insurance companies and which appeared on Whipple's map keys. Agents, underwriters, or anyone familiar with the insurance industry could move between Whipple's reports, its maps, and its insurance policies with great ease.[32]

As much as they helped to reconceptualize the city's susceptibility to fire risk, inspections provided Whipple an avenue to intervene in the physical landscape.

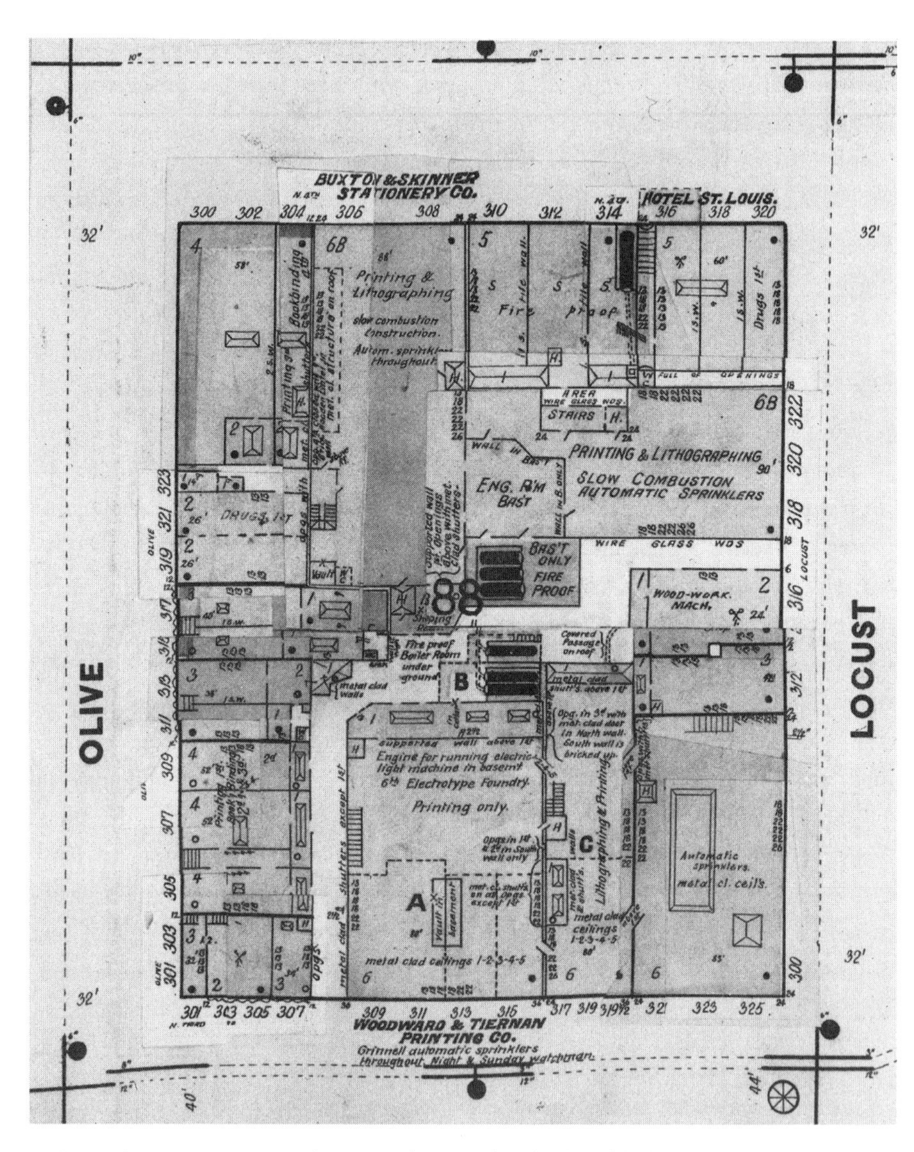

Whipple Fire Insurance Map, plate 65, volume 2, detail of city block no. 88 (Locust to Olive, North 4th and North 3rd), 1897. By the last decades of the nineteenth century, insurance maps cataloged and represented many elements of the urban landscape, reflecting the growing complexity of underwriters' efforts to control the problem of fire. Many firms even used the oversize portfolios to manage their risks, recording policy numbers and coverage amounts directly onto the atlases, helping them to avoid taking too many risks within small or contiguous areas. Courtesy, Missouri Historical Society, St. Louis

Each time surveyors discovered a material danger of fire, they brought it to the attention of the occupants and owners of the building. In addition, the next morning the *Daily Fire Reporter* publicized which properties had passed inspection and which had not. Although on some occasions there were no hazards found in a block, usually Whipple's surveyors found and enumerated dangers—on which the *Reporter* would comment. An entry from December 1882 under the heading "Inspection of Buildings" noted the hazards that gasoline and rubbish posed: "Block 6. At Nos. 514 and 518 South Second Street Gasoline is used. Ashes and rubbish are deposited against fence and outhouses at Nos. 508, 510, 514, & 518 South Second Street." Strict reinspection buttressed the effectiveness of Whipple's program—especially when it came to pressuring property owners to accede to Whipple's recommendations. Every Monday, Whipple employees reinspected the risks they had reported the previous week. If dangers had not been removed, the building and its occupants were once again reported: "All dangers of fire reported for the week ending Thursday, January 19, 1882, have been removed, with the following exception: At No. 623 Locust Street (Read's Restaurant), a cooking range is in use, the floor beneath which is not protected." Reporting unsafe conditions for the first time appears to have resulted in the removal of the fire hazard in about three-fourths of the instances. Announcing the presence of fire dangers for a second time brought even more pressure on the occupants (and/or owners) to remove the dangers of fire.[33]

Municipal officials confirmed and legitimized Whipple's authority. In the first half of 1882, the *Daily Fire Reporter* frequently reprinted a small notice from the chief of the St. Louis Fire Department offering assistance to Whipple. The note from Chief H. C. Sexton, dated February 1876, commended and approved of Whipple's plan for inspections and further advised that "should parties refuse to comply with your request to have dangers of fire remedied, I will render you such assistance as may be needed." In August 1882 the company announced that it had received a letter from John Beattie, commissioner of public buildings, regarding its system of inspection: "I have investigated your system of inspecting buildings (for the purpose of abating fire dangers), and in approving the same, I will say, that wherever dangers of fire exist and the parties responsible refuse or neglect to comply with your requests in the premises, I will render you all necessary assistance." At regular intervals and in their year-end assessment of risk in the city, the company reiterated its relationship with the building commissioner and the municipal government. If after notice of a continuing hazard a property owner refused to act, Whipple informed the owner that the company had notified the city building inspector of the variance. As the firm explained in the *Daily Fire Reporter:* "[We sent the] party a notice signed by Mr. John Beattie, the City Building Inspector, who is legally authorized to enforce the removal of dangers by fire. We have a volume of blank notices in our office made in duplicate—when we send a notice to the party we also send a duplicate to the above named official so that he is informed of what use we are making of his name." In cases of particularly acute danger, Whipple skipped the process of reinspection and immediately informed Beattie's office. The *Daily Fire Reporter* observed that the notices sent to the building commissioner "usually" had their desired effect; further, when no attention was paid by a property owner to a notice of hazard, Beattie "never failed to enforce our recommendations."[34]

Qualitative analysis of the *Daily Fire Reporter* suggests that Whipple rarely had to report property owners to the city's building commissioner. Why was this? Was it because the St. Louis municipal government's enforcement of building codes was that much more effective than other cities of the same era? Or did Whipple's authority in matters of inspection derive from a source outside the local city government? How, indeed, can we explain Whipple's apparent success at getting property owners to remove hazards?[35]

Whipple's ability to coerce compliance with its suggestions developed out of the company's complex relationship to the fire insurance industry. The *Daily Fire Reporter* supplemented Whipple's maps, diagrams, and surveys with information about particular risks at a site. Most of the risks identified by the *Daily Fire Reporter*, such as holding ashes in a wooden container, were strictly prohibited in most insurance contracts. As a result, if a fire occurred where such hazards existed (or, perhaps, had previously existed) underwriters could potentially contest a loss and refuse to pay. After the 1880s, Whipple's maps, like other insurance maps, provided underwriters with a "warranty" on the property. In other words, property owners could not alter their property (introducing new fire hazards) without potentially invalidating the insurance contract. Evidence that Whipple was aware of these relationships appears in its 1883 explanation of its inspection process. The company advised that it is not necessary for "insurance agents with a line on the risk [which did not pass inspection]" to visit the property owner. Under close scrutiny from Whipple, property owners faced the prospect of losing their insurance coverage or not receiving payment for a loss if they did not remove fire dangers.[36]

The threat of losing insurance coverage may not have been the only motivation driving property owners to remove hazards threatening their property. Whipple's inspections also provided crucial information to property owners in that it identified potential hazards *and* recommended solutions. At a time when insurance contracts paid only two-thirds or three-fourths of the value of goods lost, property owners bore a substantial portion of the risk. Consequently, inspections gave them the ability to diminish their financial liability. They also provided evidence that owners were abiding by the strict terms of the fire insurance contract— information that may have been useful when entering into disputes with underwriters about the terms of their contract. As a result of this process of inspection, reinspection, and publishing the results, Whipple found most of its recommendations for removing hazards had been heeded. The threat of government intervention, then, served as an important supplement to, rather than a substitute for, the authority that the market conferred upon the Whipple Insurance Protective Agency.[37]

Through its program of inspection, Whipple established itself as an intermediary between underwriters, agents, and the insured. Having a third party survey and analyze property eased conflicts between the parties to the fire insurance contract. Whipple performed the inspections that agents were encouraged to do on a regular basis, thus lightening the load of agents and giving underwriting companies more control over how the agents conducted business. If Whipple's maps or inspections consistently indicated that a particular "risk" was especially dangerous, insurers could advise agents not to renew that property. Rather than relying on agents to assess the fire risk of a particular property—a decision which some in the industry thought could be affected by agents' desire to get business—underwriters used maps and inspections to assess risk independently of an agent. Indeed, during the 1920s, young men entering the insurance business could be employed as "map clerks" whose primary responsibility included writing "lines" on maps "so that the company's liability can be seen at a glance." Maps could embolden companies to enter markets distant from their "home office" because they could be assured quality representations of fire risk in those places.[38]

The efforts of mapping agencies such as Whipple were a crucial component of the fire insurance industry's increasingly sophisticated and detailed production of knowledge regarding the risks of fire. As often as possible the *Daily Fire Reporter* quantified or evaluated everything about a blaze from the total financial loss to insurance loss to the origins of a fire. For instance, on January 24, 1882, the *Daily Fire Reporter* announced that on the previous day a fire had been caused when a child lit a match to determine how much gasoline had spilled in a barbershop. At other times, entries were more speculative about the origins of a blaze. The Cass Avenue conflagration of July 21, 1882 "originated in the cellar [of the Cass Avenue Planing Mill Company], which contained the main line of shafting, and was probably caused by a hot box." Even so, over 20 percent of the time in 1882, the Whipple Insurance Protective Agency could not determine the origins of a fire.[39]

Whipple transformed its daily evaluations into monthly and yearly reports of the causes, origins, and dangers of fire in the city. In its "Monthly Fire Report" for July 1882, Whipple noted that over $120,000 in fire loss had been recorded in the city; of that, approximately $115,000 had been covered by insurance. There were forty-two fire alarms, three false alarms, a second alarm, and nineteen "still" fires. Of the fifty-seven "causes" of the month's fires, nineteen were unknown, and fireworks had caused ten fires—not a surprising fact given the July Fourth holiday. Moreover, just as insurance companies tabulated statistical records of losses yearly, Whipple created a year-end statistical portrait of fire loss and risk in St. Louis—a record often placed within the context of statistics gathered in previous years. For instance, in 1882 the company analyzed the previous year's fire loss according to a number of categories, including "Causes of Fires in the Past Year," "Occupancy of Buildings in which Fires Originated," types of buildings where fires originated, a month-by-month accounting of "Loss," "Insurance," and "Loss to Insurance Companies," "Recapitulation for the Past Thirteen Years," and the "Building Inspections." When it enumerated the causes of fire Whipple not only advanced insurers' knowledge of the problem, but also it legitimized its own and the industry's authority on the subject.[40]

Additionally, the company offered evidence of the importance of insurance and helped to demystify and comprehend fire in rational terms by dissecting previous incidents of fire. In fact, the *Daily Fire Reporter* often provided a dramatic marketing message for insurance when it underscored the dire economic costs of fire. The publication gave each fire a cash value, thus clearly identifying the severe economic consequences of the fire problem (to both the community and property owners). Even relatively minor fires were reported. An early morning blaze in January 1882 caused $100 of damage to a barbershop and another "small loss" to a perfumer. It also recorded large blazes, such as the fire that destroyed an entire block of Cass Avenue and the Cass Avenue Planing Mill Company. The *Daily Fire Reporter* listed all twenty-seven insurance companies that had insured the Planing Mill Company, for $4,500 on damage to the building and $23,500 on its "contents"—both stock and machinery. The same blaze also destroyed a furniture factory that was insured by firms in London, Paris, Cincinnati, and Philadelphia. Just days after the loss on Cass Avenue, other unrelated fires in the city during the early morning hours caused over $100,000 in damage and substantial loss to insurance companies.[41]

The *Daily Fire Reporter* also promoted insurance as the best method to protect property. The newsletter emphasized that fire danger could be combated, and it distinguished between business owners who purchased fire insurance on their property and those who did not. In January 1882, it contrasted the uninsured barbershop owner who lost $100 with the perfumer who suffered a small loss but had "$600 in the Watertown." In addition to encouraging consumers to purchase insurance, it also provided object lessons to insurers—especially those firms that varied from prudent business practices. For example, the *Daily Fire Reporter* listed large fires in other cities and closely detailed heavy losses that resulted from fires in St. Louis. This practice warned insurance companies against taking risks that were too large or too closely concentrated.[42]

As the *Reporter* enumerated the horrors of fire, it recognized the benevolence of the underwriting community. For instance, it documented the role insurers played in reducing fire losses by highlighting the work of the Underwriters Salvage Corps. Periodically, the *Reporter* praised the corps, though it was expensive to maintain: "The Salvage Corps of St. Louis have made a splendid record for 1881, and have saved to the insurance companies and property owners many times the cost of their maintenance." Subscribed to by businesses, insurance agents, and underwriting companies, the salvage corps, or "insurance patrol" as it was sometimes called, used tarpaulins to cover items on the lower floors of a building to prevent water damage. Though not formally part of the fire department, the salvage corps sometimes helped extinguish fires, as happened on the morning of September 4, 1882, when they put out a fire in a three-story brick building.[43]

Insurers' surveillance of the built landscape had grown remarkably comprehensive by the early twentieth century as surveyors mapped more cities and maps grew more detailed and uniform. As maps became standard, the business of mapping was consolidated into the hands of fewer companies, especially as the Sanborn Mapping Company absorbed its competition. By the beginning of the twentieth century Sanborn dominated the industry and published atlases for nearly five thousand U.S. cities and towns. In 1905, Sanborn published a *Surveyor's Manual for the Exclusive Use and Guidance of Employees*, which became the de facto industry standard. In 1908, the Fire Underwriters' Uniformity Association adopted a "Key to Plan Notations" that explained the colors, symbols, and notations used on fire insurance maps. In 1914, the NBFU appointed a special committee to determine the utility of setting up its own map-making operation. Although the NBFU did not create its own company, its map committee worked closely with businesses like the Sanborn Company (which held a virtual monopoly on insurance surveying by 1920) to produce standard maps for over twenty thousand cities and towns throughout the nation.[44]

Trade Associations and Schedule Rating

Creating standard routines, a common visual language, and similar approach to categorizing risk helped the industry to define the problem of fire more clearly. In particular, insurers began to develop a network of local and regional underwriting associations that filled the vacuum left by the NBFU's decline in the 1870s and 1880s. These organizations created a platform from which insurers began to remake the industry's priorities, including intervening in the built landscape to promote fire safety. Likewise, the industry began to experiment with new methods of rating risks that corresponded to the developing interest in preventing fire and to the intensive process of gathering and assessing data. The expanded cooperation of underwriters, though imperfect, began to remake the industry, helping it to find more cohesion and to take the first tentative steps in a new direction during the 1890s.[45]

Beginning in the 1870s, underwriters established a rapidly expanding network of local and regional associations dedicated to supporting everyday routines and industry regulatory activities. These trade associations fostered the development of standard practices by providing crucial services to companies doing business in their region or locale, including mapping, inspection, rating advice, and loss adjustment. For example, in 1879 a number of national (and international) insurers founded the "Western Union" to control rates and practices in western states. Such associations multiplied in rapid succession. In the Southeast and West, fire insurance companies formed the Southeastern Tariff Association in 1882, the Pacific Insurance Union in 1884, and the Rocky Mountain Fire Underwriters Association in 1889. About the same time, in the Northeast and Midwest, insurance companies organized the New England Fire Insurance Exchange, the Underwriters Association of the Middle Department, and the Underwriters Association of New York State. These decentralized organizations began to receive broader coordination from the Eastern Union, established in 1893, with similar arrangements developing nationwide between local and regional underwriting associations.[46]

This loosely affiliated network of associations provided the platform from which underwriters hoped to re-create industry priorities. Through the activities of local, state, and regional associations, underwriters acquired a new appreciation for the complexities of the built landscape and fire protection. The associations operated at multiple levels—national (i.e., the NBFU or, later, the National Fire Protection Association), regional (e.g., the Western Union or Southeastern Tariff Association), and state and local (e.g., the Ohio Inspection Bureau or the Philadelphia Underwriters' Tariff Association). And, although they were not formally linked and they performed functions that often overlapped, trade associations cooperated more often than not. In fact, unlike previous attempts to regulate the industry, the plethora of loosely affiliated underwriting organizations mirrored local and regional differences within the industry as well as the nature of public regulation, in which power was dispersed to the states.[47]

Though many insurers insisted that they were not in the business of fire prevention, underwriters became accustomed to meddling in the urban environment. If insurers' focus on simple profitability and expanding their business remained central to their endeavor, their core mission gradually began to expand as they developed a taste for controlling the built landscape. In 1884, for instance, the St. Louis Board of Fire Underwriters—led by James Waterworth, C. T. Aubin, and Western Bascome—assessed the city's fire protection and argued that the infrastructure was inadequate. It recommended nearly $500,000 in improvements, which included $190,000 for several new engine houses and apparatus and over $270,000 for new water mains and fire hydrants. Within a year, nearly all the changes had been made or at least acted on. The fire department located four new engine companies in the city's commercial and industrial districts and expanded by eighty-three men, a 24 percent increase from the previous year—the largest single increase in the department's history.[48]

Underwriters' associations performed many mundane, albeit overlapping tasks; they inspected and rated fire risks, and especially sought to regulate local (or state or regional) practices. For example, the Philadelphia Fire Underwriters' Tariff Association (later the Philadelphia Fire Underwriters' Association, or PFUA), formed in 1884, worked closely with the Hexamer Mapping Company and its proprietor C. J. Hexamer to survey and rate risks throughout the city. The organization compiled minimum rates for dwellings, churches, and other small or noncommercial risks. It established a rating system for mercantile hazards such as factories, warehouses, and department stores. Each such property required a special survey, usually conducted by Hexamer, in addition to the already extant company map. During its first year of existence a tariff committee—which included nearly every member—met four days per week and set rates on hundreds, if not thousands, of properties or property classes throughout Philadelphia. As a way to circumvent the 75 percent majority needed to approve each rate, the association created rating schedules for various districts and types of properties. Using many such schedules, the PFUA surveyed and inspected each risk underwritten—especially those of large commercial hazards. In the process, the association suggested improvements in construction and design. It reported that "hundreds of risks have been permanently improved and many fires prevented." Just as the surveys of the

Whipple Mapping Agency aided the St. Louis Board of Fire Underwriters, Hexamer's maps and board-sponsored inspections provided the Philadelphia underwriters' association an avenue to alter the built environment in a manner that would prevent fires.[49]

As underwriters developed an increasingly pervasive surveillance of the built landscape, of their clients, and of themselves, leading insurers experimented with new approaches to setting rates. Many firms began to utilize a rating method known "schedule rating," which contrasted dramatically with the methods that the NBFU had proposed for establishing common rates. Inaugurated in 1893 with the publication of Francis Cruger Moore's *Universal Mercantile Schedule*, this mode of setting rates depended upon a complex formula of charges and credits based on the internal and external hazards of a risk as well as a structure's protective features. Rating in this manner rapidly became tied to the inspection and categorization performed by fire underwriting associations, and insurance companies that used the schedule for determining premiums assigned a "base rate" to a building of particular construction, type, and use. After establishing a base premium, insurers referred to the complex schedules and added or subtracted value based on a variety of factors. Dangerous fixtures, design, or construction materials would increase the premium, whereas aspects of construction deemed fire preventive would result in the lowering of a premium. Using schedule rating techniques to figure premiums involved far more effort and care than underwriters had previously shown. When they evaluated risks using such schedules they assessed more than categories of construction, occupation, and (to a limited extent) immediately external hazards. This method inventoried a wide range of information, including the material produced and/or warehoused in a building, minute details of construction that might render a building more or less safe, in-depth examinations of external hazards, and consideration of fire extinction conditions in the locality. Schedule rating reflected the intense manner in which the industry gathered knowledge and represented an escalation in the use of information in daily practices.[50]

The industry hotly debated the merits of schedule rating. To many critics, it hearkened back to the mode of establishing premiums that was used by the industry early in the nineteenth century. Although schedule rating developed from the categories produced by the industry's intense scrutiny of the landscape, it did *not* strictly depend upon the actuarial record. In spite of the fact that these systems, such as Francis Cruger Moore's *Universal Mercantile Schedule*, involved a high level of rigor and care and classified risks in a more detailed fashion than previous rating schemes, many within the industry criticized the schedule as "unscientific." Critics argued that assessments made by the schedule contained arbitrary valuations of fire hazard. Indeed, the *Universal Mercantile Schedule* and other such programs deviated from the objective loss record because the assessments they made (debits and credits) did not reflect the actual statistical record of fire experience. However, according to its proponents, schedule rating produced premiums that roughly approximated the actuarial record. In addition, according to one authoritative source, this method resulted in rates that were exactly alike provided that a risk had "exactly the same construction, occupancy, fire protection and environment."[51]

Although underwriters remained sanguine about the potential benefits, the use of schedule rates faced many of the same difficulties that attended rating activities more generally. Competition for business usually manifested itself in debates about the price of insurance—an issue that had plagued the industry from before the Civil War. Adjusting prices as a result of competition, then, repeatedly thwarted underwriters' efforts at establishing standard rates and even hindered the ability of individual firms to follow their own rules. Additionally, schedule rating involved so many factors that many critics believed this method of setting premiums actually spurred price competition, further undermining the industry's ability to create uniformity. Moreover, these schedules depended upon agreements between companies and local underwriting associations on "base rates." Not only were such accords difficult to maintain, but the task of coercing nonmember companies to comply with base premiums was nearly impossible. As a result, the *Universal Mercantile Schedule* produced rates that differed significantly from city to city and town to town but not for any objective reason. According to the *Weekly Underwriter*, market conditions accounted for price differences because the schedule was not often followed. That is, if a locale could bear higher insurance rates, then it paid higher premiums regardless of features particular to that place. The *Weekly Underwriter* further argued that the use of schedule rating systems tended to discriminate according to events of the recent past, which ran contrary to actuarial principle. In addition, the practice of using schedules often resulted in additions and subtractions to the base rate that were not contained in the actuarial record, increasing the possibility of large financial losses. For instance, in 1892 the PFUA recommended a rate hike of over 20 percent because fire losses so significantly deviated from its schedule of premiums.[52]

Despite its limitations, schedule rating conformed to the changing character of everyday underwriting practices—especially the industry's intensive program of gathering information and the growing strength of local underwriting associations. It also hinted more broadly at underwriters' increasing interest in preventing fires. Many underwriters began to believe that schedule rating provided them with a tool that would help to coerce property owners to build a safer urban landscape because this system discounted or increased premiums according to a variety of risk factors. According the PFUA, "schedule rating results in the removal of articles likely to cause fire, because, if not removed, the rate is increased by reason of their presence. It also results in the introduction of fire defenses, because the absence of such defenses is charged for as a deficiency under the schedule." Similarly, James Waterworth, former President of the St. Louis Board of Fire Underwriters, argued in his memoir that schedule rating had helped the St. Louis board improve construction in the city's central commercial district and to pass legislation mandating safer building practices. Waterworth pointed to the appearance of several fireproof structures during the late 1880s and into the 1890s as evidence of the board's success. The board had urged passage of a "slow-burning construction ordinance" and its adoption, according to Waterworth, "was the signal for beginning the equipment of St. Louis with suitable buildings for a more vigorous business life." Many obstacles remained, however. Waterworth and the board faced continued resistance from some in the architectural community, and often the "public moved faster than the board" in seeking safer construction practices.[53]

The spread of schedule rating during the 1890s suggested that a new mentality had begun to emerge in the fire insurance industry. Emboldened by their intense surveillance and common practices, many insurers began to advocate the active removal of obvious fire hazards from the landscape in order to reduce annual losses. The 1890s marked the industry's first full-fledged attempts to intervene in the built environments of urban America. After its dormancy in the 1880s, the National Board of Fire Underwriters began to revive itself and to reestablish its leadership role by becoming more active in fire preventive matters. In 1892, the board sponsored drawing up a national electrical code to supervise the installation of electric wiring and lighting. Furthermore, when the NBFU appointed a committee to help frame or to recast the Universal Mercantile Schedule, it abandoned its previous singular focus on setting uniform rates determined according to a national actuarial table. Rather than embracing the notion that it could establish premiums that would be followed by the industry's many firms, the NBFU responded to the desires of its membership. It began to assist them in the process of determining categories and rating valuations for a variety of "internal" risks, as well as "external" environmental dangers.[54]

The NBFU also took its first hesitant steps toward an activist prevention agenda by following the leadership of local underwriting associations and firefighters. In the 1890s, for instance, the NBFU joined the National Association of Fire Engineers, along with other professional organizations, to recommend modest regulations for building construction. The organization also resolved to "bring the insurance people and the National Association of Fire Engineers into closer relationship, and unite in reducing the enormous fire waste." In addition, underwriters' aggressive inspection of fire risks generated a critique of fire departments, water systems, and other aspects of fire-protective infrastructure. Beginning in 1889, the NBFU recommitted itself to inspection of fire departments and cities across the nation—including all aspects of urban infrastructure. The comprehensive inspection regimen, which sought to upgrade, standardize, and rationalize fire departments and infrastructure, derived from the surveillance that local underwriting associations had established in their respective jurisdictions.[55]

The NBFU's initiative in the 1890s only increased the scope, depth, and intensity of the fire insurance surveillance of cities. The development of a systematic, national approach to the urban and rural fire protective infrastructure had long been an interest of the fire insurance community. Ever since fire underwriters critiqued volunteer firefighting in the 1850s and advocated the use of steam fire engines, they had consistently pressured cities to improve their fire-protective infrastructure. However, the program initiated in 1890 differed from earlier efforts in several important respects. First, the effort was systematic. Spending over $5,000 annually, the NBFU hired a full-time inspector, paid for his travel expenses, and printed the inspector's reports. In 1893, the NBFU inspected and reinspected over 150 cities—a pace that it sustained through the early twentieth century. Secondly, the effort was as comprehensive as it was deep. By 1904, the NBFU developed an extensive and detailed set of "Standards for Grading Town Public Fire Protection." The standards covered water supply, waterworks, fire department organization, and departmental equipment. Lastly, after some wrangling, the NBFU forged common ground with the National Association of Fire Engineers to upgrade the nation's fire departments.[56]

During the 1890s, the NBFU drew upon the work of local associations as it began to develop a program to erect standards of prevention against fire. The Underwriters' Laboratories (or UL), established in 1894, would become one of the most significant aspects of the fire insurance industry's twentieth-century war against "fire waste." Established in collaboration with the Chicago Fire Underwriters' Association, the Underwriters' International Electrical Association (as UL originally labeled itself) supervised electrical safety at Chicago's Columbian Exposition. Directed by William Merrill, UL helped to make sure that the showcase "White City" electrical exhibition would live up to expectations as it began to tackle the larger issue of the safety of electricity.[57]

Like Underwriters' Laboratories, the National Fire Protection Association

(NFPA) developed independently, but over time it too was drawn under the umbrella of associations that the NBFU used to popularize its program of fire safety. Formed in 1896 by men primarily associated with the Underwriters' Bureau of New England and the Boston Board of Fire Underwriters, the NFPA's membership consisted of stock fire insurance companies interested in improving their surveillance of "risks." The organization's first project involved formulating standards relating to automatic sprinklers. In subsequent years, the NFPA established committees to report on other fire-protective devices, including fire doors and shutters, hoses and hydrants (in collaboration with the National Association of Fire Engineers), fire alarms, fire extinguishers, fire retarding materials, and fire pumps. In 1900, as the NBFU refocused its agenda, it sought a cooperative relationship with the NFPA to produce standards for fire-protective devices. Recognizing the overlap in the two organizations, the NBFU and NFPA agreed to work together to formulate and promote standards for fire-protective devices. The NBFU reproduced the NFPA's standards as its own, assumed the expense of publishing them, and printed the annual proceedings of the NFPA.[58]

Conclusion: Disorderly Men, Competition, and the Problem of Fire

For all their rhetoric about standardized practices, unity, and discipline, the insurance industry failed to bring the problem of fire under control. On the eve of the twentieth century the insurance industry could not boast of any systematic program that had made American cities, much less the industry itself, safe from fire. More significantly, the industry's financial status was not improving. Between 1860 and 1896, over eight hundred companies went bankrupt; profitability remained below 4 percent. Moreover, the industry remained subject to the vagaries of conflagration, much as it always had. A study conducted in 1905 confirmed that over 33 percent of the industry's losses were produced by conflagration. Only a small segment of the insurance industry—New England's Manufacturers' Mutuals—had developed a coherent program of fire prevention and safety by the twentieth century. Their research, begun in the 1880s, was led by Edward Atkinson and would provide inspiration for future work in fire prevention. For example, sprinkler systems, a fixture of late-twentieth-century construction, were developed in New England textile mills. Although the experience of the mutuals could not be easily replicated outside of the tightly controlled economic and built environment of New England textile capitalism, Atkinson and his cadre of engineers became agents of change as they encouraged the spread of fireproof construction and pushed the main body of the fire insurance industry to become more cognizant of safety. Even though stock insurance firms seemed to be awakening to new possibilities, they remained hesitant to become active in promoting public fire safety, and perhaps the most energetic voice for improving public fire safety was the National Association of Fire Engineers.[59]

In the end, the insurance industry's attempts to make business practices and manhood more rational appeared to have fallen short on many fronts. If urban fire insurance maps diagrammed danger and offered suggestions for safety, they had not yet become the underpinnings for a electrical, building, and fire department standards. If urban cityscapes sometimes included an increasing number of "fireproof" or slow-burning buildings, there was *no* systematic pattern to such construction, nor had the NBFU succeeded in its initial attempts at developing a uniform building code. If society demanded that middle-class men behave with order and regularity, such admonitions did not help the men of the insurance industry to transcend the "creative destruction" of American capitalism. Despite warnings against taking excessive risks by seeking protection from standard practices, middle-class businessmen embraced the individualistic, rough-and-tumble world of laissez-faire capitalism. As a result, their firms went belly up in large numbers. Yet, stunningly, as late as 1894 American fire underwriters remained unconcerned about public fire safety. Quite simply, it was not their responsibility, which—paraphrasing the National Board of Fire Underwriters—was to make profit from conflagration, even if this meant charging consumers higher rates. Although the seeds of change had been sown, insurers nonetheless were not succeeding in their effort to arrest the problem of fire. On the eve of the twentieth century, the marketplace, rationalized business practices, and middle-class manhood had thus far failed to bring discipline and order to the American environment.[60]

CHAPTER SIX

Becoming Heroes:

A New Standard for Urban Fire Safety, 1875–1900

When Truck Company D arrived at St. Louis's Southern Hotel in 1877, it found the building engulfed in flames. Above the pyre, almost a dozen people dangled from windows. The company hurriedly maneuvered its ladder truck along Fourth Street—impeded by streetcar tracks and blocked from the hotel's upper reaches by porches. Phelim O'Toole, a former sailor and twice commended fireman, scurried up the ladder only to find himself five feet away from the ledge where six stranded people anxiously balanced. Drawing upon years of experience, O'Toole demanded their only clothing—the bedding they had hastily wrapped around themselves. One man—a professor of English—balked, demanding, "What do you want with them?" Above the din of flames and wind, O'Toole yelled, "You pass them down and I'll save your lives." True to his word, O'Toole fashioned a rope and lifted himself to the ledge. Battling choking smoke and blinding heat, he methodically lowered the professor and his family, one by one. Then, moving to another sill, O'Toole removed another man and two women to safety. He recalled, "It got pretty hot and smoky up there, but I did my best."

If the firemen thought that their night's work was finished, the blaze had other plans. Fire Chief Lindsay spotted a man hanging in another window about to be consumed by the raging flames, and he ordered Truck Company D to reposition its apparatus. But, again, the ladder could not reach the stranded man. As O'Toole worked frantically to get to him, the man, frightened by the impending flames, threatened to jump. O'Toole implored him to stand pat, until the company maneuvered its truck closer. As O'Toole recounted, "We ran the truck into a shape that a truck never did work in this country, or any other, and never will again, though it did that time. It had nothing to support it; so we threw her against the wall some distance below the window." Without solid support, O'Toole ascended the ladder. He commanded the man to dangle from the sill. As the window crackled and exploded, O'Toole took firm hold of the ladder with his legs and feet, and leaned forward to gain better purchase. He grabbed the man's feet, and yelled "drop." Down he went, but O'Toole grasped him, struggling against gravity for nearly a minute. O'Toole remembered that the man "was very much excited, and we were hard set to get him off the ladder." Only moments after the company withdrew, the Fourth Street side of the Southern Hotel cascaded onto the street.[1]

Phelim O'Toole's heroism came at a time when firefighters in American cities were reorganizing their work and establishing the boundaries of their occupation. O'Toole's story illustrates some of the skills, tools, and apparatus that firefighters adopted as they refined their work techniques in relation to the physical and cultural dimensions of rapidly changing nineteenth-century cities. Firefighters remade the labor and the mission of urban fire departments when they fought fires more aggressively and faced an increasing amount of hazard as they penetrated deep inside the shakily built urban infrastructure. Firefighters also discovered a framework and rationale for their occupation as they began to rescue more and more people trapped by flames. Encouraged by a popular press that celebrated their work, firefighters became icons. Thus narratives of firefighting heroism defined their occupational boundaries, providing firemen considerable social and political authority, and justifying their interventions in the landscape. Firemen made possible the headlong rush of urban development even as it was routinely threatened by environmental catastrophe.

Repeated failures of public officials, insurers, engineers, and capitalists to protect urban America from fire fostered disorder in cities and provided the space in which firefighters established heroic credentials. Further accentuated by the headlong rush of capitalist development, this crisis was felt perhaps most acutely in working-class communities, which often had to face harrowing choices between economic survival and safety. These neighborhoods shouldered a disproportionate amount of the risk of industrialization and urbanization, including the problem of fire. At a time when insurers and engineers made great strides in developing safety for isolated industrial and commercial properties, in other areas firefighters stood as the first and only line of defense. As a result, firefighters' provision of safety resonated among their neighbors, and in popular culture more broadly.

Just as concerns about fire danger were written into insurance policies and drawn onto maps, they also were represented by the bodies of firemen. Firefighters established an occupation that became an integral part of municipal bureaucracies and urban life. By protecting both people and the infrastructure against environmental devastation, they preserved the system of social, economic, and political power. In the process, firefighters transformed themselves into heroes, but occupational formation occurred unevenly and incompletely. Following the Civil War, firefighters began to identify and to champion new work techniques and skills that had particular efficacy in the dynamic built environment, characterized by taller, larger buildings that magnified the danger of fire to citizens as well as property. The changing urban landscape and its ever-shifting dangers concerned firefighters greatly, and in 1872 they formed the National Association of Fire Engineers to represent their interests. The NAFE conceived of firefighting as a profession, and it used yearly conventions to establish the jurisdictional boundaries and work activities of the occupation. These meetings also became forums, like firehouses themselves, in which firefighters—from pipemen to chief engineers—implicitly, and sometimes explicitly, discussed how gender, race, and class mattered in the performance of their work. Over time, firefighters adopted increasingly similar work practices and asserted their claims as professionals. Even so, by the last decades of the century, firefighting culture varied remarkably according to local differences *between* and *within* cities, as the stories of firemen in Philadelphia and St. Louis make abundantly clear. Uneven training, insular male work groups, and machine politics inhibited the spread of professionalism and the full realization of a common occupational identity.[2]

The Built Landscape, Fire Danger, and Occupational Authority

Firemen created their occupation within the rich social, cultural, and political milieu of American cities, and they formed their identities as professionals in a complex relation with the urban built landscape. Compared to factories and other work environments that employers often controlled tightly, urban landscapes were more dynamic and complex. The process of urbanization and capitalism continually altered and transformed American landscapes. Firefighters discovered that their workplaces constantly were being reconstructed. New spatial arrangements not only complicated their labor, but also placed firefighters in ever greater danger. If firemen observed that the built environment imperiled themselves and their comrades, they especially expressed concern about the hazards faced by ordinary people. As firemen spoke out about the dangers facing their neighbors, they discovered a central organizing principle for their occupation.[3]

After 1872 the National Association of Fire Engineers set the standards for firefighting. To fire engineers, the firefighter was obligated to preserve his community's general welfare in a disinterested fashion. Such general welfare extended beyond mere control over firefighting labor; it included the broadest authority over the problem of fire. At the NAFE's first meeting, fire engineers outlined the new organization's agenda. Divided into two broad categories—"fire prevention" and "fire extinguishment"—the agenda covered eleven topics in great detail, many of which would reappear regularly at future meetings. Eventually, the nation's leading firemen initiated a program that centered on five components: (1) scientific firefighting, (2) standardization—of tools, fire department organization, and urban infrastructure, (3) codes for construction and design of buildings, (4) removing politics from fire protection and making the public's welfare a priority, and (5) organizing and directing the collective action of other specialized communities with an interest in fire safety, such as fire underwriters, engineers, and architects. The NAFE asserted its jurisdiction over urban fire protection by disseminating its *Annual Proceedings* to city governments, fire departments, and the National Board of Fire Underwriters, which regularly sent participating delegates. The NAFE later named as official "organs" of the association several prominent independent publications—*Fireman's Herald*, *Western Firefighter*, and the widely read *Fireman's Journal* (which later changed its name to *Fire and Water*)—through which the NAFE kept the pressure on others in the fire protection field to adopt its recommendations.[4]

Fire engineers observed that the shifting structure of the urban environment produced different, if not entirely new, dangers to urban residents. Increased use of new construction techniques, materials, and technologies—such as masonry and, later, steel-frame construction—changed the structure of cities, especially their downtowns. Buildings grew ever taller and spread horizontally, which also transformed warehouses into structures of remarkable size. Although the increased size of urban structures is difficult to measure, and was not uniform across the nation, buildings of major cities nonetheless gradually extended upward from between four and six stories before the Civil War to routine heights of eight, nine, ten stories, and beyond, by the turn of the century. This extension in scale is perhaps best symbolized by the appearance of skyscrapers in the 1880s, such as the

Home Insurance Building in Chicago, and the expansion of building scale in warehouse districts, such as those structures still present in St. Louis's Lucas Avenue Industrial Historic District. Not surprisingly, these rapidly expanding landscapes remade the danger of fire. Fires in such large structures burned differently depending on a variety of factors, including both construction materials and techniques, as well as building contents. Blazes often burned hotter and longer, producing intense heat and smoke that made fighting fires difficult. Sometimes fires lasted long enough to cause structural elements to lose their strength, triggering massive building failures. Towering walls toppled, falling onto firefighters or bystanders. The dangers even extended underground as deeper, larger, and more elaborate basements and subbasements warehoused goods and became smoky tombs for firefighters.[5]

Although these dangers were visible only to ordinary citizens at moments of crisis, firefighters were acutely aware that cities were becoming more hazardous. Experiences in burning buildings made them sensitive to how changes in construction methods, design, and building size altered the problem of fire, creating diverse "fire environments." Not surprisingly, fire engineers used their annual convention as a forum to discuss the new dangers, especially the materials and techniques which were used to increase the size of cities and which were often labeled as being "fireproof." Though these new materials were fire resistant, firefighters argued that fireproof construction techniques did not make buildings safer nor did they inhibit the spread of fire. Beginning in the 1880s, with the appearance of skyscrapers and the spread of fireproof construction, firefighters routinely raised important questions about the techniques. The editors of *Fire and Water*, for instance, asserted that fireproof structures "are somewhat of a delusion. They may be constructed throughout with non-inflammable materials, such as stone, iron, and plaster; but as long as the building contains material sufficient when on fire to cause the stone to split and fly, the iron to become redhot and deteriorate to a point below its breaking strength, and the plaster to crumble and fall, the building cannot be considered fireproof." The NAFE argued that most fires originated in a building's contents, not its frame, and that once weakened by fire so-called fireproof buildings became death traps. Thus, fire engineers argued, such structures constituted "one of the chief dangers to life, which firemen have to contend against." Adding to this problem was the fact that during the latter decades of the nineteenth century many fire departments did not have ladders long enough to reach upper floors of buildings and many cities did not have water supplies up to the task of propelling water into high-rise buildings. Even as new modes of construction altered dangers, older problems remained—including the common use of wood.

The NAFE remained concerned about the persistence of such construction methods and linked building practice to the larger structure of the North American natural environment: "The superabundance of wood mainly affords the principle [*sic*] of excuse in this country for its use in building purposes, and the troubles will continue as long as the supply lasts." The NAFE even conceived this sort of "tinder box style of construction so universally practiced" to be a problem of national character.[6]

Gaining a measure of control over this shifting built landscape increasingly became an issue that helped to set the boundaries of firefighting as an occupation. Firefighters agitated to diminish the growing danger, demanding that builders use construction techniques that made buildings safer or, at least, produced spaces that were easily susceptible to firefighters' physical intervention. At one of the NAFE's earliest meetings, reducing fire hazards in the landscape became a topic of intense attention, as the title of one discussion suggested: "The limitation or disuse of combustible material in the structure of buildings; the reduction of excessive height in buildings, and the restriction of the dangers of elevator passages, hatchways, and mansards." Such themes recurred over and over at subsequent meetings, and with each passing year discussions grew more expansive, nuanced, and detailed. However, the engineers' principal concerns remained focused on inhibiting the transmission of fire between buildings or within them rather than preventing ignition altogether. Through the 1870s, the NAFE received reports from its membership regarding the specifics of the relation between fire and particular structural features of the urban environment, and after 1880 the NAFE routinely solicited comment from experts outside the organization. Guest speakers included prominent underwriters and engineers such as George Hope, president of the NBFU, Edward Atkinson, perhaps the leading expert among civil engineers (and employed by New England Factory Mutuals to reduce their fire risk), Charles Knowles, president of the Southeastern Underwriters Tariff Association, and C. C. Hine, publisher of the *Insurance Chronicle*, and one of the leading industry experts on fire insurance practice. As early as 1876, again in 1884, and repeatedly at meetings through the 1890s, the NAFE transformed its discussions of environmental dangers into increasingly detailed recommendations for building.[7]

The NAFE examined risk in an exceedingly detailed manner, and, like others concerned with fire danger, it used a progressively more common and specialized language. For instance, the organization's members discussed the hazards posed by construction materials, both inside and outside buildings, openings within and outside buildings, and firewalls. The NAFE found fault with multiple and seemingly irrelevant points of construction. It challenged the use of aesthetic detailing fashioned of wood as well as other wood features on buildings. The organization noted that even buildings with brick or stone exteriors proved exceptionally vulnerable if internal features had been made of wood. It advocated use of iron shutters instead of wooden shutters, building "fire walls" between buildings that would extend five feet above roofs, and "compartmentalizing" buildings to help isolate fires. Along the same lines, the NAFE repeatedly expressed concern about how the structure of shaftways and other gaps within buildings—such as those for elevators and stairway—facilitated the spread of fires. The language also grew more specialized as the organization categorized and named fire risk. For instance, in 1874, one fire commissioner distinguished between dangers "of origination" and of "communication and extension"; what he identified as the "danger of communication" later came to be called "exposure."[8]

The NAFE's discussion of familiar hazards also became increasingly tied to considerations of how the new technological innovations were incorporated into the built environment. Indeed, as the NAFE critiqued the landscape, it realized that practicalities of construction potentially hindered progress on safety. For instance, noting that using metal in construction was expensive, the NAFE urged industry to develop fireproof materials that were inexpensive and practical. Likewise, the organization discussed the manner in which electricity, tall buildings, structures covering a greater area, and even fireproof construction techniques affected the danger of fire. Firefighters focused on a plethora of new threats that appeared in the landscape, including the presence of fumes and deadly gases that could collect and explode. In particular, the introduction of electricity became a new source for fires, even as the network of wires that crossed the urban landscape hindered efforts to extinguish fires. According to a report in 1889, "the present system of aerial wires presents the most difficult obstructions to gaining access to the upper portions of burning buildings."[9]

Fire engineers especially attended to the increasing height of cities and area (square footage) of buildings, which they feared were outstripping the ability of fire departments to reach. In 1887, New York City's fire chief reported to the NAFE, "We find mammoth structures being steadily erected to a height that the longest portable ladders fail to reach, and to which the most improved pumping machinery fails to deliver effective streams of water. . . . Even those constructed of slow-burning, or so-called fire-proof materials, are liable to be seriously damaged or even destroyed by intense heat from the burning contents, where water, the only known combatant of open fires, cannot be applied." Such fires developed uncommonly high heat (aided sometimes by slow-burning construction) that made extinguishing them all but impossible and facilitated a fire's spreading to adjoining structures, even those as far as sixty feet away—sometimes beyond the width of a street. Moreover, unless buildings had properly constructed "anchoring" walls, they were liable to collapse, thus spreading fire to adjoining structures. According to many experts, including influential insurance journals, more blazes were caused by exposure to fire in adjoining buildings, despite the introduction of fireproof construction.[10]

Firefighters were a skeptical group that, as discussed above, questioned the latest innovations in construction technique. Recognizing the variable nature of fire risk, a committee of fire engineers argued that many of the building technologies labeled fireproof merely altered the dangers of fire; they did not eradicate them. For instance, instead of burning like wood, these materials—such as stone, marble, iron, and brick—could be heated to the point of losing their strength altogether—causing massive structural failures and building collapses. Moreover, common firefighting practices, such as the use of water, affected certain building components in dramatic fashion, sometimes causing them explode into shards. If engineers' insights showcased shortcomings in fireproof practices, they did not reject them altogether. Instead the committee advocated more research and the extensive use of multiple layers of fire-resistant materials and/or proven design strategies. The committee suggested such redundancy, which it obsessively detailed: "I would divide [warehouses] into compartments as small as circumstances will admit with brick division walls, and openings in same to be protected by metal covered door, that all floors be made solid by concrete or plaster, and if not feasible then all floor timber to be lined with fire-proof material such as magneso calcite, plaster, terra cotta or any material that will insure the greatest amount of fire resistance." The NAFE repeatedly argued that fireproofing could help reduce an external hazard, but that it did not necessarily negate the internal hazard caused by heat from extremely flammable contents.[11]

Just as the NAFE sought to modify the urban environment to reduce danger, it also sought to alter the protective infrastructure to gain an advantage in the war against fire. For instance, from its inception through the end of the century, the NAFE promoted its vision of safety by recommending the improvement of, among other things, urban water systems. The NAFE contended that urbanization and industrialization had caused cities to grow beyond the abilities of fire departments to protect property and human life. It argued that many urban water pipes were too small to provide sufficient water for more than one or two steam engines at a fire, causing engines "connecting further away from the source" to "find only a vacuum in the pipes." In addition, the NAFE recommended erecting "standpipes" in all large buildings. Standpipes ran from a building's base to its roof, creating

Henry Braun, John Toensfeldt, and Louis Kauffman in pompier uniforms (St. Louis Fire Department Pompier Corps), 1877. In 1877, the St. Louis fire department established the nation's first pompier corps, led by Christian Hoell, who taught climbing and rescue techniques to firefighters throughout the nation. In this photo, three members of the corps wear the climbing and rescue gear that rapidly became common in fire departments nationwide. Courtesy, Missouri Historical Society, St. Louis

hose connections that allowed firemen to play water on a fire efficiently and quickly. Also, firefighters developed "siamese connections"—which forced water from two hose lines through a single nozzle—to recover water pressure lost from friction as it traveled through worn leather hose lines. Along with other recommendations, including cautious support of automatic sprinklers, fire engineers urged cities to build an infrastructure that assisted them in preserving public safety. The official publication of the NAFE even changed its name to *Fire and Water* in 1886 (earlier *Fireman's Journal*, later *Fire and Water Engineering*) to emphasize the important role that water, engineers, and the "waterworks man" played in improving urban fire protection.[12]

Firefighters advocated for adding inspections to their responsibilities in order to help them to become more familiar with the dynamic landscape, and therefore more effective at fighting fires. As one fire chief stated in 1893, "There is no more important feature connected with the fire department than the inspection of buildings." Systematic inspection of the urban landscape by firefighters contained the problem of fire across all its domains. It would serve as an important way of preventing fires because building owners could be informed about deficiencies in construction and dangers in the structure, which would result in the removal of hazards. Fire department leaders supported the creation of specialist fire inspection bureaus, but they also underscored the importance of firefighters themselves getting into the field, in that this provided firemen with critical data about the environments where they might someday work. Inspection, they explained, makes "the firemen acquainted with the building; shows him how best to enter and what to do in case of fire; discloses dangerous openings and faults in construction, which may save life if known and avoided." Through inspections firefighters became more effective at "subjugating" a blaze, and knowledgeable about the dangers they faced in each particular structure in their district. Ultimately, however, examination of urban hazards allowed firefighters to do more than help builders or themselves; they could become "instructors of the public in the prevention of fire."[13]

Indeed, believing that safety could be built into the landscape, the NAFE took an early leadership role in promoting a national municipal building code more than two decades before the insurance industry published its model building code. In 1876, the NAFE recommended an eleven-point plan for safe building construction. In 1877 the discussion grew in scope when engineers discussed "The enactment, by all State Legislatures, of good and wholesome building laws to be applied to the larger cities and towns." Legislating safety received attention yearly through 1882, when the NAFE established a formal committee to consider a legal code. By 1884, the committee on laws made recommendations that it hoped would become the basis for "preparing a code of laws which might be justly applicable to the different sections of the country." The code's seventy-four sections included the vast range of topics considered by the NAFE since its inception and included provisions for enforcement. According to the "practical" code, a coterie of city-appointed building inspectors would supervise new construction, and "reconstruction," of urban environments. All buildings would include relatively simple design characteristics meant to arrest the spread of fire. Moreover, the law required that fire escapes be provided in all buildings of two or more stories in which workers were employed above the second floor. The code represented the state-of-the-art in building construction in the 1880s.[14]

At least three factors limited the systematic implementation of the NAFE's proposals. First, although the NAFE found the insurance industry sympathetic to its concerns, firemen found underwriters to be tepid allies. Insurers supported proposals regarding firefighting infrastructure such as fire hydrants, water pipes, and telegraph systems, but they appear to have been relatively uninterested in systematic urban building codes. Secondly, even if underwriters were supportive, their ability to effect change was questionable. The insurance industry remained internally fractured and does not appear to have possessed an exceptional amount of cultural sway. At a meeting of the NAFE in 1894, an NBFU representative explained that the insurance industry sought to use the marketplace to coerce better construction but lamented that the NBFU's lack of authority over nonmember insurance companies limited its effectiveness. Even the activist Boston Manufacturers Mutual Insurance Company, which was created by a partnership of New England textile mill owners unable to obtain commercial insurance, primarily focused its research activities on textile mill construction, and only secondarily on other issues related to industrial design. The scope of its ability to intervene in the built environment was relatively limited and derived mostly from intimate economic ties to its clients in the textile industry. Mill mutuals, after all, were organized to improve the profitability of textile manufacturers, not to improve the general public safety. Additionally, they comprised a tiny portion of the insurance industry and held limited authority. Nonetheless, the mill mutual companies' laboratories produced research that could be transferred beyond mills, although their insights did not automatically generate a systematic program of safe construction in urban settings. The unwillingness and/or inability of the insurance industry to actively support a comprehensive legislative agenda was compounded by the fact that efforts to transform building practices remained mired in local regimens of political, social, and economic power.[15]

If the NAFE's jurisdiction over matters of building practice was limited, its efforts gradually had an impact as the organization repeatedly challenged Americans to build more safely. In particular, the NAFE repeatedly criticized the insurance industry because it believed that underwriting practices promoted fires and did not prevent poor construction practices. In a pointed exchange at its 1894 meeting, one fire engineer goaded an NBFU representative to admit the industry's failures to regulate building activities using the marketplace. Despite such frictions, by 1891 the NAFE had helped to organize a coalition with the American Institute of Architects, the National Association of Builders, the National Association of Building Inspectors, and the National Board of Fire Underwriters to establish a "Model Building Law." Although it is unclear if the coalition ever produced a model building law, the impetus for framing a code was gaining momentum. In 1892 the New York State government proposed a "model building law" for medium sized cities which the NBFU reprinted in pamphlet form and distributed to various local insurance boards as well as fire chiefs. By 1896, the NBFU began to consider the tenets of a code, which it published in 1905 and which became the national standard shortly thereafter.[16]

The NAFE's attempt to frame a systematic national building code during the 1870s and 1880s failed, but the principles behind it became central to the reconstruction of firefighting identity. In particular, firefighters increasingly associated fire danger with its human cost, whereas fire underwriters saw fire as a primarily economic risk. Even in the realm of economic matters, firefighters argued that the insurance industry behaved with a foolish abandon that threatened lives. *Fire and Water* reported, "There are any number of large buildings in this city, wherein dangerous occupations are pursued, that the firemen would hesitate to enter them in case of fire occurring for fear that the walls would fall upon them or that the fire would spread so rapidly that retreat from the interior would be impossible; instead of striving to improve the character of the structures in the city, the underwriters virtually offer a premium for inflammability and poor construction." To the NAFE, the NBFU's unwillingness to promote safe building and its apparent lack of concern about human safety in the face of profit were reprehensible. The NAFE was especially critical of imprudent underwriting practices that placed firemen at risk. It reported, "Firemen's lives, which are exposed to danger every moment from the time a signal is received until they return to quarters, should not be thus placed in jeopardy, because careless or dishonest men hold indemnity from insurance companies."[17]

Indeed, the NAFE increasingly justified its professional jurisdiction according to matters of human safety. In the 1870s, Henry Clay Sexton urged a "balcony system" of safety; he advocated that all buildings have outdoor balconies on high floors as places where people could congregate during a fire and be reached for rescue. By 1884, safety features like Sexton's balconies and fire escapes became a key component of the NAFE's suggested building code. Of course such lifesaving concerns also colored the NAFE's view of building height and of new technologies, such as electricity. As electric wires gradually crisscrossed the landscape, fire engineers identified them as dangers to firemen performing rescues. Firefighters' personal vulnerability to poor planning and/or construction also became a rationale for the NAFE's intervention in the landscape. In 1889, as a prelude to a discussion about appropriate wall construction, the NAFE argued that "of the great number killed each year by falling walls not a few were firemen."[18]

Ultimately, matters of human life gradually emerged as a consistent theme—both explicitly and implicitly—running through firefighters' dialogues about their occupation. This concern developed directly from their experience working inside buildings. As firemen labored to put out blazes, they discovered more and more that people were trapped in the upper stories of buildings, sometimes beyond the reach of their ladders. They also encountered falling walls, collapsing floors, hidden gaps in floors, and sudden explosions caused by built-up gases or fumes. Underscoring the unstable urban environment, fire department leaders, firefighters, and firefighting publications often repeated heroic narratives, like that of St. Louis fireman Phelim O'Toole, which was repeatedly used as an example by the influential St. Louis fire chief Henry Clay Sexton in the 1870s. During the 1880s and 1890s, *Fire and Water* routinely reported on large fires, fatal fires, and rescues. For instance, in 1886, *Fire and Water* noted, "New York and Chicago are running a close race this year for the honor of saving the greatest number of lives." Rescues became integral to defining firefighting as an occupation, and firefighters justified their claims to public authority by emphasizing the danger faced by themselves and their neighbors. The ever-changing fire danger, and firefighters' efforts to protect their neighbors, affected how they organized their daily work activities in the closing years of the nineteenth century.[19]

Eating Smoke, Saving Lives, and Professionalism

Ironically, when firefighters redefined their occupation around human safety, they made their labor more acutely dangerous. As their workplaces changed along with the urban environment, firefighters faced more dangers because they adopted new tactics; firemen channeled their aggression more directly toward blazes themselves. Mastering fire no longer meant pumping water onto it from afar, or simply checking its spread. Rather, fighting fire became an intimate and personal experience. Firefighters saved victims by scaling buildings whose heights extended beyond the reach of their ladders. They also penetrated deep into burning structures whose mammoth scale and varied construction created ever changing, complicated, and more hazardous fire environments. Urban journalist Jacob Riis described this new ethos, which was fully articulated in the NYFD and other departments nationwide by the 1890s: "New York firemen have a proud saying that they 'fight fire from the inside.' It means unhesitating courage, prompt sacrifice, and victory gained all in one." In the 1870s, this strategy had emerged from the daily work of firefighters in many large cities. It expanded through urban fire departments via the meetings of the NAFE and through the earliest manuals of the incipient occupation, which emphasized the importance of battling fire up close. Instructional guides urged firemen take action from a position "as near the fire as the heat and condition of the building will permit." Company foremen were told to place "pipemen" strategically, in such "position that water will strike directly upon burning materials." This aggressiveness engendered a host of new attitudes, work techniques, and tools, such as hooks, ladders, axes, and other equipment—which, as they became routinely used, distinguished professional firefighters from their predecessors.[20]

In the 1870s, as firemen moved their work to the inside of buildings, they encountered many new hazards, and chief among these was a primary byproduct of fire. Smoke became almost a singular concern, even an obsession—at times obscuring the emphasis on performing rescues. Smoke emerged as a more prominent hazard in part because the built environment grew in scale and durability, thus trapping gases and literally shrouding victims, as well as the fire itself. Indeed, smoke—not fire—would turn out to be the principal reason that people died in burning structures. Against this intensifying threat, firemen adopted the strategy of "ventilating" buildings with axes, hooks, and other tools. As one chief told colleagues at an NAFE meeting in 1895, "My first action on arrival is to ventilate the building." Removing the smoke and heat allowed firefighters to locate the blaze more quickly, and it facilitated the rescue of disoriented victims. Additionally, it prevented disastrous "backdrafts," which occurred because superheated smoke and gases that had collected inside confined spaces exploded when oxygen was introduced to the closed area, usually by firefighters opening a door. It is perhaps not surprising, then, that ventilating buildings became so critical, or that smoke became a defining aspect of firefighters' culture. As Jacob Riis explained, "The fire is the enemy; but he can fight that, once he reaches it with something of a chance. The smoke kills without giving him a show to fight back. Long practice toughens him against it, until he learns the trick of 'eating the smoke.'" Eating smoke would become almost as central a marker of a fireman's prowess as saving lives.[21]

Confronting the consuming element, according to observers, demanded a stunning combination of gifts—of physical strength, intellectual vigor, and of character. It was these attributes that energized firemen in this grand drama to control nature; indeed, the NAFE demanded that firemen possess the "inventive genius, tireless devotion, ceaseless watchfulness, unfailing wisdom, strength and endurance to show that we are able, in God's strength to meet every instrument of destruction with superior skill, alacrity and tact, until . . . fire shall no longer be a terror to mankind." Listed among the skills of firefighters were a host of physical characteristics—including dexterity, strength, technique, teamwork, and precision—all of which are evident on the pages of firefighting manuals and in the discussions of fire engineers. As Jacob Riis described it, "Firemen are athletes as a matter of course. They have to be, or they could not hold their places for a week, even if they could get into them at all. The mere handling of the scaling ladders, which, light though they seem, weigh from sixteen to forty pounds, requires unusual strength." Likewise, the work of the fire department, according to the NAFE, demanded firemen with "the skill necessary . . . to reach dizzy heights, or to apply recently invented machinery." Yet, to the NAFE, skill implied more than operating equipment or being technically knowledgeable. It also included "courage, endurance, quick perception, [and] common sense." Additionally, being a fireman required "that mental vigor, that intellectual keenness, without which man cannot vie to deeds of bravery, without which he is unreliable." And, finally, none of these attributes mattered if a fireman did not possess innate personal qualities that predisposed a man to being brave; according to the NAFE, "a mind that receives inspiration in time of public calamity and danger [reflects] the natural qualities that form the foundation of an ideal fireman's character."[22]

Phelim O'Toole's actions at the Southern Hotel Fire in 1877 embodied the physical abilities, technical skills, and daring that were emerging as the hallmarks of the fire service. O'Toole, foreman of the city's Skinner Truck (a type of ladder vehicle) Company, saved over a dozen people from the burning hotel. As the press reported, "It was Phelim Toole [*sic*] who carried Charley Kennedy, the New Yorker, down the Skinner ladder from a fourth story window when the walls of the building on that side were in danger of giving away. . . . The window from which Kennedy had been taken five minutes before was buried in a mass of smoldering ruins below." O'Toole and his fellow laddermen frequently received commendations for saving buildings and people from destruction through their exceptional skills. Using conventional ladders, ropes, and various climbing techniques, they also took hose directly to the base of fires located on buildings' uppermost floors. If O'Toole and his colleagues demonstrated high levels of skill, they displayed even more bravery. During one fire, O'Toole directed the hose nozzle at the fire's base until he fell unconscious and had to be carried from the building. At another fire, he saved an adjoining building from damage by dangling several floors in the air to "remove a firebrand" from its roof.[23]

O'Toole's legend grew with the telling, reaching an apex when he died and became an icon of heroism and professional duty in the STLFD, and throughout the nation. According to a report after his death, O'Toole performed countless acts of "unusual heroism" at over "one hundred" fires. His heroism embodied the ideal of disinterested public service characteristic of professional firemen. Henry Clay Sexton eulogized O'Toole as "one of the bravest men who ever lived—the bravest of the brave." And, several years later, Sexton related the O'Toole incident to the NAFE as part of an appeal for stricter building laws and pensions for firemen. Even after twenty years, when the NAFE's annual meeting again was held in St. Louis, a keynote speaker recalled O'Toole's "daring deeds," offering them, as well as those of the city's other martyred firemen, as a model for the fire service. For in 1877, shortly after the Southern Hotel fire, O'Toole became the fourth member of his company to have been killed in three years. Although O'Toole's legend grew with each successive generation, such narratives reveal the very real danger that lurked at every fire.[24]

Firefighters' headlong rush into burning buildings depended as well on their ability to command technology, the particulars of which changed as firefighters recast the priorities of their occupation. For example, when steam engines were the focal point of fire department labors, the "practical machinists" who operated them were the most highly compensated of rank-and-file firefighters. By 1900, however, the premium placed on this particular skill gradually diminished as departments redefined the basic lexicon of firefighting work and expertise. Traditional impulses among firefighters to arrive early at fires and to play first water eventually reappeared in innovations such as the quick-hitch harness and chemical engines. More critically, though, the drive to save lives and to fight fire up close generated renewed interest in the importance of hose, as well as rarely used tools such as hooks and ladders. The development of specialized apparatus that facilitated rescues—such as Hayes, Skinner, and Leveritch trucks, or the Cronin portable platform—underscores the emphasis that firefighters placed on human life as their work techniques shifted along with the environment. Furthermore, the fact that these innovations were developed by (and frequently named after) men who served as, or later became, fire engineers reveals the link between firefighting competence and innovation. *Fire and Water*, as well as other firefighting publications, routinely published the latest in firefighting technology, providing a service to firefighters, waterworks employees, and business interests alike. By the late 1880s, some issues of the journal contained more than twice as many pages devoted to advertising—

Fireman Martin Tierney on ladder, roof of Engine House No. 6, September 27, 1892, photograph by O'Reilly. Like other St. Louis firefighters, Martin Tierney learned work techniques, such as ladder skills, at his engine house. Around the turn of the century training became more regimented and common, sometimes involving specific drills and formal instruction. Courtesy, Missouri Historical Society, St. Louis

classified and placed ads—as it did to pages about fire protection. In particular, regular features on "patents," books for firemen and engineers, "contracting news," and "business specials" kept the field abreast of the latest technologies. Likewise, demonstrations, displays, and conversations about new workplace innovations occupied a significant part of the NAFE's convention.[25]

The importance of ladders—as well as associated work techniques and rescue tools—cannot be overstated. Indeed, fire engineers routinely challenged firefighters to scale new heights of innovation, goading them with pointed questions, such as: "Who will be ingenious enough to find the ability to get to 130 feet?" At its second convention in 1874, the NAFE bemoaned the fact that "truck companies"—also termed "hook-and-ladder" or "ladder" companies—had received little attention among firefighters. It further argued, however, that such companies and their equipment were increasingly necessary as "the constantly increasing height to which buildings are now carried requires a radical change in this respect, to bring the departments up to their former comparative standards of efficiency." The NAFE recommended that all departments should have at least one truck company and that cities with populations greater than 25,000 should have two. In 1877, William Lewis, then publisher of the influential *Fireman's Journal*, recommended a more dramatic increase in the proportion of truck companies in his *Manual for Volunteer and Paid Fire Organizations*, advocating that every department have at least two "truck" companies for every three "engine" companies. By the 1880s, as most urban fire departments began to increase the number and use of ladders at fires, the ratio of engine companies to hook and ladder companies in American cities declined. In St. Louis, the ratio went from 10:1 in 1870, to 5:1 in 1880, to almost 3:1 in 1900. In Philadelphia, the ratio of engine to truck companies remained steady at about 6:1 from the department's inception in 1871 through the 1890s, when it began to decline; by 1906, it had lowered to less than 3:1. More broadly, in American urban centers with populations greater than thirty thousand the ratio of pumping apparatus to ladder apparatus had declined to nearly 2:1 by 1917.[26]

The formation of specialized life-saving units—the pompier corps—in American fire departments accompanied the proliferation of trucks. In 1877, following the devastating Southern Hotel fire, two German immigrants to St. Louis, Zero Marx and Christian Hoell, developed a plan to rescue people trapped high up in buildings. Having learned scaling techniques in his native Germany, Hoell instructed a corps of gymnasts, German turners, on the use of the simple apparatus, which they demonstrated to the fire department, city leaders, and a crowd estimated at fifty thousand. The St. Louis fire chief, Henry Clay Sexton, quickly incorporated the plan into the department because of these units' extraordinary ability to gain access to buildings regardless of structural impediments, such as electric wires, narrow alleys, and high buildings. Using scaling ladders, a special climbing belt, climbing techniques, and rope, pompier corps could carry hose to a "commanding position" or lower people to safety. Hoell and his pompier corps became a model rescue unit, but according to Sexton and the NAFE, ordinary firemen also possessed the required skills. Even so, employing pompier techniques required great dexterity and discipline. Firemen lifted the iron-hooked end of the pompier ladder, caught it to the sill of a first-story window, and then scurried up the scaling ladder's narrow rungs. Next, the firefighter straddled the sill, and once again lifted the hooked end to the next story. Followed by other firemen, the corps formed a chain of ladders from which they could use ropes to pull up hose or to perform rescues. Fire departments sometimes used hose guns to propel a line of rope onto a building's roof into the hands of a waiting pompier fireman. As one observer noted, "It stirs one to see a 200 pound man run up a wall of a four-story building by no other means than a skeleton ladder, twelve feet long—a device that appears to be unable to bear the weight of the average youth." Creating a unit of specialist rescuers dramatically reflected the new emphasis of firefighting late in the nineteenth century, as removing people "out of the jaws of death" came to define the fire service.[27]

The rapid spread of pompier techniques throughout American fire departments underscored not only the importance of lifesaving in the new occupation but also the growing influence of the NAFE's emphasis on professionalism. Shortly after its formation, the St. Louis pompier corps traveled to a fireman's tournament in Chicago, where one of the original pompier corps, Thomas Mayer, won first prize for rapid climbing and Phelim O'Toole and Patrick Conway won the prize for pairs climbing. Fire departments in New York, Chicago, Boston, and Cincinnati quickly emulated the innovation, and by the 1890s pompier corps had become standard in American fire departments. In 1882, at the request of the New York Fire Commissioners, Hoell trained New York firefighters on pompier technique, presenting what was perhaps the nation's first formal training program. Soon pompier training became a significant part of firefighting drills across the nation—in Kansas City, St. Louis, Chicago, and New York. The training varied, but the instruction techniques helped to make pompier skills commonplace, if not widely used in firefighting. In Chicago, for instance, firefighters learned about the techniques as city instructors toured engine houses; in New York City some thirty-five thousand firemen had passed through the city's sixty-day training course by 1903. The spread of Hoell's innovation reveals the paths through which new firefighting techniques were transmitted.[28]

The pompier corps reflects, too, the internationalization of firefighting late in the nineteenth century. Although pompier techniques journeyed from Europe to the United States with waves of immigrants, they would return in a new form, as part of the NAFE's agenda for professionalism. Firefighters and fire chiefs took a more expansive interest in global firefighting during the 1890s as part of the organization's broader agenda to spread the science of firefighting to the world. For instance, in 1893 fire engineer George Hale led an American delegation to the International Fire Congress in London. Recognized as a great inventive genius among firefighters, the Kansas City fire chief developed a swinging harness that cut the time required to hitch horses to steam engines, a "tin roof cutter," the "Hale electric wire cutter," and the "Hale water tower," which *Fire and Water* described as being one of the "most important additions to fire apparatus of the nineteenth century." Hale's inventiveness and the efficiency of the American team not only brought home prizes for their performance, but also thoroughly stunned the world audience. According to the *Fire and Water* correspondent, "The American fire team, by their display in London of the American method of fire craft, have surely entered the leaven into the British loaf that will in time leaven the whole foreign system."[29] As the firefighting trade journal would make clear in successive issues, American firefighting techniques proved far more advanced than those practiced elsewhere around the globe, in no small part because U.S. firefighters had to respond to a decidedly more dangerous urban environment. Just as American capitalism would infect the globe, so too would the professionalism of American firefighters.

In 1896, the NAFE transformed itself into an international organization, becoming the International Association of Fire Engineers. Recasting firefighting also entailed the transformation of the individual and collective bodies of firefighters, and the occupation's leaders began to codify the component tasks and skills of firefighting. One of the first such efforts, Lewis's *A Manual for Volunteer and Paid Fire Organizations*, published in 1877, revealed the outlines of the new occupation just as it emerged as a distinctive line of public service in large American cities. Lewis emphasized that firefighters battle blazes up close and provided specific instructions on how best to put water on a fire, including more than ten pages of instruction, with over forty specific recommendations regarding the use, handling, and care of fire hose. Lewis especially urged foremen to strategically place "pipemen" in such a "position that water will strike directly upon burning materials." He advocated fortitude on part of foremen, who, like pipemen, would take action from a position "as near the fire as the heat and condition of the building will permit." The manual, however, devoted little attention to saving lives or to innovations in the use of ladders, offering only two, relatively vague, instructions about their operation. This omission underlines the degree to which the occupation of firefighting was a work in progress during the 1870s. Although Lewis recognized the importance of ladders and other tools, the techniques and tactics for using them had not yet become part of the lexicon of firefighting. Nonetheless, the manual reflected the shifts in firefighting work.[30]

A decade later, in 1887, the NAFE became interested in establishing standard work practices for firefighters, and it commissioned Andrew Meserve to illustrate and describe firefighting work. Meserve's *The Fireman's Hand-Book* introduced firefighters to a host of new skills and practices that had emerged in American fire departments and that had been publicized at NAFE meetings in the 1880s. It discussed techniques for rescues, the use of ladders and pompier equipment, and innovations in departmental management procedures, including "organization," "tactics," "drills," and "funeral honors." Most significantly, Meserve devoted nearly one-third of the *Hand-Book* to the techniques needed to use, move, and climb ladders. Well illustrated, the volume discussed how to use various "respirators," detailed the proficiencies needed to get a hose close into a fire, and taught knot-tying techniques as well as the competencies required to "rescue insensible persons." In addition, it even documented cultural practices, such as parading, competing, funerals, and the "pleasant drill" of how to give and to receive a kiss. Later instructional books expanded on the lessons taught in Lewis's *Manual* and Meserve's *Hand-Book*. In 1898, Charles Hill published a manual modeled after firefighter training in New York City that updated firefighters on the latest innovations, especially those competencies used to perform rescues or to fight especially difficult blazes. Fifteen years later, in 1913, New York fire chief John Kenlon continued this trend when he publicized the culture, history, and training routines of the NYFD. Along with the NAFE's regular attention to firefighting labor, these training manuals would become the basis for courses of fire department instruction, which emerged late in the nineteenth century and appeared more formally in department training schools in the twentieth century.[31]

As firefighters defined their service, they established the boundaries of their occupation, and many began to describe firefighting as a calling worthy of professional status. Such claims began with firefighters' recognition of the seriousness of fire as a national problem and gained momentum when firefighters emphasized the characteristics and skills of the job itself. The NAFE argued that their occupation was part science and part craft, acquired only through disciplined training and years of experience. Firefighting demanded precise knowledge of hydraulics and engineering, but the application of that knowledge varied at every fire. Only through experience, forged in direct contact with the flaming landscape, did firefighters' become experts. Yet, expertise was gained only through discipline. Recognizing the role that formal education played in defining the boundaries of other professions late in the nineteenth century, some firefighters sought to establish a college for firemen, so that firefighting would be recognized as a "science." At the same time, others wanted to take the process further, seeking to "establish under the law that fire extinguishment is a profession and can only legally admit those of proper qualifications." To be sure, transforming firefighting into a profession conferred an elevated status on firefighting, but it also became a panacea for many of the issues that firefighters faced: incomplete training regimens, poor wages, widespread differences in organization, management, and work both between and within departments, and the persistence of rough cultures in engine houses. A college for firemen would transform the experience of lifelong firefighters into practical and meaningful lessons—at once science and craft; it would elevate and make standard the practice. Mostly, though, it would help firefighters combat political intervention in fire departments by elected officials. As one proponent explained, until firefighting became a profession "one of the most humane, ennobling and self-sacrificing professions will be looked down upon as a degraded, unsystematized and uncertain service." Not only would firefighting become free of outside interference if it were elevated to an independent and objective profession, but also firefighters would acquire social and political capital, helping them to become more effective advocates for fire prevention.[32]

Although obtaining professional status received wide approval, the NAFE and firefighting leaders typically fought political intervention and perceptions of lax discipline with a more easily obtainable solution: training firefighters themselves. Using instructional guides, such as Meserve's *Hand-Book*, fire departments used drills to inculcate discipline and to establish common work practices in all departments, regardless of whether they were volunteer or professional. Departments embraced this strategy because they discovered that lack of "thorough and systematic drill in the different branches of the fire service" rendered advances in firefighting equipment useless. The NAFE noted, "When the extension ladders were first introduced, it was almost impossible to get men enough around a seventy-foot ladder to raise it in any reasonable time and with safety." Additionally, manuals taught physical drill at a time when many fire department leaders wondered whether contemporary firemen possessed the endurance of their predecessors who acquired manly power working "the brakes of the hand engine." Despite making bold claims about firefighters' innate qualities, the NAFE recognized that firemen were made, not born. Drilling produced discipline, which made men amenable to following orders and placing themselves in danger for the good of the company and the public. Training further helped to distinguish between those firefighters who could and could not handle the difficulties of firefighting—revealing those who "lost their nerve." Training regimens also developed a framework for the occupation, at once encouraging the creation of common bonds and challenging the informal aspects of local cultures. The NAFE did not want men to "trifle away their time in games and story-telling." Training firefighters kept them focused; it allowed fire department leaders to answer the NAFE's concerns about their technical and physical competence—"Do we keep our men up to the line of progress with the machinery?" By disciplining themselves, firefighters defined the tasks and expectations of their occupation, which bolstered their claims to possessing a "calling," if not a profession.[33]

Gender, Race, and Class among Firefighters

As firefighters brought discipline to the chaotic process of urbanization, they established an occupation characterized by their particular understandings of gender, class, race, and ethnicity. Most obviously, firefighters' beliefs were expressed in the preeminent symbol of their occupation—the heroic body of the life-saving fireman. This dashing figure physically and metaphorically restored order at every fire scene—a point not lost in a popular culture that celebrated the heroism of firemen regularly. The coordinated elements of firefighting provided security to urban residents in a highly visible way, and in clear contrast to the individualistic culture and sense of manhood that characterized late-nineteenth-century businessmen and capitalism. As performing rescues elevated firefighters in the eyes of their neighbors, it also became connected to popular understandings about manhood, race, and social class.

Artists, local newspapers, and national magazines furthered this identity when they depicted the omnipresent dangers of fire and the manly heroism required to preserve lives and the social order. In the 1860s, Currier and Ives produced a popular print series, *The American Fireman*, which depicted professional firefighters at work. As early as the 1870s, firemen also graced the cover of *Harper's Weekly* and appeared in other national magazines in poses that replicated firemen's understanding of their manhood. Fire chiefs formed a professional association and firefighters read publications printed exclusively for them beginning in the 1870s and expanding during the 1880s. Local newspapers regularly reported nearly every fire and featured the heroic exploits of firemen on front pages. Crowds of thousands gathered at fires. Sheet music publishers recorded the heroics of firefighters, and vaudeville performers incorporated firefighting heroism into their productions. By the 1890s, firemen made their first appearance in the movies, as Thomas Edison selected the spectacular fire exhibit at Coney Island as an early subject of his new motion pictures. By 1900, stories of firefighters heroics had become standard in their presentation, much as the NAFE hoped firefighters actions would become when faced with harrowing situations. The press valorized firemen who demonstrated physical strength, technical competence, and public spirit by vanquishing fire or saving a damsel in distress. The fireman became an icon of heroism at a time when, in the face of industrialization, manhood itself seemed to be in decline.[34]

Although the nineteenth-century media played a critical role in constructing firefighters' heroic culture, it was ultimately firefighters who authored this cult of manhood. News accounts almost always began with the actual work of firefighters, who often described their labor with a remarkably self-effacing personal style. Constructed as self-consciously as the other elements of firefighters' culture, this unassuming quality seemed to encourage journalists. As Jacob Riis noted, "I have sometimes wished that firemen were not so modest. It would be much easier, if not so satisfactory, to record their gallant deeds." Although journalists often exaggerated, the day-to-day activities of firefighters did not really need much embellishment. For example, it does not seem to be a stretch to believe Riis when he says, "The first experience of a room full of smothering smoke, with the fire roaring overhead, is generally sufficient to convince the timid that the service is not for him. No cowards are dismissed from the department, for the reason that none get into it."[35]

Firefighters put such modesty aside, however, when they deliberately placed themselves into a grand life-saving struggle—a battle that they transformed into a test of their mettle as men. When the environment went amok, *firemen* rushed into buildings. They saved people, especially women and children, with all the cultural resonance that that possessed. As Andrew Meserve recommended in his firefighting manual published at the behest of the NAFE, "Always give precedence to women and children in rescuing lives." Not only did firefighters adopt strategies that stressed protecting the "innocent," but they put themselves at significant risk in the process. The intensity and danger of their labor heightened the contrasts that firefighters relished—between their steely nerves and muscled physiques and those of the people they rescued, who generally were soft, weak, and frightened. As one firefighter recalled about saving a woman, "I wish that she hadn't fainted. It's hard when they faint. They're just so much dead weight. We get no help at all from them heavy women."[36]

As firefighters forged their identities in the battle with fire, they seemed intuitively aware of the metaphorical dimensions of their struggle. Long the object of dread fascination, fire has evoked a complex series of contradictory symbolic associations between the physical and spiritual, good and evil, and destructive and regenerative forces. Control over fire has been perceived, at least metaphorically, as a marker of manhood. As historian Robin Cooper has argued, conquering fire was critical in order that man "ensure mastery over himself and his property (including human property) and to secure general stability and order." Urban journalists

Firefighter using pompier techniques, 1898, illustration from Charles A. Hill, *Fighting a Fire* (1898). Ladders played an increasingly important role as firemen prioritized saving human life. Charles Hill's *Fighting a Fire,* one of the earliest guides to firefighting, illustrated pompier techniques. A series of four prints, including this image, showed how to use a specialized scaling ladder to ascend a building, one window frame at a time.

such as Jacob Riis picked up on this drama with a patterned story that always began with a moment of disorder—the outbreak of fire. Professional firefighters of the 1880s, though, changed that narrative fundamentally when they dashed to do battle with the devouring element up close, inside buildings or saved human lives. As they saved the day and rescued innocent victims, firefighters returned as heroes. Tapping into the conventions of heroic myth underscored firefighters' singular purpose and asserted their control of the elements and cities. As Riis suggested of firefighting, "It was the fancy of a masterful man and none but a masterful man would have got up the ladder at all."[37]

If struggle to contain fire underscored firefighters' ability to command nature as well as themselves, winning that battle—and especially rescuing people—frequently demanded that firemen behave with reckless abandon. Becoming the bravest of the brave, as O'Toole had done, often meant taking chances that no reasonable, or even altruistic person would; being a noteworthy firefighter demanded that firemen put themselves at high risk. Such impulses certainly derived from the pressures of the occupation. As a reformist firefighter in Philadelphia recalled, "All wanted to be heroes in the eyes of their associates or who ever [*sic*] was in a position to observe them, with the consequence a great number received serious injuries or lost their lives." Yet, as firemen redefined their occupation in the second half of the nineteenth century, the heroic character of the work may itself have attracted men predisposed toward danger. As Riis described, "Doubtless there is something in the spectacular side that attracts. It would be strange if there were not." At times, journalists even suggested that firefighters took special pleasure in the adrenalin rush of their job; according to Riis, "The one feeling that is allowed to rise beyond [their occupation] is the feeling of exultation in the face of peril conquered by courage."[38]

Recent social psychological research on heroism, including that of firefighters, would appear to confirm Riis's conclusions. James McBride Dabbs has found that hotshot firefighters—those most likely to perform rescues or take dangerous assignments—possess personality traits associated with skill mastery *as well as* an abundance of testosterone, the hormone linked both to male physiology and high levels of aggression and competitiveness. In fact, mastery of self and firefighting skills do not appear to be enough to inspire firefighters into heroic acts. According to Dabbs, "Testosterone is needed to translate that motivation into action." Of course this is not to suggest that there is some biological basis to excellent firefighting work or to heroism; rather, it reveals the critical importance that both aggression and competitiveness have played in the creation of firefighting as an occupation. In the nineteenth century, when firefighters attacked fires up close and placed a premium on performing rescues, they had laid the groundwork for a high-risk culture. Firefighters created an occupation in which an aggressive recklessness coexisted uneasily with mastery of skills and self. On one hand, the job demanded athleticism, technological skill, and self-control; on the other, success sometimes depended upon excessive risk taking, thus drawing men who had either technical skill or courage to the point of foolhardiness—and in some cases men possessed both qualities simultaneously. Firefighters of the 1870s and 1880s had redefined what it meant to be a fireman; not only had they given their occupation a well-defined boundary but they forged a transcendent and frequently contradictory identity as heroes.[39]

Peering yet further beneath the veneer of heroism reveals a more complicated story about firefighters' definition of manliness, one colored by broader understandings of class, race, and ethnicity. In fact, the professionalism advocated by firefighters carried with it a somewhat contradictory message, with appeals to characteristics of male behavior commonly associated with both working-class and middle-class manhood. On one hand, when the NAFE adopted the rhetoric of professionalism as the best means to discipline firefighters and cities, it asserted the hegemony of middle-class cultural mores. Not only was the language of professionalism increasingly a cultural construct, but the NAFE advocated middle-class values of sobriety and self-discipline as a way to discipline the working-class expressiveness of firefighters' everyday work cultures. Professional behavior demanded that firefighters replace rough working-class activities with more decorous behaviors befitting self-sacrificing heroes. Moreover, the NAFE clearly intended virtuous professional firemen to become examples in their communities. Fire engineers fashioned themselves as "representative men engaged in one of the grandest professions in the land." As sober and efficient firemen, as well as "good husbands" and "good fathers," fire engineers influenced urban residents better than "ten ministers."[40]

On the other hand, fire engineers injected elements of working-class populism into their vision of professionalism. According to advocates of professionalism, firefighters should possess strong physicality, and be able to engage in extremely dangerous and intense labor. Moreover, accounts of heroic firefighters rescuing the middle class, which was not accustomed to the rigors of physical activity, provided an object lesson about the values of a strenuous life. The stories of O'Toole's rescue of a university professor and his family echoed the language of the middle-class sporting community, which itself sought the physical "virtues" of working-class life. The NAFE argued that that physical virility could inculcate moral character, thereby assuaging concerns that contemporary life had sapped men's vitality. Drilling and regular exercise provided discipline that invigorated firemen "physically and morally." Such recommendations predicted the way that the middle class appropriated the rough physicality of the working class to create a new norm of manliness that would ultimately gain ascendancy during the Progressive Era in figures like Theodore Roosevelt.[41]

Firefighters' calls for safety carried not only the evangelical spirit of the middle class but also the reformist zeal of working-class agitators. Firefighters believed that their manliness could counteract the corrupting values of unfettered laissez-faire capitalism. The NAFE contrasted firefighters' heroism with the dangers that builders and underwriters created as they built cities higher and higher. "Let the engines of destruction multiply; let the skill of men continue to invent material to send life into death . . . we will endeavor by our inventive genius, tireless devotion, ceaseless watchfulness, unfailing wisdom, strength and endurance to show that we are able, in God's strength, to meet every instrument of destruction with superior skill, alacrity and tact, until . . . fire shall no longer be a terror to mankind."[42]

Furthermore, firefighters' occupation benefited from and utilized the racial stereotyping common to late-nineteenth-century society to define heroism as the exclusive domain of white men. Such a definition was reflected in nineteenth-century popular culture and was buttressed when the NAFE promoted and perpetuated those images at meetings and in publications. Stereotypical caricatures of African Americans appeared in firefighting imagery during the 1880s. Currier and Ives, which had produced the heroic print series *The American Fireman* and *Life of a Fireman*, issued nearly 550 comics during the firm's seventy-five year history. Approximately half of those prints belong to a subcategory known as the "Darktown Comics." Located in a tradition of "comic" prints ridiculing African Americans, dating from the 1820s and 1830s, one set, titled *The Darktown Fire Brigade*, derided the notion that African Americans could serve as effective firefighters. Whereas *The American Fireman* and *The Life of a Fireman* valorized the heroic white bodies of volunteer and professional firefighters, *The Darktown Fire Brigade* infantilized African Americans as simpletons, incapable of skilled labor or coherent action.[43]

Ostensibly set against a backdrop of rural volunteer firefighting in the South, *The Darktown Fire Brigade* carried significance beyond those work and regional communities. Thomas Worth, who made the prints, depicted African Americans as bungling circus clowns incapable of performing tasks that had become routine among urban, professional firefighters. In fact, he portrayed African Americans failing to use tools, such as hooks and ladders, which were essential to firefighters' occupational identity. The point could not have been lost on urban and rural audiences throughout the nation: African Americans lacked the physical and mental dexterity necessary to be firefighters. Just as had been the case fifty years earlier when Philadelphia volunteer firefighters pressured African Americans not to create an independent fire company, professional firefighters excluded blacks from their ranks. Furthermore, firefighters routinely used racial caricatures to depict inefficient, inelegant, and "primitive" firefighting activities. Such drawings appeared in the preeminent fireman's professional journal, *Fire and Water*, which in 1891 published a pictorial history of "old time fire apparatus" tracing the development of hose from ancient Egypt through early modern Holland. The drawings of Egyptian firefighting, titled "The Friendly Serpent," depict sambo figures strikingly similar to those shown in Currier and Ives's *The Darktown Fire Brigade*. The figures display such ineptitude that only after a snake hiding in a tree assists them, by becoming their hose, can they extinguish the fire. In striking contrast, *Fire and Water* showed several models of European firefighting traditions under the title "Some Types of Old Fire Engines," which portrayed heroic European (white) firefighters using evolving technology. These forebears of American firefighters are shown as innovators, initiating a cycle of technological progress.[44]

Such prejudices, which reflected the racial politics that underpinned firefighters' occupational culture, sometimes exploded at the NAFE's annual meetings. In 1885, for instance, during a session on "politics," Richmond's Chief (G. W. Taylor) turned the discussion from party politics to racial politics. According to Taylor, the issue was not a political fight but a war between the races. He explained, "We don't wage a war with the Republican Party; we wage a war against the black man. Unfortunately the black element in our country is allied with the Republican Party, and we are heralded forth as ku-klux, simply because we want to maintain our rights." In the era immediately following Reconstruction and during the early years of Jim Crow, Taylor asserted that much more was at stake than simply firefighters' skills and proficiency. White male hegemony, especially in southern states where blacks often outnumbered whites, depended upon controlling the machinery of the state. As Taylor and his peers sought to bar African Americans from social and political power, they denied them the opportunity to protect the public interest. They connected firefighters' manly service to white racial identity by negating the possibility of black participation.[45]

Advocates of firefighting professionalism may also have used the notion of whiteness to limit ethnic identification among firefighters. By underscoring the firefighters' common background—their shared "race"—whiteness added another layer to the brotherhood of firefighters. And, indeed, overt expressions of anti-immigrant or ethnic bias were rare occurrences. Interestingly, at the 1920 IAFE meeting, in the wake of the expression of anti-Irish and anti-Catholic sentiments, Chief John Kenlon of New York City resigned his post as the organization's president. However, when he appealed to his fellow firefighters "on the broad grounds of Christian unity and firemanic brotherhood," his colleagues insisted that he remain head of the IAFE, and his status in the occupation remained untarnished. Likewise, it would appear that ethnicity and immigration rarely became issues within local fire departments, which were heterogeneous institutions. In St. Louis, for instance, about 15 percent of all firefighters who served in the department in the nineteenth century reported having been foreign born. Moreover, if surnames are any indication, there was a diverse mix of firefighters of Irish, German, Eastern European, and Southern European heritage—likely the children of immigrants, if not themselves recent transplants—in the departments of both St. Louis and Philadelphia. Undoubtedly, these backgrounds found expression in engine houses, and may have tightened the bonds between firefighters of certain companies and/or their commitment to the fire service more generally. For example, in St. Louis firefighters who served more than thirty years were more often foreign born, predominantly immigrants from Ireland or Germany. In ethnically diverse fire departments, the language of a common brotherhood may have helped to ease any ethnic differences between firefighters, without completely overshadowing firefighters' links to their local ethnic communities or their families. Being heroic white firemen harnessed them more tightly to the occupation, and certainly fomented brotherhood.[46]

The notion of being part of a white brotherhood also may have buffered firefighters from being identified with either working-class militancy or middle-class capitalism. In the nineteenth century, especially as they developed a clear occupational identity, firemen remained relatively separate from both their working-class neighbors and middle-class professionals. First and foremost, they were white male heroes, dispassionate in their service of the community good. Heroism offered firefighters a seemingly neutral position that stood outside the class conflicts that so cleaved Americans during the Gilded Age and Progressive Era. Even so, the tension between being a professional and living and working within chaotic and diverse urban neighborhoods could not be resolved easily. In fact, balancing community, career, and occupational identity in engine houses and at fires proved precarious for many firefighters as they sought to preserve order.

Politics, Professionalism, and "Bully-Boys"

How did the tensions between professionalism, working for municipal governments dominated by urban political machines, and day-to-day work activities play out in the lives of firemen? What were the everyday experiences of firefighters, battling a changing problem of fire in American cities? Answering these questions, peering beneath the hyperbole and dust of time into the work lives of nineteenth-century firefighters, suggests that their quotidian experiences differed from the cult of manly heroism touted by advocates of professionalism. On the surface firefighters appear to have created a powerful occupational identity, judging simply from the *average* length of their careers. Firemen entering service in St. Louis in the nineteenth century averaged more than twenty years; in Philadelphia they averaged over fifteen years. Additionally, in cities elsewhere in the United States firefighters appear to have commonly worked careers of this length. According to a survey published by the NAFE in 1889, over 36 percent of the nation's professional firefighters had at least fifteen years of experience in their departments. As the NAFE and American popular culture more clearly defined firefighting as a heroic service, successive cohorts of firefighters worked longer careers. Nevertheless, although such longevity suggest firefighters' affinity for their jobs, their careers varied greatly. Differences in experience existed between firemen in the different cities as well as firemen within the same department. This variability suggests that local working conditions mattered as much as, if not more than, the NAFE's vision of firefighting as a professional calling.[47]

Evidence of frequent dismissal from departments underscores the instability of firefighting work. In St. Louis, for instance, of the firefighters that exited the department prior to 1880, nearly 75 percent were discharged, and 11 percent resigned by choice. Over the next twenty years 46 percent were dismissed, and a paltry 17 percent resigned. In Philadelphia, about 40 percent of all firefighters exiting in the nineteenth century were discharged; by the same token, nearly half resigned. In contrast to firefighters in St. Louis, then, Philadelphia firemen would appear to have exercised a greater degree of control over their employment. More broadly, comparative data suggests that firefighters' work patterns in the nineteenth century seem to have resembled those of industrial workers more than they differed from them, except in terms of *average* career longevity. Although precise comparisons are difficult to make because of the paucity of the data and differences in its collection, there can be little doubt that during the nineteenth century firefighting was capricious and volatile employment, although the level of arbitrariness varied by locale.[48]

The NAFE's goal of making firefighting standard across departments was not achieved by the end of the nineteenth century. Indeed, the technical aspects of firefighting advocated by the NAFE—and represented by the heroic proficiency of Christ Hoell and Phelim O'Toole—filtered very slowly into fire departments, and even more gradually into local engine houses. For example, when James Gilbert joined the PFD in 1885 he reported that he received little training from his colleagues about proper work techniques. Perhaps worse, few companies used the same nomenclature for standard equipment. Gilbert noted that "there was no preliminary instruction given . . . fire service tools that were carried on the different apparatus would be known in many instances by names coined by the men of the different companies." Although Gilbert's reformist agenda may have led him to hyperbolize, a report by Philadelphia's mayor in the twentieth century repeated similar sentiments, and a *Fire and Water* decried the inefficiency of the PFD in 1892. Just as common training methods developed slowly (with the NAFE not issuing a standard manual for drilling until 1889), common practices and rituals made their way into engine houses very gradually. Indeed, with the exception of New York City, firefighter training typically occurred at individual engine houses, with relatively little centralized authority or instruction in technique. The Chicago Fire Department, for instance, described its training as a "school system" in which instruction took place at individual engine houses: "New firemen are initiated and old ones skilled in the dangers and necessities of their calling at the engine houses where hook and ladder companies are stationed." To the degree that training regimens remained centered in engine houses, they reinforced intense local connections as much as they provided a vehicle for creating common work practices. Not surprisingly, on the eve of the new century, firefighters' experiences varied widely, even in the same city.[49]

Firefighting may have had heroic cachet, but everyday work conditions tell another story, perhaps accounting for some of the instability in firefighters' careers and experiences. When compared with other nineteenth-century jobs, firefighting did not offer especially appealing salaries, work, or benefits. Firefighting publications noted the salary disparity when compared to other "workmen whose work calls for special knowledge, skill, and training." Yet, firefighters earned slightly more than laborers, and they experienced little seasonal change and were relatively unaffected by layoffs resulting from economic fluctuations. Even so, the terms of firemen's employment were daunting. Firefighters worked twenty-four-hour shifts, six days per week, with only two or three one-hour breaks for meals. Additionally, during the nineteenth century firefighters had not yet received many of the benefits—such as pensions and civil service protection—that would become common among municipal employees in the twentieth century. On top of this, firefighting was dangerous labor, in which the possibility of death constantly loomed and injuries were common. Indeed, in a profile of the Pittsburgh Fire Department in 1896, *Fire and Water* reported that the city's 303 firefighters suffered 54 "serious" injuries that resulted in over a thousand lost workdays during the previous year. All of these factors—low wages, extreme working conditions, localized regimes of power, and intense danger—diminished the likelihood that a man might remain a firefighter for a long period of time. Paradoxically, these very same conditions may have reinforced the singularity of firefighting culture, by winnowing department membership to men with particular social or cultural predilections—and those who possessed particularly aggressive qualities. Relatively poor working conditions may have fostered employment of men inclined toward excessive risk taking or physical activity.[50]

Work conditions reinforced the localized pattern of power in fire departments;

Fire at 717–721 Arch Street, Philadelphia, 1886. As the twentieth century dawned, firefighters attacked blazes ever more aggressively, using ladders to gain access to upper floors and dragging hoses deep into buildings.
Courtesy, Fireman's Hall Museum, Philadelphia

intense connections to community, neighborhood politics, and small all-male work groups dominated the life of most firemen, making life difficult for outsiders or men with different cultural backgrounds. The insular work cultures evident in engine houses grew from a competitive, rough notion of manliness characteristic of working-class life and evident in the hurly-burly of machine politics. To a large degree, the strength of these cultures buffered firemen from the NAFE's message of professionalism. Such strong work groups possessed their own customs, making it difficult for the NAFE to disseminate training techniques. As a consequence, daily life at engine houses only rarely evoked the heroism portrayed in the popular press; it was more characterized by a visceral physicality, jocularity, and community parochialism.

The structure of firemen's work groups facilitated localism and insularity by enforcing long-term, close contact with the same group of men, as well as the same urban neighborhood. For example, St. Louis firemen who entered the department before 1880 worked and lived with the same colleagues, and within the same neighborhood for much of their working years, as they were only rarely transferred away from the firehouse to which they were initially appointed. This pattern of workplace immobility did not change significantly until about 1890. After that, each subsequent cohort (by decade) grew increasingly likely to be transferred, with the mean number of transfers increasing to over four per career for firemen appointed after 1930. Additionally, the nature of firefighting work, with men laboring together for twenty-four hours per day, six to seven days per week, intensified firefighters' relationships to their colleagues—for better or worse. Each day, firefighters received three one-hour periods to return home for meals. Such scheduling made it difficult for them to spend much time with their families, unless they resided very near the firehouse. Some lived so far away that an hour was not sufficient time, although in some cases firemen were allowed a single three-hour period for meals and home visits.[51]

The insular, all-male world of the engine house encouraged rough behaviors commonly attributed to working-class male sociability—such as drinking, fighting, and gambling. Qualitative assessment of STLFD personnel files reveals that drinking was an endemic part of firefighters' culture. Although one fireman was dismissed for "canning beer next to the engine house," it does not appear to have been very common for firefighters to be dismissed for intoxication. It appears that firemen accused of drinking on the job usually received reprimands or fines. In Philadelphia, reformist firemen—such as James Gilbert, who would later head the city's fire training school—complained of endemic insobriety at the city's engine houses. According to the Gilbert, the gravest danger faced by firemen was their own drunkenness; he even suggested that most deaths in the line of duty occurred because a man had been drinking.[52]

Gilbert further described the rough culture of the PFD as the "bully-boy" system, in which firefighters coerced a common behavioral code through physical violence or collective recrimination. In his memoir Gilbert wrote, "Each company was a law unto itself." Each work group had a pecking order and diverse cast of characters. Being hired into the department was the first step in becoming a fireman; the second required negotiating the complexities of the work group. When Gilbert joined Engine Company No. 36 in Philadelphia's Holmesburg neighborhood, a fellow fireman promptly told him "that I could not remain as a member of the company . . . he informed me that the station was in the 35th Ward and only men of that Ward could serve in it. As I was from the 33rd ward, the inference was that I would have to get assigned to some Fire station located in the ward that I lived in." Though prohibited by rules, Gilbert did not back down and fought with his antagonist, who it turned out had been backed into a similar corner several years earlier. By besting his opponent, Gilbert gained the respect of his fellow firemen and remained active in the department.[53]

A decade later, Gilbert took the same approach to assert his command authority after he was promoted to replace a foreman who had been driven out of the department. The company to which Gilbert had been appointed had systematically challenged the previous foremen's authority by not obeying commands. Eventually, one fireman set fire to the hay in the engine house; the incident drew attention to the company and the foreman resigned. As a result, the leadership of the PFD appointed Gilbert to lead the troubled company. In order to best the collective strength of the men, Gilbert used existing departmental rules to divide the members against themselves and to defuse their collective power. He assigned each man a particular responsibility within the engine house and recorded that responsibility in the appropriate "watch book." In this way, any incident created by the company to embarrass and disgrace Gilbert could have harmed their fellow company members.[54]

The competitive culture of the engine house never relented, and the rough manliness carried over to fire scenes as well. Gilbert's success at taming tumultuous fire companies had placed him at odds with many of his fellow firemen, and he felt the consequences firsthand at a blaze. During 1912, after taking command of another disorderly company, Gilbert reported that he earned the enmity of his fellow officers in the process of making it one of the city's most efficient companies. He recorded in his memoir that because that he had shown up his fellow officers in the department, they placed him in an untenable "position" at a fire, which severely tested his abilities as a man and leader of men. As he had done two decades earlier, Gilbert took the challenge and fought his enemies on their terms. Within a month, he returned the favor. Gilbert placed one of his earlier antagonists in an equally difficult situation. When the man could not "hold" his position and "retreated from the fire," Gilbert achieved his victory. According to Gilbert, "It was the day of the Bully Boy tactics in the Fire Bureau and the Lord help those that could not hold their end up on the Fire ground."[55]

Late in the nineteenth century, fire departments gradually introduced bureaucratic rules to their departments, but those rules appear to have had little impact on the daily experiences and careers of firefighters. By the 1880s, most Philadelphia fire companies kept a ledger into which foremen chronicled the daily roll call, requisitions for materials, the care of hose and/or apparatus, and fire runs. Rarely very detailed, these records suggest a simple level of accountability and do not appear to have been scrutinized closely by departmental leaders. Moreover, although it seems almost certain that fire officials, especially the fire engineer, issued commands to the entire department, it is not clear how those orders were transmitted or if they were recorded. Judging from company watch books, the level of departmental coordination appears to have been minimal, perhaps helping to explain why fire companies in Philadelphia used different names for common tools. Likewise, recording disciplinary incidents does not appear to have impeded promotion, as evidence from the New York Fire Department (NYFD) suggests. As with other departments, the NYFD rewarded exemplary courage and the experience gained from length of service with promotion, but being disciplined or charged with insubordination did not preclude career advancement. By the first decade of the twentieth century, all but one officer in the NYFD had at least fifteen years experience, and more than two-thirds had received a commendation, being placed on the "Roll of Merit." Additionally, a striking number—almost 40 percent—had at some point been brought up on charges of having violated departmental rules.[56]

The intransigence of local work cultures was deeply connected to broader understandings of manhood and politics. Advocates of reform—such as Gilbert—believed that training and discipline could transform local engine houses but worried that the connections between politics and everyday life in American cities posed a more serious threat to effective professional firefighting. Indeed, the importance of political connections and the influence of local regimes of power on departments were of special concern to advocates of firefighting professionalism. In fact, to firefighting professionals, local regimes of power—such as Philadelphia's bully-boy system—were interconnected with political practice and manliness. Reformers frequently criticized the coupling between politics and firefighting, which resulted in ineffectual firemen. The NAFE identified the connections between politics and engine houses as perhaps the most powerful inhibitor of the dissemination of regular, routine work habits. At meeting after meeting, it considered the negative impact that political intervention had on efficient, professional firefighting.[57]

The intimate connections between politicians and fire departments are perhaps most evident in the hiring and promotion of men to the position of fire chief. Most late-nineteenth-century fire chiefs had worked in their departments, generally in increasingly more responsible posts, for many years. However, that experienced firefighters became chiefs does not necessarily indicate a "triumph of professionalism." Indeed, although most chiefs were veterans who had the requisite skills and savvy to be chief, the vast majority nonetheless acquired the position through political considerations. Fire engineers came to head the department precisely because of their political connections, although many chiefs fancied themselves as representative firemen and the firefighting press typically described chiefs as men who had risen through the ranks. In fact, most chiefs appear to have been political appointees, including a large number that championed the professionalism of the NAFE. For instance, the STLFD's reformist chief, Charles Swingley, experienced just such a rise through the ranks, earning his first four-year term as chief when his predecessor, John Lindsay, "was retired." According to department personnel files, Lindsay was appointed to the department in 1867 and "dropped" in 1895. Being dropped by the department was not much different from being discharged. Lindsay's being dropped coincided with a political change—it occurred two years after St. Louis residents elected a new mayor in 1893.[58]

More striking than his replacement of Lindsay was Swingley's extraordinary rise through the ranks. Biographical statements contained in Swingley's scrapbook noted his rapid promotion but insinuated that the promotions had come over a long period in a career marked by his "devotion to his duties, his skill and intrepidity." In actuality, after being appointed in 1869, Swingley languished as a pipeman—his position of entry—for his first twenty-four years in the department. Twice transferred, Swingley questioned his future prospects at least once. In 1882, he resigned for several months before returning to the department. Although he later earned his reputation as a proponent of merit and efficiency, Swingley's rapid promotion resulted from his close political connection to the city's pro-business Republican business elite, represented by Cyrus Wallbridge. Swingley did not receive a single promotion until *after* Wallbridge won the mayor's office. After that date, he was promoted through the ranks with extraordinary rapidity. He received an appointment to assistant foreman on May 1, 1893, and was promoted to foreman December 14, 1893. Then in February 1894, he became assistant chief. Swingley succeeded Lindsay on May 15, 1895—moving ahead of other men such as assistant chief Eugene Gross, who was widely lauded for his long and meritorious career. Gross had been appointed to assistant chief in 1881 (after being appointed in 1869 and promoted at regular intervals). According to prevailing practice, as "first" assistant chief in 1885 Gross had been first in line to succeed Lindsay (whose term had started in 1885).[59]

If the rough world of turn-of-the-century urban street life and the individuated nature of job competition in industrial society nurtured the bully-boy system, urban politics provided the glue that bound it together. The paternalism of machine politics tied firemen to ward leaders and structured their relationships to their local communities. In return for assistance during elections, ward officials rewarded firemen and other municipal workers with employment. In Philadelphia's fire department this sometimes resulted in hiring and promoting men with severe physical disabilities, such as blindness or consumption, or men who had little experience. Even Gilbert's promotion to assistant foreman had depended upon his connections to the mayor; otherwise he may have been passed over. Being promoted because of favoritism so haunted Gilbert that he proclaimed that he never again would support corrupt machine politics, but his close connections to reformist politicians certainly aided his career in the long term.[60]

At the heart of the relationship between local politics and the fire department was the patron/client relationship between the political boss and his community. In the context of the broader labor market, the exercise of such connections to gain employment conformed to common job acquisition strategies. At a time when most men gained their first job through personal or family links, it is not surprising that being hired and promoted in both St. Louis's and Philadelphia's fire departments depended upon relationships to ward politicians or the political machine. For instance, in St. Louis, a man became a fireman after "a man, or his friend or friends" presented an application at the chief's office. Even though Swingley boasted of his disinterested hiring practices, all applicants required "persons who know him and can vouch for him."[61]

The benefits of possessing political ties went beyond finding a decent-paying job. The fortunes of firefighters often rose or fell with that of their patrons. Just as James Gilbert achieved considerable status—and promulgated reform—when a reform mayor took office, firefighters supported their benefactors with gusto. According to Gilbert, on election days few firefighters showed up at the engine house because most served as foot soldiers at the polls. In fact, a man's ability as a political operative sometimes counted more than firefighting skill. Gilbert reported that "there were some of the members of the Fire Department that had no aptitude at all as Political workers, they were out of luck and had to watch their step unless they were the relative of some politician." Whereas men without vote-getting skills had short careers, those with campaign ability found themselves beyond their superiors' discipline. Company leaders, even those who had acquired positions through favoritism, could not reprimand such men without being subject to potential political backlash. Ironically, even those firefighters who lacked personal connections supported the machine indirectly. Well into the twentieth century, Philadelphia's Republican machine extracted regular contributions from municipal employees' paychecks. In Philadelphia, the Republican political machine coerced campaign contributions from municipal employees amounting to as much as $2 million dollars annually in the first decades of the twentieth century.[62]

The connections between firefighting and politics, between firefighters' work cultures and urban political machines were not accidental. Both shared certain fundamentals, especially a common understanding of what constituted manliness. Firefighters and political leaders identified themselves according to similar rules of dominance, physicality, and competitiveness. Machine politics demanded an effervescent personality, charisma, and a rhetoric of public service, and it depended upon a competitive spirit. Its participants strove to dominate rivals in the frequent sparring and jousting of campaigns, backrooms, and council floors. Indeed, most political bosses considered themselves as *the* man among men. Likewise firefighters competed incessantly; whether they struggled for company leadership, control at a fire scene, or departmental hegemony, firemen who could not stand up to such contests found themselves discredited, dismissed, overlooked for promotion, or simply embarrassed. Perhaps more viscerally, there was a certain romance to firefighters' conflict with nature that appealed to the popular culture and elected officials. As urban father figures and nominal fire department chief executives, mayors frequently attended fires and basked in the heroism of the city's firemen. Symbolically (at the least) they claimed credit for having restored order through harnessing firemen's vitality. As much as firefighters and machine politicians shared a common sense of manliness, a common struggle to order and control the urban environment especially bound them together.[63]

Conclusion: Manliness, Order, and Public Safety

By 1898, when Jacob Riis celebrated the "Heroes Who Fight Fire," firefighters had already given shape to their occupation and defined its organization and labor. Firefighters had become life-saving heroes, prominent in the public imagination. Of equal importance, firefighters had refined their work techniques; they no longer battled blazes from the outside, but from inside buildings. Firefighters also sought status by calling their work professional, and they participated in the dramatic expansion of governmental bureaucracies in the United States. As cities grew and their residents demanded improvements in the urban infrastructure, municipalities negotiated with private corporations, engineers, and other professionals to build waterworks, sewer systems, and public health organizations. Like the development of other nascent municipal bureaucracies—such as health, police, and water departments—the development of fire departments occurred gradually and as a result of constant negotiation between capitalists, firefighting professionals, and the state. The expansion of municipal fire departments paralleled the growth of other urban services, but it differed in important respects.[64]

When firefighters and department leaders used the rhetoric of professionalism to organize their efforts to restore order, they claimed status as one of many communities of experts that began appearing in the United States following the Civil War. Firefighters never achieved the formal institutional authority bequeathed to the professional societies established by engineers, lawyers, and physicians, however. Yet, even without formal legal sanction, they wielded nearly as much influence as engineers, and were invited to weigh in on the development of the technological infrastructure. Although firefighters' claims to be professionals blended into broader efforts to reform municipal employment and preceded attempts to make police departments professional, such influence distinguished firefighting from other blue-collar pursuits, helping to elevate the occupation. By clothing themselves in the values associated with middle-class professional organizations—discipline, rationality, and sobriety—firefighters obtained broad social respectability, despite their working-class and immigrant origins.[65]

Firefighters' claims to stature often came at the expense of the fire insurance industry, against which firefighters contrasted their service. Firefighters' ability to expand their authority occurred, in part, because of the complex role that insurers had in public fire protection. Unlike the expansion of other municipal services, such as sewers, electricity, and transit, where the role of capitalists in providing those public amenities was usually quite clear—usually as a contractor, partner, or innovator—insurance firms were significantly less visible in fire protection. Underwriters offered little direct economic pressure on or assistance to firefighters in the nineteenth century. They chose instead to leave the operation of fire departments in municipal hands. Additionally, insurers almost completely excused themselves from any role in public fire safety. Even setting aside insurers' flippant remarks that conflagration could be managed, the differences between firefighters and insurers could not have been more stark.

Firefighters and underwriters generated two very different visions of safety. Underwriters' labor existed as part of the larger capitalist endeavor, in which entrepreneurs transformed American cities and brought striking and intense new dangers. Insurers created an economic means by which industrial society could control the problem of fire by protecting and preserving capital, if not the material infrastructure. Underwriters—like corporations, professional organizations, and the growing middle class—sought to manage the economy and risk. Fire danger became an abstraction, represented on maps and in statistics. Insurers removed the hazard from the physical world even as they wove it into the fabric of industrial society. Danger became a commodity that could be bought and sold. At the close of the nineteenth century, insurers' efforts to control the problem of fire remained marginal and relatively ineffective. A significant number of Americans did not or could not purchase fire insurance to protect their financial interests. But, more significantly, at a time of dramatic conflagrations, most Americans understood the danger of fire in terms of its physicality and material meaning to their lives—a concern to which fire insurers devoted little public attention. Most importantly, however, the fire insurance industry also was failing to provide basic economic protection to policyholders. Adopting the ethos of laissez-faire capitalism, underwriters seemed unconcerned about the large numbers of firms going out of business—even encouraging the failures in the belief that this made the industry stronger. Fire losses mounted yearly in the decades following the Civil War, and "the engines of destruction multiplied"—to borrow firefighters' phrasing. The problem of fire remained one of the most potent dangers facing American cities.

Into this vacuum firefighters stepped, often casting themselves as heroes. Yet, firefighters remained rough and flawed characters, as life in their engine houses suggests. In truth, the ability of firefighters to discipline cities, much less themselves, was limited—perhaps as imperfect as that of fire underwriters. Firefighters strayed far from the ideals expressed by the life-saving fireman. Their work cultures remained expressive, tied to local communities, and relatively unfettered by bureaucratic procedure. Despite their rhetoric, they were anything but trained professionals. Local political regimes and the work cultures of local engine houses frequently prevented firefighters from developing standard work habits or tools. At the same time, evidence suggests that the heroic fireman was sometimes a bully and occasionally drunk.

Nonetheless, urban Americans appear to have forgiven firefighters' foibles because they offered protection at a time of mounting danger. Urban dwellers watched as firefighters raced into fires, saved lives, and prevented conflagration. Although they certainly offered an icon of elevated white manhood, firefighters claimed a place in the popular imagination mostly because they so visibly struggled to control the problem of fire when nobody else seemed able to do so. Indeed, the fireman rushing from a burning building carrying a woman or child distinguished firefighters' interest from the economic concerns of middle-class underwriters, providing authority derived from moral, not economic power. Although firefighters' legacy contained no small amount of hyperbole, there can be no doubt that firefighters placed themselves in great danger as they struggled against fire. *Fire and Water* captured the degree to which firemen's bodies literally embodied the legacy of their struggle when it described the Chicago fire chief who "escaped death on several occasions by the narrowest of margins, and, as a consequence bore on his body the marks of an infinity of burns and cuts." In much the same way, firefighting as an occupation came to be marked by tales of valor, danger, and personal sacrifice. In the nineteenth century, at a time of great instability in American cities, firefighters provided a sense of security, though not absolute safety, which reassured urban dwellers and capitalists alike. By the time fire engineers discussed "a practical plan for mutual identity" in 1898, firemen had already defined their occupation as heroic and selfless—so successfully, in fact, that they continued to be judged through the late twentieth century as disinterested figures of security. Well before progressive reformers dreamed a rational city or the insurance industry sold fire prevention, firefighters offered a vision of urban order. Firemen established the first systemic standard of fire safety in the United States when they made themselves into heroic icons.[66]

Part IV / Paper

CHAPTER SEVEN

Consuming Safety:

Fire Prevention and Fire Risk in the Twentieth Century

What had been unimaginable a decade earlier became a reality during the first decades of the twentieth century. The fire insurance industry, led by the National Board of Fire Underwriters, dramatically changed course and embraced fire prevention, which would lead to fundamental and rapid improvements in urban fire safety. As Harry Brearley recalled in his history of the NBFU, fire prevention "would never have entered the consciousness of the constitution makers in 1866, at least it seems to have occurred to no one. In those days, the business of underwriting was underwriting—neither more or less." For the first time, the industry systematically promoted safety in the public realm when it began to propagate legal and behavioral standards aimed at preventing fire and minimizing its economic consequences. Underwriters drew these new principles from a variety of places, especially decades of company loss tables and decades of research by engineers and architects into construction techniques. Yet, building codes represented but the first of several steps designed to mold the urban landscape and consciousness into a form more favorable to the industry. Insurers also created guidelines for organizing and planning the network of water and electrical systems in the twentieth-century city. Emboldened by early successes, they helped municipalities to organize fire defenses as well as the spatial layout of cities. Perhaps most critically, in

1916 the NBFU embarked upon a massive public relations campaign that targeted individual Americans. National Fire Prevention Day, established in 1911, and National Fire Prevention Week, organized in 1922, represented the most visible results of this effort. Arguing that a careless public was to blame for "fire waste," insurers added new dimensions to their message as they worked to inculcate the notion of individual responsibility deeper into the fabric of the nation's cultural life.[1]

On the eve of the twentieth century, the insurance industry faced a puzzling situation: as it became increasingly important to the nation's economic health, the industry continued to stand on the brink of failure, facing what seemed to be perpetual crisis. Although it is difficult to say just how much of the nation's property was insured, there is little doubt that fire insurance had become indispensable to the expansion of capitalism and the economy. When the Merritt Committee investigated the insurance industry for New York State, it bluntly stated that the system of credit so integral to the economy, and to most financial transactions, was "founded on the institution of insurance." It went on to note that just "as the welfare of society is founded on the free operation of credit, by so much is the institution of insurance of importance to the public, quite aside from its value in actually distributing loss." A leader of a major department store echoed these sentiments, and emphasized that fire insurance figured in the most basic aspects of everyday commerce. He reported, "It would be impossible to carry on business without insurance against loss by fire. It would so disturb values of all property that it would materially interfere with the loaning of money; credits which are such a vast aid would be impossible. . . . In extending credit to merchants, I am constantly considering questions concerning a customer's fire insurance."[2]

At the same time that the industry became so critical to economic development, it remained remarkably unstable—hardly the reliable backbone to commerce that business leaders desired. The industry faced heavy and unpredictable losses from fire, as hundreds of bankruptcies and repeated conflagrations demonstrated all too well. Indeed, some industry experts estimated that nearly one-third of the industry's losses resulted from conflagration, causing dramatic fluctuation in the fortunes of individual firms. In this context, the industry searched for ways to improve profitability and stability. Only reluctantly—in the context of continued industry insolvency, public pressure, and competition within the industry—did underwriters become advocates for public safety. If insurers' motives changed little, their strategies did. Underwriters began to focus on stopping fire before it started, reducing the possibility of conflagration and making fire losses more predictable. Fire prevention became the industry's new rationale and represented a dramatic departure from the era in which the industry had argued that the marketplace could coerce safe behavior.[3]

Public fire safety did not enter the popular consciousness systematically until the NBFU began to engage the issue late in the nineteenth century. Of course, firefighters as well as a limited number of mutual insurance firms, engineers, and architects had advocated fire prevention and better construction practices as early as the 1870s, but the bulk of the fire insurance industry did not become interested in this problem until the 1890s. After several decades of practice centered on observing, recording, and ordering knowledge about incidents of fire, a coalition of the most powerful fire insurance companies in the United States became proactive in their approach to controlling the problem of fire. The National Board of Fire Underwriters used its extensive network of member institutions to create, foster, and disseminate an extensive and systematic program of public safety. In 1901, the NBFU rewrote its statement of purpose to reflect this new emphasis. For the first time, the NBFU vowed to fight against what it called "fire waste." Adding a fifth clause its mission statement, it committed "to influence the introduction of improved and safe building construction, encourage the adoption of fire protective measures, secure efficient organization and equipment of fire departments, with adequate and improved water systems, and establish rules designed to regulate all hazards constituting a menace to the business." No longer content to confine debates about the problem of fire to their own industry, insurers expanded their discussions and expertise about the hazard to include the broader society. In so doing, underwriters explicitly made preventing fire and enhancing public fire safety a centerpiece of their work.[4]

Fire prevention signaled a new mode of thinking about fire risk, not to mention profitability and solvency. Previously, underwriters believed that standardized business practices and rational decision-making processes would produce fiscal health for their industry. By imposing an unvarying economic calculus on their industry, underwriters had reasoned that they could also coerce customers and society more broadly into behaving with safety in mind; the secondary consequence of standardization would be better building practices and fewer fires. The NBFU's revision of its mission statement repudiated that approach and inverted it. In the twentieth century, safe building practices and better fire defenses lessened the total fire amount of fire loss, and in so doing became key avenues toward profitability. To a certain extent, this shift actually had represented the next logical step for the industry. By 1900, it had developed routine practices, had instituted the mechanisms for standardization of rates, and had placed the built landscape under unprecedented surveillance. As these systematic initiatives produced new understandings of the problem of fire, the social and political landscape of Progressivism provided a window of opportunity for dramatic reform. In this context, insurers transformed knowledge about hazard into programs of safety. Advocating and seeking enforcement of standardized building codes, electrical codes, and fire protection defenses indicated the industry's move away from conceptual action to physical management of the landscape.[5]

Building codes represented a first significant step in this campaign, but they were only part of the fire underwriters' larger efforts to improve safety. To a large degree, the shift in the industry's philosophy intersected with changes taking place in the everyday practices of underwriters. Already in the late nineteenth century, insurers had begun to transfer the order implicit in their manuals, statistics, and maps to the urban landscape. Likewise, the industry debated its rate-making procedures, considering alterations that might encourage fire prevention. It also began to publish its knowledge about the problem of fire and to finance and support the research of safety engineers and architects concerned with developing safer construction practices. By the first decades of the new century, the industry had transformed a mixture of hodgepodge initiatives of fire prevention into a carefully orchestrated and systematic program of prevention.

In addition to building codes and industry practice, fire underwriters evangelized new behavioral standards with a ministerial zeal. The insurance industry did more than offer advice about safety; it prescribed a way of life. Underwriters told insurance workers that the promotion of safety was connected to a standard of middle-class manhood that had increasing resonance in an expanding consumer society. In a business that was about taking economic risks, insurance men had to balance the expression of individuality with standardized management practices; they learned to embrace the uncertainty of the marketplace by rationalizing themselves according to prescriptions of prudence and safety. Moreover, underwriters hoped to make all Americans behave more rationally by connecting their safety campaign to another societal standard—the middle-class consumer family. The insurance industry established the middle-class household—led by a man responsible for his wife, children, and a home mortgage—as the basis of its safety campaign in the 1920s. Companies urged men to behave with deliberate care, to be guardians of their families' futures by purchasing insurance. Among other things, they even implemented curricula in schools that taught children to become "responsible" adults and to inspect their homes.[6]

Ultimately, the industry dramatically affected the American landscape, although its efforts to promote safety remained largely invisible to the public, in stark contrast to the public safety provided by heroic firefighters. Nevertheless, by the 1950s fire prevention was a self-evident public good, as building codes and cultural standards became embedded deeply in the urban infrastructure and into the fabric of American life. These legal and behavioral canons became more pervasive with the extraordinary suburban development following World War II. Although most urban dwellers remained marginally aware of how this broad initiative affected their lives, they nonetheless supported its expansion whenever they purchased insurance to protect their property or when they purchased consumer products marked by safety standards, such as the UL (Underwriters' Laboratories) label. The insurance industry was transforming itself and the nation. In the twentieth century, underwriters sold more than insurance; they trafficked in safety, making it into a commodity. As Americans purchased insurance, they bought security, and implicated themselves evermore deeply into a new behavioral discipline. Ultimately, the rapid spread of legal and behavioral codes had a lasting effect on the physical and cultural landscape, arguably making it the most impressive and successful of all Progressive Era reforms.

Building Codes

If its mission statement signaled a shift in direction, the building code issued by the NBFU represented a first step toward systematic fire prevention. In 1905, the NBFU drew upon decades of experience to draft guidelines for safe building construction. Compiling the code into a single piece of legislation, the NBFU hoped cities would adopt its recommendations as written. As a palimpsest for cites and towns of all sizes, the code folded fire safety concerns into prescriptions for standard construction practices, producing a city-planning strategy structured around controlling the problem of fire. The plan included the most mundane of recommendations, such as fire limits on wooden construction, as well as relatively new ideas, such as mandatory standpipes in certain types of structures. The building code revealed the industry's continued belief that responsibility for preventing fires should become a matter of public responsibility, and marked a departure in how the industry viewed its own role in stopping the problem of fire. Underwriters no longer focused exclusively on making their own business more disciplined, but they now hoped to make the American built environment more and more standard, at least in regards to fire danger.[7]

The *Model Building Code* proposed by the NBFU was the systematic realization of decades of intense but often fragmented research into the problem of fire. Such ordinances were not new in 1905, nor had the information in the proposals been uncovered recently. Indeed, in the preceding fifty years the insurance industry had fervently embraced a program of intense observation of fire and its dangers as part of its pursuit of scientific underwriting. Additionally, organizations outside the industry, including the International Association of Fire Engineers (IAFE), actively pursued establishing municipal fire codes as a way to diminish the dangers that firemen faced. And the mutual fire insurance companies that wrote insurance on New England textile mills experimented with different strategies for preventing fires in industrial mills, which they disseminated to and installed in the factories of their clients. Perhaps most importantly, in the 1890s the state of New York worked to pass a code for medium-sized cities, which was disseminated by the NBFU.[8]

In addition, during the 1890s, two new voluntary organizations—the Underwriters' Laboratories (UL) and the National Fire Protection Association—began to explore how fire could be prevented. Established in 1894 in collaboration with the Chicago Fire Underwriters Association, the Underwriters' Laboratories (originally the Underwriters' International Electrical Association) accomplished an agenda tacitly approved by the NBFU: incorporating new technologies into the landscape without increasing fire risk. In the context of the expansion of electricity in the 1890s, UL anticipated the particular concern growing throughout the fire insurance industry—the rapidly growing fire waste that resulted from improperly installed electrical systems. In the 1890s, for instance, the NBFU, which had long examined how new technologies altered fire danger, expanded the functions of its "Committee on Heating and Lighting," formed an "Electrical Bureau," and hired "consulting engineers." By the early twentieth century, UL served as the NBFU's primary testing facility.[9]

Like the Underwriters' Laboratories, the National Fire Protection Association (NFPA) developed independently, but over time, it too was drawn into the pool of associations that the NBFU used to popularize its program of fire safety. Formed in 1896 by men primarily associated with the Underwriters' Bureau of New England and the Boston Board of Fire Underwriters, the NFPA consisted of stock fire insurance companies interested in improving their understanding of fire danger and minimizing it. The organization's first project involved formulating standards relating to automatic sprinklers. In subsequent years, the NFPA established committees to report on a number of other fire protective devices, including fire doors and shutters, hoses and hydrants (in collaboration with the National Association of Fire Engineers), fire alarms, fire extinguishers, fire retarding materials, and fire pumps. In 1900, as the NBFU refocused its agenda, it sought a cooperative relationship with the NFPA to produce standards for fire-protective devices. Recognizing the overlap in the two organizations' membership and purposes, both agreed to work together on this. The NBFU reproduced the NFPA's standards as its own, assumed the expense of publishing them, and printed the annual proceedings of the NFPA.[10]

Building on and combining previous research into fire safety, the *Model Building Code* represented the first systematic attempt—and certainly the most comprehensive effort to date—to standardize the nation's urban environment according to the dictates of fire safety. To a large degree, this systematization appropriated and combined the disparate previous research on fire protection. For instance, the proposed law reflected the previous decade of collaboration between the NBFU and NFPA on standards for fire protection apparatus, and referred to no less than eighteen previously published NBFU or NFPA standards. Similarly, many of the recommended construction techniques were already used by builders who were conscious of fire safety. Other components, too, such as recommending fireproof construction methods and improving water delivery systems, were not unfamiliar to many urban political leaders. Although these techniques and strategies may have been well known, they had not been systematically organized into municipal building laws, especially in smaller cities and towns. The NBFU's proposal also included elements of the National Electrical Code, rules for storing fuel oil, and advocacy of automatic sprinklers. Additionally, the NBFU drew support and information from the IAFE, which had long argued that construction methods should be changed to promote safer building practices. The proposed legislation united what had been separate avenues of research into a unified and systematic manifesto of fire prevention.[11]

The NBFU's program was broadly conceived, and as the organization stated elsewhere, it intended its legislative suggestions to be expedient and not onerously difficult to implement. The NBFU created a "practical" code, which meant that reform was attainable and even commonsensical. Its proposals did not demand a radical revision of everyday construction or legislative practices. In fact, the recommendations contained references to any number of measures long championed by fire underwriters, such as provisions for fire limits within which wood structures could not be built. The code not only represented the codification of existing safe practices but included recommendations from new lines of research. For example, the NBFU sought to update previous municipal ordinances on flammable substances with the latest research, and the building code recommended new standards for using flammable liquids in lighting and heating as well as chemical fire extinguishers. Again, the industry did not seek to transform cities overnight. Its recommendations did not purport to offer the panacea to the problem of urban fire by advocating that new technological protective features be built immediately. For example, the code did not require all buildings to have standpipes. Neither did it exclusively favor a single technique for minimizing fire danger, such as fireproof construction for all structures. The NBFU wanted to create a meaningful legislative guide for building safety that would effectively transform the landscape over the long term. Despite this gradualist approach, the NBFU hoped that its proposed legislation would not be taken piecemeal. It hoped that the entire set of proposals would be taken as a unit; it wanted its code to supersede existing municipal fire codes.[12]

The code was as comprehensive as it was flexible; it urged a multilayered approach to reducing fire danger within cities and accounted for a variety of building types, styles, and methods of construction. Even as the code recommended that new structures be built according to the basic principles of fireproof construction, it also included more limited suggestions, such as how to construct elevators and stairways in order to prevent the transmission of fire. The code also identified construction techniques that could make wood structures relatively safe, even as it instructed builders about safe construction with iron or steel. Although underwriters had long known many of the hazards lurking in the built environment, the national building code demonstrated underwriters' greater sensitivity to the complex relationship between fire and the landscape, especially the effect that technology could have in altering fire danger. For instance, the NBFU took account of how electricity altered the urban environment's susceptibility to fire and introduced new hazards for firemen battling blazes. By referring legislators to its National Electrical Code, the NBFU began to create the connections so necessary to establishing an interlocking legislative safety net to minimize fire risk.[13]

The NBFU's model building code reflected the fire insurance industry's interests, practices, and financial concerns. The code devoted much attention to structures used by large numbers of people, especially hotels and office buildings, but implementing it was not driven by the same passion for saving individual lives that characterized firefighters' culture. When the NBFU reserved a special section for theaters and other "public" buildings, which included manufacturing facilities, churches, department stores, and any building "where large numbers of people are congregated," it targeted cultural edifices with great symbolic value. By obtaining fire safety in the construction of such structures the industry would enhance its own prestige and gain social capital that would be invaluable in implementing its fire preventive agenda elsewhere in society. There was also great motivation in the realization that as buildings became increasingly large and expensive, they posed a potentially devastating financial exposure to insurance companies.[14]

In the early twentieth century, the industry remained most concerned with those areas of company portfolios that showed the greatest volatility and/or especially large individual economic exposures. Not surprisingly, the model building code devoted little attention to individual residential structures. By according private, freestanding homes precious little attention, the code reproduced the industry's overwhelming concern about preventing large economic losses. Although dwellings remained the most consistently profitable section of many companies' risk portfolios, creating special measures to prevent fires in homes remained secondary, and would remain so for fifteen more years. Targeting residential construction may well have been a culturally and politically difficult proposition—after all, early in the twentieth century, few Americans had yet become conscious of safety as an issue in the home.[15]

The model building code also reflected the concerns implicit in everyday underwriting practice, especially rate-making procedures. Far from advocating a systematic guideline out of some sudden altruistic public spirit, insurers saw an advantage in making the urban environment more uniform. If the primary advantage accrued from reducing losses, the industry also believed that standardization of construction would improve the ability of the insurance community to classify fire losses and to improve the quality of statistical data. In turn, this would help to make accounting for and assessing risks much simpler and more efficient. Indeed, the proposal meshed with the underlying principles that governed the process of evaluating risk within the industry. The code was organized according to the two criteria that most affected categorizing danger and setting rates: the use of a property as well as its construction. On one hand, it distinguished between basic property uses such as private dwellings, apartments, tenements, hotels, and office buildings. Although a distinction between private dwellings and public buildings structured the code, its major organizational sections broke according to features of construction. Each section outlined a particular aspect of building a structure. For example, part 17 covered "roofs, leaders, cornices, bulkheads, scuttles and tanks," and part 6 considered "walls, piers, and partitions." By emphasizing construction methods so strongly, the plan reflected the shifting way in which the industry was evaluating risks. Namely, while still paying attention to property use, underwriters focused an increasing amount of attention on construction methods and fire preventive devices when setting rates.[16]

The national building code further revealed how the industry chose to participate in public fire protection; it especially indicated an unstated conviction within the industry—that fire prevention and fire protection were neither the fiscal nor the moral responsibility of fire underwriters. Even though insurers had begun to urge fire prevention and the use financial incentives to coerce safe behavior, most continued to disavow any direct accountability for public safety. The national building code developed at the nexus of this discussion within the industry and represented a compromise position. Fire insurers resolved to take an advocacy role, but aside from paying for inspection and research, the industry limited its cost outlays for prevention. By regularly updating and disseminating the legal code, the fire insurance industry placed the responsibility for acting on fire safety squarely within the public sphere.[17]

Providing intellectual expertise to public officials offered tangible benefits to the insurance industry. Becoming an expert resource on fire danger did not cost the industry a great deal. The NBFU's standards capitalized on the industry's already well-developed program of research; not only were the costs of subsidizing the NFPA, UL, and underwriting associations minimal, but much of the activities—and expenses—involved in gathering knowledge about fire were already part of daily business routines. In addition, graciously providing data and opinion at no charge to the public improved the industry's image—which may well have needed boosting, given underwriters' nineteenth-century pronouncements. Underwriters came to act as consultants on improving public safety, rather than greedy businessmen most interested in making money. The bottom line, however, was the bottom line. Preventing or minimizing fire loss by upgrading fire defenses promised to improve profitability by reducing losses—especially the likelihood of conflagration—and by making fire loss more predictable.[18]

The NBFU's proposals had an immediate and profound impact on American cities. Available evidence indicates that municipalities adopted the codes in large numbers. Distributed to all cities and towns with populations larger than five thousand, the code was remarkably successful. At its annual conference in 1906, the NBFU's committee on construction of buildings reported that it had distributed over four thousand copies to public officials, underwriters, and fire chiefs. In addition, the committee reported that "the Code has met with general approval . . . we have been informed of many cities and towns where it has been made the basis of a new building law or of an intelligent revision of existing ordinances." Progressive reformers of every ilk uncritically adopted the NBFU's agenda. For instance, in 1919 the Women's League for Good Government in Philadelphia cited the insurance industry's leadership in the war against fire waste. The association recommended more building inspectors to enforce existing codes and an improvement of the city's water supply—both of which had been recommended by the NBFU when it inspected fire conditions in the city in 1911.[19]

It is clear that many municipal governments adopted NBFU reforms, but charting their enforcement is another matter. Although building codes contained provisions for inspectors to monitor new construction, as well as alterations to existing structures, there is scant evidence with which to assess the extent of their enforcement. Usually, failures to apply building laws were discovered only when they resulted in a spectacular fire that made newspaper headlines. Even so, the dissemination of the NBFU's proposals into municipal legislation undoubtedly had the positive effect of making builders more conscious of fire safety—whether they conformed to the rules or not. Of course, the insurance industry recognized the importance of enforcement and encouraged municipalities to make sure that the new laws were followed. For instance, in 1940 the NBFU supported the simple practice of having zoning laws include a requirement that all building permits be viewed by a building inspector familiar with fire safety legislation.[20]

Early in the twentieth century, the NBFU had transformed its knowledge-gathering practices into a program of fire prevention that it pursued relentlessly. Buoyed by widespread acceptance of its proposals, and recognition of its expertise, the NBFU updated the code regularly. As the twentieth century progressed, these standards increasingly structured fire-safe building practices across the nation. As cities adopted its recommendations, the NBFU came closer to achieving its goal of securing "uniform building laws throughout the country." Although gratified by the rapid implementation of building codes, insurers remained realistic about the obstacles facing them. Accordingly, the industry recommended patience with its fire prevention strategy. Underwriters frequently reminded themselves that the strategy would not become effective until "the distant future." Nonetheless, the insurance industry had begun to reconstruct—literally and permanently—the urban environment according to its assessment of the problem of fire. Insurers' system of classifying risk—represented on statistical tables, maps, and management charts—was being transformed into a new physical order.[21]

The Committee of Twenty

As the NBFU released its model building code, the organization also led the fire insurance industry in a more intensive examination of the conflagration hazard in the nation's cities. "The Committee of Twenty" was formed to make recommendations, and it issued reports that focused on a number of environmental, structural, climactic, administrative, organizational, and technological factors associated with municipal fire defense. The committee distributed these reports to local officials, commercial interests (such as boards of trade), fire departments, and throughout the insurance industry. Of course, the NBFU as well as local underwriting associations had made recommendations regarding the problem as early as the 1890s, but following the Baltimore and San Francisco conflagrations of 1904 and 1906, respectively, the NBFU's examination of urban fire defenses grew more comprehensive, systematic, and widely publicized. In its first two years, the special committee used its annual budget of more than $100,000 to inspect forty-six cities, and reinspect eleven others. Over one hundred insurance companies subscribed to its engineering reports for a yearly fee of $500, and all NBFU members financed the committee by paying a small assessment in addition to their yearly dues. The board further dispersed reports to "educational institutions, technical journals and prominent individuals interested in fire protection engineering," as well as to "commercial bodies, local underwriting authorities, and heads of municipal departments" in inspected cities. Insurance capitalists did not passively urge cities to implement legal codes that mandated better building practices; they increased pressure on municipalities to create a more systematic and comprehensive fire protection infrastructure.[22]

When the NBFU sought to curtail the threat of conflagration, it did not just seek to address the problem in the nation's largest cities, although that is where it began its efforts. If, by 1920, the NBFU had produced over two hundred detailed surveys of the nation's largest cities, the committee of twenty especially targeted smaller cities, arguing that small cities and towns could not afford investigations of the size and scope of those conducted by the NBFU. Nor could these municipalities secure the expertise necessary to make a thorough evaluation of their fire defenses. The NBFU would help growing cities mature through their current "stage of development." The committee would draft a systematic program of municipal fire defenses that the town could grow into. It believed that by providing such critical services across the landscape underwriters were "in a position to place themselves in a proper light before the public and, by the expenditure of a nominal sum in each city now, save much in the future."[23]

Through the committee of twenty, the NBFU exhaustively studied every facet of everyday urban life in the cities it inspected. Subscribers received an overview of a locality's civic affairs, population and growth, topography, street layout, winds, temperatures, and fire record. The committee created a historical and contextual view of an urban place, which individual companies used to "fix lines"—to determine rates appropriate to a city's fire defenses. Other information, such as "property valuation and tax rate," provided equally important intelligence that helped to foster its agenda. By evaluating a city's income from property values and taxes, the committee ascertained a municipality's fiscal health and its ability to implement reform. It even dwelled upon such mundane items as what sorts of fuel most urban dwellers used to heat their homes. The soft coal used in Pittsburgh, Chicago, and Cleveland, for instance, caused tin roofs and ironclad structures to deteriorate quickly and increased the fire danger.[24]

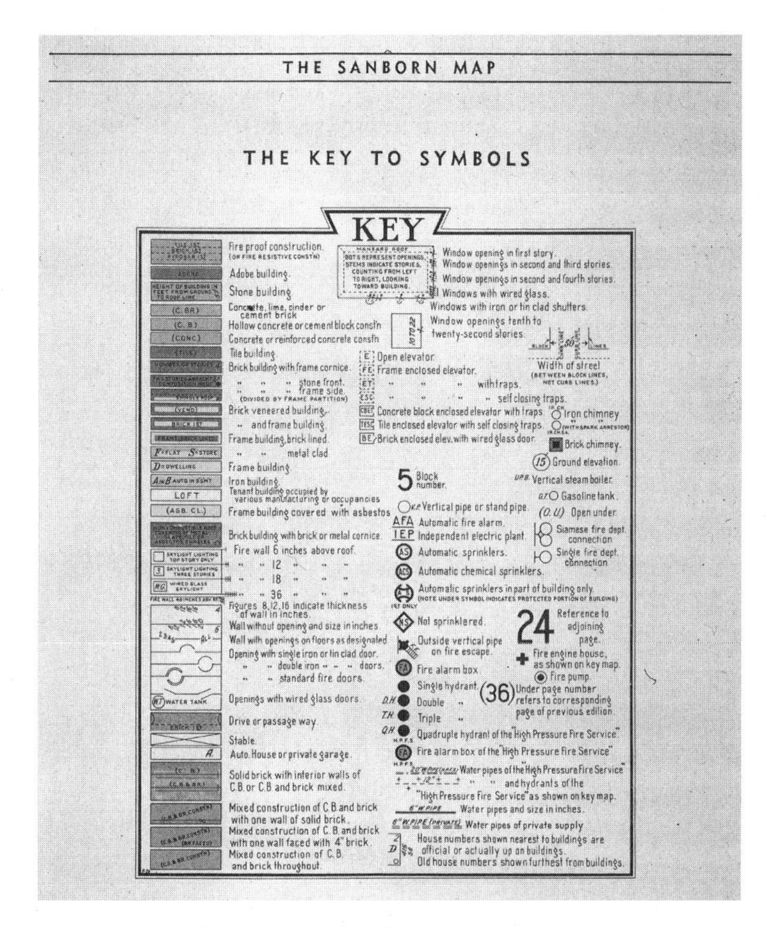

"The Key to Symbols," from the Sanborn Fire Insurance Company, 1940. Sanborn maps continued to account for increasingly detailed features of the built environment well into the twentieth century, although the maps became gradually less significant in the daily practice of fire underwriters by the 1920s. Courtesy, Geography and Map Division, Library of Congress

The attention given to the above matters paled in comparison to the committee's examination of urban water supply, firefighting, and fire alarm systems. In keeping with the NBFU's general strategy, the committee recommended that municipalities should obtain direct control over every key aspect of fire protection, but especially in the case of municipal water supplies. It reported that supplying water for fire protection "involves so many elements widely different in character from those affecting domestic supply that it is difficult to draft a satisfactory binding agreement [with a private owner]." Ironically, even though many within the insurance industry recommended economic self-interest as the best way to prevent fire, the NBFU's committee of twenty complained that a similar economic rationale was inappropriate in the provision of urban water supplies, especially for fire protection. According to the committee, private owners of waterworks could not appreciate the complexities or expenses associated with the fire-protective aspects of water supplies. It reported that "a water system, very poor from a fire protection viewpoint, might be, and often is, a profitable investment to its owners." As with so many other elements of the fire-protective infrastructure, the committee recommended public ownership of this crucial utility as a condition for receiving the best fire insurance rates.[25]

As it agitated for municipal ownership of water supplies, the committee also advocated simple (and economical) solutions to water supply problems. For instance, it recommended using water meters, combined with restrictive legislation, to lower water waste and boost the pressure in water pipes. And it counseled cities to increase the amount of water in reservoirs. Such changes often cost less than adding to water systems. In addition, the committee urged cities to adopt a high-pressure water system to protect commercial and industrial districts, especially downtown areas. It believed that high-pressure water pipes, though expensive, significantly reduced the conflagration hazard.[26]

Equally intriguing were the NBFU's recommendations regarding fire department matters. The board broke with its practice of allowing the IAFE to take the lead in setting department standards as it had done in the past. Rather, it utilized the committee of twenty's inspections as the basis for its own proposals. Then, using its considerable economic and political clout, it worked to implement those recommendations, hoping to make the provision of firefighting services more rational. Each published report examined departmental organization, including districting, engine location, and officer assignments; discussed firefighters' salaries, duties, training, working conditions, and qualifications; and proposed apparatus, tools, and other innovations. The suggestions demonstrated the insurance industry's interest in gaining more authority over fire department composition, technology, and organization. In an anecdote that subtly ridiculed the vaunted New York City fire department, the committee argued that ineffective firefighting could cause significant losses to the insurance industry. It noted that the NYFD let a fire in a small brick building spread to nine surrounding structures, arguing that "there was nothing else left for this fire to feed on, otherwise it might have continued its destruction." In short, inadequate firefighting could cost the industry millions of dollars in losses, and thus its responsibility should not be left solely to firefighters.[27]

The NBFU pushed hard to rationalize firefighting practices without alienating firefighters, all the while courting the support of business and municipal reformers. The industry rarely criticized firefighters directly. Even when it drew a portrait of them as a rough-and-tumble fraternity, the NBFU suggested that firemen were not the problem; lax discipline was to blame. The committee argued that firefighters would be more effective if they engaged in regular training, followed strict work rules, and routinely drilled. Such reasoning allowed the NBFU to advocate greater professionalism without criticizing the heroic smoke-eating firemen that stood so prominently in the popular imagination. Moreover, by challenging departmental patronage and unprofessional conduct, the NBFU echoed the demands of local political reformers and the IAFE. Such recommendations involved walking a fine line, and the NBFU also allied itself with both the IAFE and rank-and-file firemen by accusing municipalities of not funding departments appropriately. The committee's typical report almost always included a proposal to increase departmental expenses for manpower, apparatus, and engine houses. Such recommendations met with the approval of the IAFE, business leaders, and even local political officials. Merchants saw that the city's beefing up fire defenses offered them not only better firefighting protection, with little cost to them, but potentially lower insurance rates as well. Local political leaders who may have been hesitant about the increased expenditures likely viewed the NBFU's demands as a way to improve services and expand the municipal bureaucracy with relatively low political costs.[28]

By publicizing engineering reports and reinspecting cities, the NBFU pressured municipal leaders to heed its recommendations—a strategy that appears to have met with a remarkable degree of success. Within eighteen months of making its first report (on Norfolk, Virginia), cities inspected by the committee (forty-three total) collectively spent $5 million on improving their fire defenses. Over $1 million of that sum was allocated to fire departments and nearly $4 million to water systems. Moreover, many municipalities often adopted more than 50 percent of the specific proposals mentioned in reports. St. Louis, for instance, received its initial inspection in March 1905, and addressed 55 percent of the committee's suggestions within one year. Among other changes, the STLFD began regular drills, located three additional engine companies near to the recommended locations, and replaced two aging steam engines. Even so, firefighters may not have welcomed all the changes. For instance, following the NBFU's recommendations, the Boston Fire Department retired all "chief officers" over sixty years old, replacing them with younger "more energetic and up-to-date men." If such changes angered firefighters, they nonetheless would have met with the approval of reformers, including the leadership of the IAFE, and more often than not, firefighters benefited significantly from the recommendations made by the NBFU and its committees.[29]

In Philadelphia, both the insurance industry and the committee of twenty exercised similar power over municipal fire defenses. In fact, even before the committee first examined the city in 1906, the Philadelphia Fire Underwriters Association (PFUA) pushed for reform by taking dramatic action. The PFUA raised rates by as much as 50 percent in 1900, causing panic among many of the city's commercial interests. The Trade League of Philadelphia convened an insurance committee to study the infrastructure of other cities and urged the city to adopt a high-pressure water system to cover its financial center. Such intense pressure laid the groundwork for the NBFU's study of the city in 1906, after which Philadelphia's municipal government adopted many of the committee's recommendations. For instance, the fire department purchased twenty-two new fire apparatus, including seventeen steam engines, during the first decade of the twentieth century. Even with such swift action, the condition of Philadelphia's fire defenses remained substandard, at least in the eyes of insurers. When the NBFU revisited Philadelphia in 1911, the city's insurance community expressed little surprise that it "discovered shocking things," and hoped for "strenuous action." Although the city responded too slowly for the tastes of its underwriters, the municipality again upgraded fire defenses according to the report. Among other things, the fire department purchased fifteen more motor-driven apparatus. In addition, the city began adding to the high-pressure water system, which protected its congested district. Perhaps most importantly, the NBFU's reports and reinspections fostered an ongoing debate in Philadelphia about the state of the fire department.[30]

The program of urban inspections performed by committee of twenty supported the dissemination of the model building code and significantly affected the development of American cities. Through the committee's work, the NBFU and by extension the stock insurance industry took responsibility for transforming the built environment. Not content merely to issue recommendations, the NBFU actively engaged municipal leaders in a conversation about building codes, as well as the entirety of their fire-protective infrastructure. These interventions represented a remarkable expansion of the fire insurance industry's influence over the urban landscape in both its breadth and depth. The industry's prevention agenda no longer took back seat to managing company risk portfolios; it became fundamental to its profitability. What had been a relatively hodgepodge interest grew into a systematic program of fire prevention. And this discipline would grow ever more pervasive and invasive as the twentieth century progressed.

Schedule Rating, Underwriters' Associations, and State Regulation

The preventive agenda gained further momentum in the twentieth century when the fire insurance industry adapted daily work practices—which included rating by schedule, cooperating through "underwriting associations," and collecting of actuarial data—into instruments of prevention. Beginning in the 1890s, underwriters experimented with new methods of setting rates that were not tied to actuarial data, but rather were based on schedules that adjusted the cost of insurance according to the presence or absence of fire hazards. Although substantive debate surrounded their efficacy, the industry quickly incorporated schedules into everyday practices. As the use of schedules gained influence, the industry relied on them to coerce consumers, municipalities, and state governments to make the built environment safer. At the same time, local and regional underwriting associations expanded their efforts to standardize industry practice by connecting the application of schedules to the other services that they already performed within the industry. By the 1920s, the industry sought to adjust schedules by tying them to the actuarial record. In addition, the industry used local and regional underwriting associations to develop partnerships with state and local governments that emphasized standard rates tied to actuarial experience. Altogether, schedules, underwriting associations, and collecting loss data became integral to the industry's daily business activities and an important part of insurers' efforts to improve profitability by preventing the outbreak of fire.

Developed by industry leader and innovator Francis Cruger Moore, schedule rating gained influence steadily in the decades subsequent to its introduction in 1893. By the 1920s, rating by means of a schedule had become typical throughout the United States, although setting premiums in this manner could hardly have been less uniform. The industry used variants on the two dominant schedules: Moore's "Universal Mercantile Schedule" and A. F. Dean's "Dean" or "Analytic" Schedule. Twenty-two states used the more uniform Analytic Schedule that was developed in 1904; twenty-six states used variations of the older Universal Mercantile Schedule. Unlike the Analytic Schedule, the Universal Mercantile Schedule was not copyrighted and therefore was not applied in a standard fashion. As a result, rating varied widely across regions, within states, and even within localities. For instance, in suburban New York over fifty different schedules were used to set rates.[31]

Setting rates by schedule complemented the new preventive agenda insofar as schedules established a base rate for insurance that was modified by credits to the policyholder for specified construction methods, firefighting appliances, and other prevention efforts. Customers also received a charge for various factors, including poor construction, nearby dangers (exposure), or special risks associated with the use of the property. The dominant schedules represented a similar approach to fire danger. For example, the Universal Mercantile Schedule established a "basis rate" on a "standard building," and the Analytic Schedule developed a "key rate" on an "ordinary building" in an "ordinary city." Additionally, both schedules made adjustments for the fire protective infrastructure of the town in which the building was located. However, beyond this point the schedules differed in important respects, especially in that the Analytic Schedule sought to tie rates to the actuarial record of a particular city and region. Despite this and other differences, proponents of schedules, such as the founder of the Analytic Schedule, A. F. Dean, advocated using them to encourage property owners, and especially manufacturers, to build more safely.[32]

About the same time, the fire insurance industry also developed a method of evaluating the fire-protective infrastructure of cities and towns that, by 1940, had become a standard part of setting fire insurance rates by schedule. In 1915, as the industry codified its program of inspections, the NBFU established a schedule for grading American cities and presumably, to rate them as well. The *Proposed Schedule for Grading Cities and Towns of the United States with Reference to Their Fire Defenses and Physical Conditions* evaluated fire defenses according to a variety of categories and classified cities according to ten levels of safety. Substantively, the *Schedule for Grading Cities* . . . produced little new information for urban leaders, but it provided systematic comparative data to the underwriting community. Besides being useful to insurers, especially those firms setting rates by schedule, the proposal represented the continuing expansion and intensification of the NBFU's preventive program.

Likewise, the spread of schedule use developed alongside—as well as encouraged and benefited from—the growing power and expanding range of influence wielded by underwriting associations. Developed by the industry from the 1870s through the 1890s as a critical part of its efforts to control the problem of fire, the diverse mix of underwriting associations became instruments for helping to promote standardization in insurance practice and promised significant economic benefits. Insurance companies ceded responsibility for setting rates—as well as other practicalities of their business such as performing inspections, adjusting losses, developing standard forms, and monitoring commissions—to local and regional underwriting associations in exchange for very real cost savings. Indeed, as Robert Riegel noted in his treatise on such organizations, "Obviously each company might individually employ raters to apply the schedule, but if there were two hundred companies doing business it would be folly to employ two hundred agencies to apply the schedule in a given locality when one might do it for all. The companies therefore quite naturally co-operate." As the industry had demonstrated in the nineteenth century, cooperation was hardly "natural," but expansion of schedules and the standardization of other elements of the insurance business made it only logical that firms would make use of underwriting associations.[33]

As these organizations flourished, they became more effective, consolidating and extending the benefits of cooperation. Not only did they provide cost savings and encourage the further "specialization and division of labor," but associations also promoted standardization of rates, commissions, classificatory systems, and forms and clauses in contracts. Additionally, underwriting collectives enhanced the industry's ability to inspect large numbers of dangers, which made surveillance a more effective tool of inducing change in the built landscape. However, during the 1890s, such societies had brought the industry under the scrutiny of those who were suspicious of trusts. Although the federal government exempted fire insurance from federal regulation, state authorities began to regulate the industry, out of concern regarding the far-reaching power of local and regional underwriting associations. The expansion of schedule rating helped the industry to combat such efforts, helping to settle these critical political and social issues. By making the process of setting rates more transparent, schedules de-emphasized the cost of insurance as a point of contention between the insured and agents and between agents and companies. Of equal significance, using schedules made an increasingly contentious and unfavorable political arena into more sympathetic terrain for insurance companies. Schedules deflected criticism of high insurance rates because they shifted the responsibility for rates from insurers (and underwriting associations) to the hands of policyholders. Indeed, according to one insurance company, schedules were a "business rejoinder to the anti-compact law." The use of schedule rates, along with preventive policies more generally, made it easy for insurers to respond to critics—a strategy that the industry often used. More broadly, by making rates at least partly dependent on customer behavior, the industry reinforced the prevention message contained in building codes and recommendations for reducing the conflagration hazard.[34]

Even as the use of schedules expanded and helped to allay public fears about insurance associations, the practice remained a topic of much debate. Not only did insurers debate which schedule was superior (and as a result no clear consensus emerged in favor of a single schedule) but also they held lingering doubt about the practice for very real business reasons. Insurers expressed concern that rating practices, especially rating by schedule, did not reflect the actuarial record of fire loss. They worried that a system of rating divorced from the actual record could harm company bottom lines because it allowed the cost of insurance premiums to diminish without a concomitant decrease in fire loss. Additionally, insurers emphasized the need for an objective standard for rating risks. Thus reformers continued to agitate for creating a mortality table for fire losses, thereby modeling the fire insurance industry after life insurance. Although collecting and categorizing industrywide loss statistics was not new, this practice remained as salient in the twentieth century as it had been in the 1850s. Continued debate about rating procedures exposed fire insurers' ongoing struggle to both manage information and make it more applicable to their central business problem. Indeed, how could the process of collecting and organizing loss information matter in controlling the problem of fire? And, in particular, how exactly could rates and practices—including rating schedules—reflect this important information?[35]

Collecting, organizing, and managing information about the fire hazard had not become any less critical to twentieth-century insurers. As evidenced by the NBFU's new mission statement early in the century, underwriters continued to promote the development of standard administrative routines and business forms, and they continued in their efforts to collect actuarial data. The NBFU emphasized long-standing and banal pledges of its goals, such as "to promote harmony, correct practices and the principles of sound underwriting," and "to secure the adoption of uniform and correct policy forms and clauses, and to endeavor to agree upon such rules and regulations in reference to the adjustment of losses as may be desirable and in the interest of all concerned." Additionally, the NBFU remained committed to gathering statistical data from companies and establishing "such classification of hazards as may be for the interest of members." If collecting statistical data had been controversial because of its ties to programs of uniform rates, the NBFU removed any points of contention by formally abandoning that platform. The process of gathering and developing statistics of fire loss, aided significantly by the standardization of business routines, became critical to how insurers would remake rating practices—especially how they would use schedules—later in the twentieth century.[36]

When the NBFU established an actuarial bureau in 1915 to collect and to classify loss data, it reshaped industry practice in a subtle but significant manner. The bureau gathered loss data from member companies as well as from public sources, such as state fire marshals' offices, fire departments, etc. It then compiled a table that listed the "amounts written, premiums written, and losses paid on all classes in the National Board List of Occupancy Hazards, divided by state, and further sub-divided by construction and protection, and in addition thereto, showing the burning rate and loss ratio in each class." Although the bureau cautioned that its figures could be skewed by variations in record-keeping methods, it nonetheless produced an impressive amount of data in a short period of time. In its first year alone, the committee collected information from over five hundred thousand loss reports submitted by the industry. By the late 1920s, the committee had made remarkable progress in its effort to create a mortality table of fire loss; its records included about 98 percent of all losses. By disseminating the industry's quantitative record to members, the bureau encouraged an important but elusive reform: making rates—and hence company account ledgers—reflect actual loss data.[37]

The product of the actuarial bureau's labor—the data itself—had an impact on insurance practice, but not by direct inclusion into schedule rating. Though many insurance firms wanted to peg rates directly to the actuarial loss record, insistence on setting rates according to schedules continued to hinder the application of loss data to rates in most states through the 1940s. However, this did not dissuade the NBFU and others from urging a relationship between rates and losses. For instance, in 1916, when E. G. Richards presided over the NBFU, he expressed his expectation and hope that rates would one day be based upon actual experience. Richards continued to press for computing rates based upon using loss records instead of "estimates"—a case he made in *The Experience Grading and Rating Schedule.* He argued that loss experience should become part of the calculation of insurance rates. Concretely, he suggested applying a ratio (a state's loss record divided by that of the entire United States) to set a rate. For instance, using data collected by the NBFU, he calculated that in the United States between 1903 and 1912, the average loss rate (per $100 of insurance) was 1.125, but that it varied widely from state to state; in New York the average loss rate was .751 but in California it was 1.463. If the insurance industry followed Richards's argument, insurance consumers would pay nearly twice as much in California as they would in New York to insure the same property. Although few adopted Richards's schedule, his method clearly made inroads; by the 1940s, insurance costs on the same categories of property varied between states, according to loss records.[38]

Additionally, Richards recommended that insurers develop a more standard and exhaustive system of classifying risks in order to provide a more sensitive reading of the landscape. To do this, he recommended applying many of the same categories used in the Universal Mercantile Schedule and the Analytic Schedule. He also championed using the NBFU's proposal to grade the fire defenses of cities and towns. More significant than the way Richards wanted to categorize risks was his dramatically different method for collecting the data. He urged that the industry adopt a method of tabulating losses with which a small portion of the industry (about eighty-five companies) had begun to experiment: punch cards. Richards described the method: "A card for each writing, cancellation, reinsurance and loss is punched in the company's office and then forwarded to a well-known statistical bureau which from these cards tabulates and completes the classification of the entire business of the company." Richards's recommendations slowly made their way into practice. Within two decades, the statistical bureau handled over six million cards each year and required fewer than one hundred employees to do this massive job. By the 1950s, the use of cards to organize company risk portfolios gradually gained sway. In fact, the card system meshed so well with expanded programs of categorizing risk and schedule rating methods that they obviated the need to view fire hazards in their spatial context. Classification schemes were becoming so objective that many firms abandoned the use of fire insurance maps altogether. Richards's advice about using the actuarial data also gradually made its way into the industry via increased cooperation between insurance companies, underwriting associations, and state regulators.[39]

Setting rates in accordance with the actuarial record—at least in general terms—grew more normative as the fire insurance industry developed a cooperative relationship with state and city governments during the twentieth century. At the outset of the century, the industry had an antagonistic relationship with state regulators, with whom it had never sought to cultivate a relationship. As a result, local and regional underwriting associations found that their efforts were being nullified by a variety of "anti-compact" measures in state legislatures, especially in the Midwest and Southeast. With the writing of its new mission statement and the implementation of a fire prevention agenda, however, the NBFU began to change the industry's relation to regulators. It committed to promoting "laws and regulations as will secure stability and solidity to capital employed in the business of Fire Insurance, and protect it against oppressive, unjust and discriminative legislation." Although the NBFU adopted strong language, for the first time it explicitly sought to work with states, consumers, and reformers to develop legislation that was favorable to the industry and to society—evidenced by fire prevention standards. In addition, the industry also sought a cooperative relationship with regulators that would allow underwriting associations and other rating organizations to establish rates within localities and regions. Aided by the transparency of schedule rates and the publication of loss data, the industry helped to create a new legal climate. By 1920, twenty-four states had authorized the work of underwriting associations and/or rating compacts. By 1945 thirty-five states tacitly recognized underwriting associations, and by 1950 this approach to regulation was all but universal.[40]

The development of cooperative relationships between the insurance industry, reformers, and state regulators proved decisive in finally transforming industry practice. Most significantly, the entrance of a third party—state boards of insurance—into the equation ensured that within any given state insurance rates more closely reflected the actual cost of insurance and protected consumers against arbitrary and excessive pricing. By demanding that companies file rates, organizational data, and loss information with state boards, insurance commissioners created the mechanisms for oversight that proved critical in keeping rates in line with losses. Although the system had many imperfections and wide regional and local variations, state and municipal officials everywhere learned that they could use publicly available data to petition for reductions in rates. And the coupling of rates and loss experience had a noticeable effect. During the 1930s fire insurance rates diminished throughout the nation, with particular deep reductions—nearly 50 fifty percent—on fireproof public structures. More importantly, the new political and business arrangement served to focus the work of the insurance industry on improving public safety.[41]

In connecting actuarial data on fire loss to rating schedules the industry transformed the daily business of insurance into an effective part of its fire prevention strategy. Of course, in their own right, rating schedules encouraged Americans to build safer commercial structures. But schedules did not, of themselves, make insurance consumers especially invested in the broader battle to control the problem of fire waste; after all, they received discounts for fire prevention whether fire losses diminished or not. However, by tying rating schedules to the loss record within a city or state, the industry coerced individual consumers and their elected officials to become invested in its success, or failure, at controlling the fire hazard. Insurers, then, did not just advocate preventive behavior, but demanded that those behaviors be evidenced in the objective record of fire loss. This further shifted the burden of fire prevention to consumers who benefited most (usually in the form of lower premiums) when the material safety of the landscape improved.

During the first three decades of the twentieth century, insurance practice reflected the industry's broader prevention agenda and showed dramatic results. Just as building codes and inspections of fire departments offered concrete instructions on how to build more safely, so too insurance practices offered meaningful and real incentives for improving the built environment. The industry did not just use municipal law to minimize fire danger; it had created effective market mechanisms to control the problem. And this new regimen of prevention and practice began to yield results. Between 1910 and 1940, rates and losses diminished by almost 50 percent, and even the severity of conflagrations grew less dramatic. Of course, many factors contributed to this decline, including the Great Depression and declining use of open flames in homes and workplaces. Nonetheless, as the fire insurance industry remade its business strategy, its daily practices began to have a significant impact on the American landscape.[42]

Spreading a Gospel of Safety and Manliness

As it inculcated fire prevention into everyday practice, the insurance industry also indoctrinated employees about what behaviors made a good insurance man. Early in the twentieth century, the industry developed more formal social and professional associations to teach the business of insurance to workers in the industry. Employee training was no longer simply acquired on the job or through professional manuals. The twentieth-century insurance worker was educated in good practice and effective manhood in a variety of settings, including local and regional underwriting associations, professional societies, and local insurance libraries. A new array of organizations provided social opportunities and educational certification within the increasingly complex insurance business. Organizations such as the Fire Insurance Society of Philadelphia (FISOP), the Chicago Underwriters Club, and Hartford's Insurance Institute joined the Boston Insurance Library in educating fire insurance workers about their industry. By 1910 these organizations had initiated lecture series and collected industry literature, including engineering reports, in libraries. The mission of these societies was quite different from earlier associations. Whereas groups such as the Philadelphia Fire Underwriters Association or Fire Underwriters of the Middle Department had offered business services to corporations, organizations like FISOP, constituted their purpose as educational and fraternal and helped individuals to make sense of their profession. As the by-laws of FISOP made clear, it focused on the individual within the industry. Its purpose was to assist members in studying "the questions arising in connection with the Insurance business; to promote intercourse and the exchange of views and information among them; to give them, and through them the general public, instruction in matters of construction, fire protection, and fire prevention." The association also collected "books, essays, plans, apparatus and devices bearing upon Insurance" and established "a laboratory for chemical and physical experiment and demonstration." FISOP promoted this agenda by serving as a business club. It sponsored a lecture series, provided a regular lunch to its steadily growing membership, and published a regular newsletter, which kept members abreast of local insurance practices and industry news.[43]

As the underwriting business grew more complex and specialized, FISOP pro-

Cover of a National Board of Fire Underwriters booklet, *Safeguarding the Home against Fire*, 1918. As the fire insurance industry advocated building standards, it also began to emphasize safety in private dwellings. Through the NBFU, the industry distributed pamphlets that informed homeowners about how to protect their property and family from the danger of fire. Courtesy, ACE America Corporate Archives

vided members with crucial educational resources. From its inception, the organization sponsored regular lectures, given by local and national industry experts such as Charles Hexamer. In 1911, FISOP organized a series of sixteen lectures, which members could attend for a total cost of $3.50. Examining the topics of these addresses offers insight into how underwriters had reorganized their priorities and trained younger members of the occupation. From February to May, members attended programs about fire underwriting practice and history: Early Fire Insurance History; Conditions of the Insurance Contract; Schedule Rating; Adjustment of Fire Losses; Fire Insurance Law in Practice; Expense and Taxation. After a brief summer break, when its bulletin briefly ceased publication, member reconvened in the autumn for the remaining talks on the industry's expanding strategy of fire prevention. Those topics included Fire Insurance Regulation: Governmental, State, and Local; Public Fire Protection; Building Construction and Private Fire Protection; Causes of Fires and Methods Used in Placing the Responsibility; Inspection of Electrical Conductors with Special Reference to Fire Protection; Fire Waste. The agenda even included presentations on marine and liability insurance.[44]

FISOP's lecture series—which focused on the growing complexity of the insurance industry and its business practices—provided the industry's labor force with both the knowledge base about fire danger and the information-management skills that were so critical to their risk-taking endeavors. As the industry grew in size and intricacy, it spawned an increasingly specialized division of labor. With programs about the work of specialists—agents, brokers, lawyers, adjusters, inspectors, and engineers—FISOP provided members with the means to negotiate the network of bureaucratic relationships that characterized the complex twentieth-century insurance industry. At the same time, underwriters learned about the particularities of the basic technological tools of their trade—insurance contracts, surveys, and expense reports. Finally, the presentations instructed agents and underwriters alike about the ins and outs of calculating insurance premiums; for instance, as schedule rating became more influential, FISOP included lectures on that topic. And although schedule rating and rating associations had taken much of the judgment out of that procedure, experts nonetheless recommended that insurers become intimately acquainted with how rates were determined. After all, it was important that competent underwriters knew when to take risks. As Robert Coyle lectured in 1912, "It is a simple thing for anybody to apply a schedule. The thing is to know the rule book well enough to know when exceptions to the schedule can be made."[45]

Likewise, lectures became an avenue that industry leaders frequently used to disseminate the latest knowledge about fire prevention. In fact, from its very first lecture series in 1904–1905, FISOP members learned all about "insurance engineering"—a field focused on using engineering to prevent fires. Robert Coyle's advice to "get any kind of reports you can on fires, which show their origin and spread" suggested the importance of applying knowledge to the problem of fire. General topics, such as "fire waste," schooled members in the language and reasoning behind the industry's strategy. Meanwhile, discussion of the building construction and the causes of fires introduced young underwriters and agents to the methods and tools that the industry used to combat the problem—or as one lecture put it "Stop the Fire from Starting by the Power of Law: The State Hand in Hand with Fire Insurance Companies for Conservation of the Country's Wealth and Natural Resources, Producing the Logical Result of Cheaper Rates of Indemnity for the Public." Finally, specialized lectures provided concrete examples of how the fire prevention strategy and its methods worked in the real world. Discussions of such topics as "Safe and Sane Handling of Gasoline and Other Volatile Liquids" made the methods and principles of fire prevention real to underwriters and agents.[46]

These educational initiatives also spawned a specialized curriculum through which the industry could further indoctrinate it employees into the gospel of fire prevention. In 1909, FISOP and other insurance societies and library associations formed the Insurance Institute of America. The IIA codified industry practice and expertise into a curriculum and, according to FISOP, offered "every ambitious man the opportunity to . . . acquire knowledge which he would other wise have to acquire by rule of thumb." By offering a certificate as evidence that students had passed a comprehensive series of examinations, the institute hoped to establish an important credential within the industry. By the 1950s, the IIA joined with other organizations to develop systematic academic standards for professional designation. Though an extraordinary range of programs, the IIA offered credentials to help men and women working in the industry to gain technical competency and career advancement. As the institute's curriculum became prevalent, successive generations of underwriters and agents learned the industry's basic ethos and practices.[47]

Additionally, insurance associations facilitated personal connections and camaraderie by melding social interaction into its business functions. FISOP, for example, became a center of social as well as professional activity among Philadelphia insurers when it began to offer lunches and other opportunities from its offices, centrally located in the heart of "insurance row," which ran along Walnut Street between the Philadelphia Stock Exchange and Independence Square. Agents, brokers, and representatives of insurance companies came to FISOP to discuss and to conduct business, and to socialize. FISOP facilitated these exchanges by serving lunches to its 680 members. In 1921, the society served 33,000 meals; this meant that, on average, members dined at the society about once per week. In addition to lunches, insurance men shared cigars after meals or at smokers, a common ritual among middle-class professional men across the nation. Indeed, before the tobacco industry began marketing cigarettes to women in the 1920s, it was considered taboo for women to enter rooms where men were smoking, and smoking at professional meetings or lunches created a distinctive all-male business climate. This culture thrived at FISOP, which proffered members cigars following meals, banquets, and talks. In 1921 alone FISOP collected nearly $1,400 in donations for the cigar fund.[48]

As industry leaders taught budding underwriters about business, they also educated them about the proper basis for business. At lectures and in the lunchroom, professional organizations like FISOP instructed workers about constrained business manhood. The industry urged men to balance personal ambition against the needs of the corporation, to demonstrate individual judgment within the boundaries of increasingly codified standards of business practices, to take risks but to make certain that those risks were financially sound. Typically, these sentiments found expression when industry leaders discussed which clients to cultivate. Understandably, corporations recommended doing business with successful men, and when FISOP recommended that insurers transact business with men who were "conscientious," it used credit-reporting agencies as a guidepost. By the 1930s, the reports issued by organizations like the R. G. Dun Company (later Dun & Bradstreet) contained increasing amounts of quantitative data to help insurers evaluate the moral hazard component of the risk-taking endeavor. In 1940, for example, Dun & Bradstreet issued a list of balance sheets and operating ratios of over seventy industries, titled *Relativity of the Moral Hazard.* The pamphlet provided comparative data that was organized into quartiles that laid out the parameters of success or failure in each of the business lines. Recognizing that moral hazard—the human component of risk—was dynamic and changed with business conditions, the report provided a basis for making judgments about which men constituted the best risks. Such quantitative reports provided the basis for a more objective definition of a good risk, in much the same way that schedule rating and fire prevention used financial incentives to promote safety. Economic success made one a better risk, and it became a marker of manliness. This message could not have been lost on insurers, who realized their value as employees and men depended upon their contribution to the corporate bottom line.[49]

Even as numbers counted far more, the industry did not completely abandon imprecise measures of character when making decisions about which risks to take. FISOP, for instance, wanted insurers to do more than inspect credit reports. It wanted them to also seek out the "very large percentage of the population" that were not listed—those men whose lack of individuality, expression, and aversion to risk kept them hidden from the gaze of those societal institutions being developed to manage economic hazard. Yet, building on long-standing cultural directives regarding sobriety and self-control, underwriters prescribed discipline as a defining characteristic of appropriate business manhood. For example, FISOP urged insurers to seek out "the successful man who has attained success gradually through sound and careful business methods. Almost irrespective of the physical hazard of the risk, the business is highly desirable." Such self-control, however, mattered little without a careful assessment of the physical risks involved, and FISOP added a caveat that agents should insure property only "if the physical aspect warrants it." Conversely, FISOP discouraged underwriters from taking risks on well-situated structures that incorporated fire-preventive appliances if the customer did not behave in a careful fashion. FISOP warned insurers not to transact business with "the man whose success is due to unscrupulous methods (adulterating output, underpaying help, etc.)," to "avoid him by all means, even if his sprinkler system has three independent [water] supplies." At a time when Progressives sought to rein in the excesses of laissez-faire capitalism, insurers emphasized a more restrained capitalism as the basis for proper male behavior. Just as the broader program of fire prevention took a long-term view on industry profits, so too professional education emphasized a standard of behavior that reoriented industry practice toward a more conservative model.[50]

Urging disciplined behavior carried a weighty message in an industry that had spent decades developing standard business practices and routines, which became ever more important with the ascendancy of the fire prevention agenda. The emphasis on following rules reinforced the important activities of rating by schedule, coordinating the industry through hundreds of federated associations, and creating precise actuarial databases. Insurers also learned that cutting corners—not following schedules or fire prevention recommendations—dampened the industry's future prospects. Moreover, the emphasis on careful risk taking also offered a message to corporations about how they should do business. By disciplining their practices and their employees, underwriting firms could achieve profitability, and provide the basis for the nation's economic future. The industry's message came through clearly: the successful businessmen was one who achieved quantifiable economic gains by adhering to sound management procedures. This missive tied insurance workers to their corporations by advising them that their prosperity rose and fell in direct relation to their companies, underscoring the importance of procedural innovations in the industry.

Even as the industry created boundaries for proper behavior, many men chafed at the restrictive rules developed to govern them. For example, as schedule rating became more popular, lingering doubts about its efficacy were nurtured by a sense that it diminished the manliness of fire insurers. In 1911, FISOP published a humorous anecdote that lampooned schedule rating, capturing insurers' concerns:

> Schedule for Determining Office Boy's Salary
>
> BASIS RATE: Standard Boy, per week 5.00
> (Note: Standard boy is one constructed strictly according to specifications detailed in Volume 34, Section xxiv, issued by the National Board of Salary Equalizers.)
>
> Deduct for deficiencies as follows,
> Appearance:
>
> | | |
> |---|---|
> | Red hair | .15 |
> | Freckles | .10 |
> | Warts (according to size) | .01 to .20 |
> | Teeth out (each) | .06 |
> | Etc. etc. | |
>
> Habits:
>
> | | |
> |---|---|
> | Chews gum | .10 |
> | Interested in Base Ball | 1.50 |
> | Talks too little | .00001 |
> | Defective flue | .75 |
>
> Etc. etc.
>
> These are only a few of the two thousand items in the schedule. Everything is included, from his musical training on the mouth organ to the part of Germany his father came from.

To the degree that this satirical story reflected anxiety about schedule rating, it also exposed the degree to which white-collar workers resisted the expansion of quantification in management practices when they were applied to their own workplaces.[51]

Many underwriters argued that the more regimented procedures removed their judgment—their individuality—from the risk-taking endeavor. This concern emerged at a time when the middle class in general worried that office work had removed zeal and expressiveness from their lives. Insurers expressed anxiety that following schedules blindly—of not taking risks as firefighters so often did—emasculated them. In addition, the satire revolved around the issue of salary, pointing to a contradiction never far beneath the surface of middle-class manhood. No matter how much reformers moderated the rough edges of acquisitive capitalism, income remained a stark and unyielding measure of success and manliness. How could prudence and discipline ever be reconciled with economic success—especially when it so explicitly involved taking risks, as the insurance business did? If following rationalized corporate rules produced men of character, standardizing salaries removed essential incentives and differences between men. The "board of salary equalizers" with its tinge of anticapitalist thought implied that reason-based manhood was at odds with the emphasis on market capitalism so rooted in American culture.

Like other members of the twentieth-century middle class, insurers' concerns centered on their desire to express their individuality, to take risks that enhanced their identity as men. In the broader culture, this desire was often expressed in terms of taking physical risks, such as those written about in novels and embodied by Teddy Roosevelt's jingoistic manhood. Yet, even here, the middle class did not seek out risk simply as an end in itself. For example, businessmen saw taking economic risks as a necessity in a capitalist society, but they created managerial practices and business routines to minimize their exposure to hazard. Additionally, such men established prescriptions for behavior that balanced prudence and individuality as an antidote to business problems. More importantly, perhaps, middle-class men wanted to bring such discipline and self-control to bear on social ills by implementing what they believed to be rational economic interventions in disorderly places—especially factories and urban streets. The insurance industry, like the middle-class more broadly, saw this vision of manhood as the means for controlling the problem of fire. By acting in a disciplined and rational fashion, underwriters could conquer environmental danger and bring order to urban America.[52]

The Gospel of Individual Responsibility and Home Ownership

The insurance industry created a cultural as well as material standard for safety that placed an idealized middle-class family and home at the center of its campaign to combat fire waste. The industry reached beyond the industrial and commercial sectors, and began to prescribe building practices for homes, suggesting that individuals take personal responsibility for fire safety. This effort began in 1916 when the NBFU inaugurated a massive public relations campaign. The program matured quickly. Within two years the NBFU established a "Special Committee on Public Relations" as a permanent standing committee. The NBFU's public relations crusade built upon its earlier and continuing efforts to disseminate standards to public officials, industrialists, and professionals in the fire protection industry. In addition, the organization targeted a new audience of homeowners, apartment dwellers, and school children. It remade itself as a public service organization through a variety of activist programs, which included National Fire Prevention Day (later National Fire Prevention Week), a fire prevention periodical *Safeguarding America Against Fire* (SAAF), and a series of pamphlets and educational materials aimed at school children. Not content to advocate fire prevention solely through building codes and in daily practices, which the organization increasingly believed would take decades to be fully effective, the NBFU extended its gospel of safety directly into the nation's homes.[53]

From its inception, the committee on public relations prepared and disseminated a steady stream of publications on a variety of topics to a wide public audience. Its pamphlets covered a remarkable range of topics, including: Does Fire Prevention Prevent?; What Is Inspected Fire Hose?; What Is Causing American Fires? Fires from Small Electrical Devices; Fire Prevention and the High Cost of Living; Institution of a Spring Clean-up Week; The National Board Campaign for Conserving Grain, Cotton, and Other Staples. Along with a wall chart detailing causes of fires, the NBFU distributed these pamphlets to an audience that included public officials, manufacturers, members of the insurance industry such as local agents, and fire protection leaders (i.e., fire marshals and fire chiefs). To reach the broadest public audience, the organization sent copies of its publications to the wire services and newspapers, which published their contents as news items, special editorials, and feature stories. The campaign produced immediate results. After newspaper stories drawn from the pamphlet "Fire Prevention and the High Cost of Living," the NBFU received nearly twenty thousand requests for the pamphlet. Articles that caught the public imagination, such as one in the *New York Tribune*, generated special pamphlets. Moreover, in September 1918, it issued its first periodical meant for a general audience; each month, the NBFU distributed over forty-four thousand copies of *Safeguarding America Against Fire*. Like the pamphlets, *SAAF* served as a conduit through which the NBFU's message made its way to the general population. With issues dedicated to topics such as "Fire-Safety in the Home," "Fire Prevention Week," and "Our Vanishing Forests," the group provided news editors with copious information on the causes of fires, precise calculations about fire damage, and emotional stories about fire's dangers. For instance, in addition to informing its audience that chimneys caused over 12 percent of blazes, the NBFU enumerated the measures people could take to reduce the possibility of a chimney fire and included a diagram of what constituted a safe chimney.[54]

As the NBFU refocused its prevention work to include the home, it continued to emphasize the importance of prevention in industrial and commercial facilities, and also focused energies on issues of safety in public buildings, such as schools.

The public relations campaign, with authoritative pamphlets and *SAAF*, was coordinated with and buttressed the organization's efforts to heighten fire safety awareness among public officials and manufacturers. With exacting loss statistics and prevention information, *SAAF* provided business and political leaders with materials for fire safety campaigns. In 1922, for instance, *SAAF* issues methodically covered fire safety in hospitals, schools, churches, and hotels. Each issue delineated the most common causes of fires in each of those types of institutions and suggested solutions for diminishing those dangers. Yet, *SAAF* and the myriad of pamphlets did not supplant more bread-and-butter efforts at reducing fire loss. Rather, they complemented the expanding number of building codes and fire prevention recommendations, often promoting adherence to those codes as the principal solution to fire prevention.[55]

The NBFU specified in excruciating and vivid detail the particular hazards to which American society should be sensitized. According to the NBFU, nearly 30 percent of "all American fires" drew from "strictly preventable causes"; another 48 percent came from "partly preventable" causes; and the remainder came from unknown causes, "which were largely preventable." The NBFU boasted that its numbers came from a comprehensive study of "hundreds of thousands" of fires, which the organization privately believed constituted at least 95 percent of the nation's blazes. Along with the numbers, the NBFU offered a damning and evocative critique of "carelessness." Such irresponsibility caused significant economic loss, amounting to a yearly "fire-tax" greater than the production of the nation's gold, silver, and copper mines, plus its oil wells. Moreover, carelessness produced a staggering human toll equal to a road of devastation from Chicago to New York, on which "at every three quarters of a mile . . . [one] would encounter the charred remains of a human being who burned to death." If the numbers did not impress, horrifying accounts of the death and disfigurement accompanied them.[56]

The changes in how the NBFU used numbers demonstrates the manner in which the industry had expanded its use of the statistical record, and underscores the gradual and striking shift in the insurance industry's mentality toward fire danger and public safety. Rather than use actuarial data simply to standardize rates, the NBFU made its revitalized program of uniform accounting and statistics the basis for its program of prevention, especially emphasizing areas of public concern. As the 1920s progressed, the NBFU published the aggregate "fire record" from the previous year in *SAAF*, specifying the most frequent causes or most serious hazards. It argued that all fires were at least partly preventable and, hence, the responsibility of individuals. This message was not just for public consumption. For the first time, insurers themselves began to view fire safety as their responsibility as well; the NBFU had transformed its message. Over a one-hundred-year period, insurers' approach to the problem of fire had shifted from resignation about fire's inevitability to a conviction that they could predict losses sufficiently well to guaranty profitability, and finally to the belief that they could minimize economic losses by working to prevent fires altogether.[57]

Fire safety meant more than being cognizant of the many dangers lurking in the landscape; it implied a new way of behaving, and the recognition that prevention was both an individual and a public responsibility. This new attitude was especially evident in the NBFU's attempt to reduce fires caused by "exposure." *Exposure* referred to any fire that extended "from the place of its origin either to another building or to a quite independent unit of the same building." According to the NBFU, exposure alone accounted for between 15 and 20 percent of all fire losses. As the single largest cause of fires, it could be stopped only by closer scrutiny of fire dangers and developing a "fire consciousness." This awareness included a host of public provisions, such as better building laws, with strict enforcement, as well as better fire defenses. These suggestions raised the insurance industry's program of surveillance and inspection to new levels. Underwriters exhorted the entire nation to observe one another in order to lessen the dangers and consequences of fire.[58]

Protecting the home—as a physical and cultural space—became central as the industry redoubled its efforts to control the problem of fire. In 1916, the NBFU disseminated standards targeted at the construction of dwelling houses, thereby increasing the intensity and depth of its fire-preventive mission and at the same time protecting the most predictable and routinely profitable segment of their risk portfolios. Bringing home construction under the industry's watchful eyes and its preventive umbrella reflected only one aspect of underwriters' attempts to reform American society. The insurance industry also was still working to encourage an ideal of manly responsibility. According to insurers, it was a man's duty to protect his family and his property from fire. Fire insurance policies and the industry's program of prevention gave men a way to guard their most valuable possessions. In its first set of guidelines for "dwelling houses," the NBFU commented that homeowners' "apathy is strange, for one would naturally suppose that a man's first thoughts, would be for the safety of his family and the protection of the home which is so essential to his comfort and happiness." In *SAAF,* the NBFU hammered the message home even more forcefully: "It is a strange fact . . . that men will provide factories and offices with elaborate systems of exit and life-saving devices to protect employees from fire, and then will erect homes to shelter their own families without a single precaution for saving their lives in a similar emergency."[59]

The industry had transformed safety into a commodity around this notion of manhood. It stressed that agents and underwriters were not opportunistic hucksters but men concerned about preventing fire in the community—men taking public responsibility for danger—and helping other men to live up to a code of middle-class male family responsibility. The NBFU expected that agents would use the information provided in *SAAF* to promote safety and security, and, of course, to sell fire insurance. Fire insurance agents comprised the single largest *direct* recipients, and redistributors, of *SAAF* and other NBFU pamphlets; of the 44,000 *SAAF* issues printed each month during 1919, insurance agents received over 39,000. It did not urge a "hard sell"; rather, the NBFU editorialized that "some authority on the art of selling has declared, plausibly enough, that the most effective salesmanship often is the kind that is unaware of itself." The NBFU provided agents with statistics, stories, and other fire safety information in its publications, and it even produced a "motion picture" detailing the public service provided by insurance agents. The film, *The Keystone*, argued that fire insurance was "the mainstay of the system of Commercial Credit," and it helped to "indirectly sell" the "idea of adequate prevention." Agents knew that elevating popular consciousness regarding fire danger fulfilled their obligations to their employers and to their communities, as they became manly conservators of community by preventing fire. Naturally, in taking such public responsibility agents believed that they also would sell more fire insurance.[60]

The message of protecting children served the insurance industry rhetorically, and underwriters also employed children as important agents of safety. As an important aspect of the middle-class household and its idealized image, children symbolized the order that insurers and fathers were supposed to guard. Just as firemen exhibited manliness when they rescued innocent women and children, underwriters too wanted to protect children from the problem of fire. In addition, by targeting children in its safety campaigns, the fire insurance industry cultivated a future generation of homeowners and safety-conscious citizens. The NBFU argued that "in a few years they, themselves, will be owning property, and paying the taxes of the nation." At the same time, underwriters created a fire prevention curriculum for schools through which it placed children at the vanguard of the industry's safety message. The NBFU published over 275,000 copies of *Safeguarding the Home against Fire: A Fire Prevention Manual for the School Children of America*, which became an official textbook of the Boy Scouts of America and was used in schoolrooms in nearly every state. Endorsed by the United States Department of Education and countless state departments of education, the publication was "unquestionably the most widely circulated piece of fire prevention literature ever prepared," according to the NBFU's committee on public relations. By 1920, some state legislatures, such as New Jersey's, mandated that fire prevention become part of state curriculums at public and parochial schools. By instructing children about safety, the industry found a new route to extend its surveillance into the everyday lives of Americans. It provided them with safety inspection blanks so that they could literally carry the gospel of safety into their homes to search for hazards. Children, then, served an important programmatic role in fire prevention, even as representations of protecting them served as an organizing metaphor for the dangers that fire posed to society.[61]

When it diverted so much energy to protecting the home, the insurance industry established the white, middle-class family and its cultural norms as the behavioral model of safety, akin to building codes or the "standard buildings" of schedule rating schemes. Underwriters idealized the home, as *SAAF* editorialized in 1923: "As many are aware, it was exactly one hundred years ago this month—in the May of 1823—that an American named John Howard Payne, sojourning abroad, introduced a song which he called Home Sweet Home. The world, and America in particular, is celebrating that event today in church and school and theater. How fitting and sensible and practical a memorial it would be if every American made a resolution that his home—situated in the same land as for which Payne longed—should not fall a prey to flames." By so emphasizing the home, underwriters invoked the moral authority of the middle-class family to bring order to a chaotic society. By encouraging men to protect their dwellings from the public, external menace of fire, insurers provided a message consistent with broader middle-class gender roles and the division of male and female labor. Men managed fire safety in the public arena, especially in workplaces, and women claimed responsibility for reducing private dangers lurking within the home, although increasingly they also guarded the domesticated spaces of neighborhoods and communities.[62]

If fire insurance industry rarely ventured beyond the cultural values of the white middle-class when it proselytized safety, underwriters seemed aware that the middle class was not the only group of Americans purchasing homes and fire insurance. For instance, in the introduction to its suggested dwelling house code, the NBFU expressed surprise that so few homeowners took seriously the responsibility of preventing fire, pointing out that important differences divided homeowners as a group. Underwriters argued that working-class and poor Americans especially needed insurance support because of their "moderate circumstances" and because their dwellings represented "a large portion of the family capital." However, unlike some life insurance companies, such as Metropolitan Life, which targeted working-class and immigrant clients using customized marketing strategies, the fire insurance industry remained focused on a standardized vision of safety.

Even so, the industry intimated that fire prevention was an obligation of all citizens. The stridency of *SAAF*'s message of individual responsibility was, at least implicitly, a means of Americanizing immigrants, of erasing differences by establishing the white middle class as the standard of safety.[63]

Conclusion: Fire Prevention and Consumer Safety

By the time veterans returned home after World War II, great strides had been made in the battle. Led by the insurance industry, municipalities had reformed their legal codes to account for fire dangers and upgraded fire defenses. Also, as a matter of course, newly developing towns and suburbs adopted the NBFU's rules—including standards for housing that had previously fallen outside the framework of most construction ordinances. Similarly, as an adjunct to legal codes, fire underwriters offered standards for safe behavior. The industry demanded that underwriters and agents act as protectors of the polity, property, and people by teaching fire prevention. By arguing that the problem of fire was every individual's obligation, the insurance industry emphasized that the individualism so much at the heart of American capitalism was the best means to achieve order and safety. Although the gospel of fire safety provided an invaluable public service by heightening awareness about the problem of fire, it also improved underwriters' bottom line and reinforced dominant social values, especially the hegemony of white middle-class homeowners.

Even more striking, as the insurance industry restructured itself, it developed an approach to the problem of fire that was appropriate to an increasingly consumer-oriented society. The Underwriters' Laboratories, which had served at the vanguard of the NBFU's strategy to place new technologies under surveillance, provided a new mode of disciplining the landscape—by monitoring and inspecting industrial and consumer products. By 1915, the laboratory distributed over fifty million labels—its mark of approval—per year; by 1922 that number grew to over five hundred million. UL performed three types of surveillance, which included examination, and reexamination, of goods at factories as well as inspection of factories. The third and highest level of supervision was UL's blind tests of materials. In the early twentieth century, the laboratory inspected materials primarily used for industrial purposes, especially electrical devices and those used in fire protection. As consumer items became more central in American life, UL was among the earliest organizations to certify manufacturing and consumer goods as safe. By the second decade of the twentieth century, UL had begun to test systematically passenger automobiles at the behest of insurers. Although broadly interested in the safety of cars, UL focused on those aspects of automobiles related to prevention of theft and fire—locks, engines, and electrical devices. The UL's approach to safety—approving consumer products as hazard free before they were used—underscored the dramatic way in which the insurance industry was transforming the problem of fire.[64]

Moreover, insurers offered individuals a variety of products and services appropriate to a society in which citizenship came to be defined in terms of the ability and right to participate in the polity as a consumer. Not only had insurance policies become more effective at indemnifying policyholders against economic losses, but also the industry offered a host of services—from the Underwriters' Laboratories to updated safety information—that promised complete protection to consumers and their possessions. Through this guardianship, insurers guaranteed the durability of material artifacts, as well as the memories and cultural significance bound into each and every consumer item. As following expert standards gained social currency, safety had become a commodity, perhaps most evident in the fact that the UL mark became tradable. Through such symbols of safety, the insurance industry offered manufacturers advantages in the marketplace, and thereby insinuated its vision of security yet deeper into the nascent consumer economy.

As engineer Frederick Taylor preached a gospel of efficiency, so too the fire insurance community became evangelical about fire safety. By the 1930s and 1940s, school curriculums, fire insurance policies, fire prevention programs, and the UL label had become visible symbols of safety, alongside other less visible but no less pervasive elements of insurers' safety discipline, such as legal standards, coercive rating strategies, prescriptions for manhood, and an extraordinary surveillance of the built landscape. Yet, no matter how normative these symbols became, firefighters remained the dominant icons of fire safety. Indeed, despite underwriters' growing hegemony and their efforts to rationalize fire protection, the smoke-eating fireman and his culture of heroic manhood endured. Firefighters remained at the moral, if not institutional, center of systematic urban fire protection.

CHAPTER EIGHT

Eating Smoke

Rational Heroes in the Twentieth Century

When St. Louis firefighters arrived at the Simmons Hardware Company in the summer of 1911, no smoke or flames were visible from the street, but below ground in a basement vault a mixture of hay and excelsior sat smoldering. Firefighters entered the building looking for the blaze, and they discovered acrid smoke emanating from the basement, up elevator shafts. As quickly as they began to enter the basement, with nozzles blasting water, they began to drop, one by one, from the fumes. Thousands of the city's residents watched the "weird scene"; smoke slowly seeped from the building and waves of firefighters entered the building, only to be dragged out moments later by their comrades. Gradually, the smoke thickened and billowed; ambulances arrived to care for firefighters—ten or twelve of whom sat gasping on the sidewalk at any given time. The department's chief, Charles Swingley, became "annoyed at the failure of the men to penetrate the basement." He entered the fray. But, he too was soon carried out. Gradually, a strategy emerged from the chaos; newspapers reported, "the 'smoke-eaters' of the department, the veterans at combating such conflagrations, were sent first, and others wearing face masks held the nozzle into the vault." Firefighters worked in relays of five-minute shifts, entering and exiting the basement, where the water became so deep that they feared drowning. By the next morning, when the blaze was finally under control, nearly seventy firemen had been overcome, some more than once, including Chief Swingley. The *Post-Dispatch* reported that "until he was carried out the sixth time, and went to the City Dispensary, [Swingley] refused to admit that he had found a new thing in firefighting."[1]

Although the calculations of underwriters and fire prevention campaigns made headway in the battle against fire, firemen remained the most visible standard of safety in American cities. In fact, fire insurers sought firemen out as advocates to visit schools during fire prevention week, on the premise that "a fireman in uniform is a hero to the average child." The apparent simplicity with which children viewed firefighters belied the reality of their work lives. Firefighting was growing yet more complex in the twentieth century, as shifting urban environments continued to generate new hazards—automobiles, chemicals, and new construction materials—and contained more familiar threats, such as falling walls, electrical wires, and tall buildings. Facing mounting dangers, firefighters reorganized their work, created extensive training regimens, and introduced new technological tools and techniques. In a culture obsessed with technology, firefighters transformed themselves, at least rhetorically, into an efficient machine. By the 1930s the advent of rescue squads in Philadelphia and other cities signaled that a more constrained professional heroism was developing among firemen—one that became more rule-bound and tied to technological acumen and teamwork. Though they emphasized the rational and technological, they nonetheless continued to identify saving lives as the single most important facet of their service. Indeed, the ideal of the male hero crossing a flaming threshold, baby in hand, disappeared neither from popular culture nor from fire engine houses. And firefighters continued to define their service on a continuum—between the physical and technological, between individual acts of courage and brotherhood.[2]

With support of political reformers and the insurance industry, firefighters once again remade their service. As fire department leaders negotiated the rocky world of Progressive politics, they produced extensive changes, consistent with the recommendations of professionalism made by the International Association of Fire Engineers (IAFE.) Department leaders demanded that the urban technological infrastructure be updated, and bureaucratic structures be streamlined, and they established formal firefighting work procedures and rules. As firefighting became highly regular and standardized, firefighters thrived. They acquired more routine careers, better pay, improved working conditions—all without losing status as popular icons. Even so, the new administrative regimens circumscribed the power and eroded the cohesive culture of firemen's small all-male work groups. Wherever one looked in the 1930s—in engine houses, on fire grounds, in training schools, and administrative offices—departments emphasized technology, efficiency of action and procedure, and sober work discipline.

Significantly, the new work discipline did not eradicate the heroism or love of danger that previous generations of firefighters had made so much a part of firefighting. If anything, it pushed the contradictions so evident in engine houses to the fore, and in many respects fire chiefs came to embody the tension between discipline and daring that characterized of twentieth-century firefighting. For instance, the St. Louis Fire Engineer, long a champion of reform, personified the contradictions inherent in firefighting. Swingley epitomized the local traditions of firefighting, especially its roots in the visceral physicality of working-class life, but also represented values of sobriety and efficiency, normally associated with the middle class and ideals of firefighting professionalism. As firefighters struggled to make careers amidst departmental reform, they undoubtedly negotiated the same slippery cultural terrain as their leaders. Rather than accept a risk-averse vision of manhood, they appear to have done the opposite, embracing greater amounts of danger. For instance, at the turn of the century, the New York Fire Department developed rules about eligibility for its "roll of merit" that defined heroism more narrowly—only acts in which firefighters placed themselves in personal danger. Balancing judgment against reckless abandon, firefighters attempted to get closer and closer to fires and higher and higher into buildings to save lives; they ate smoke in larger and larger quantities. And, in the face of increasing dangers, they introduced "rescue squads" into the lexicon of their occupation. Staffed by the best and brightest firefighters, rescue squads embodied a new, more modern ideal of firefighting: the melding of men and technology into an efficient, lifesaving machine. Firefighters' sense of their manhood helped them to negotiate contradictory tensions that recurred again and again in twentieth-century society—rationalism verse expressiveness, efficiency verse passion, and modernity verse tradition. Long before astronauts personified such contradictions, firemen had created a culture of rational heroism.[3]

Reforming the St. Louis Fire Department

Providing fire protection to American cities remained contested political terrain crowded with the interests of business and insurance capitalists, firemen, and machine politicians. A common goal for years, improvements in fire protection came slowly and unevenly. Reform did not occur systematically until the cusp of the twentieth century, when its pace intensified during the Progressive Era. Although fire department reform mirrored Progressive change more broadly, there were nonetheless important differences. Unlike debates about safety and the environment, which often pitted the interests of industrial capitalists against local communities, debates about fire protection created strange bedfellows. Local communities, firefighters, capitalists, and politicians all struggled to improve fire protection because it was perceived as such an unassailable social good. Indeed, the language of fire as a bad master and good servant remained prevalent, and social and political leaders continued to use the battle against fire to assert their mastery over nature and the urban environment. Although the emphases of their reform programs often varied, capitalists, politicians, and firefighters nonetheless found themselves working alongside one another—forging both temporary and long-term alliances to suit their needs.[4]

Urban political leaders and firefighters wielded the most direct power over fire departments, but the cause of reform advanced only with the support of insurers and local merchants. Local businessmen and insurers supported calls for higher wages, better working conditions, and more rational departmental management because they did not bear directly the costs of reform; municipal governments and taxpayers (including businesses) paid for the maintenance and operation of fire departments. Additionally, businesses and insurers agreed philosophically with the ideas behind reform efforts—an increased reliance on technology, rational management, and improved efficiency. Lastly, as insurers shifted away from simply indemnifying property toward preventing fires, they became interested in improving fire defenses, especially stopping small blazes before they became conflagrations. Commercial interests owning property in cities recognized that improved fire defenses could reduce or maintain fire insurance rates, and their business costs, especially as evaluations of departments and urban infrastructure crept into insurance rating schemes.

Reform-minded firefighters seized on the insurance industry's support and the opportunity presented by the political climate of the Progressive Era. They advocated for departmental changes consistent with those championed by the IAFE. Reformist firefighters especially targeted political interference in fire departments, but also sought to improve wages, working conditions, and pensions. In their effort to circumscribe the meddling of politicians and to achieve independence, they frequently allied with insurance and commercial capital in order to gain sway in the shifting sands of urban politics. Yet, paradoxically, job survival and/or promotion continued to depend upon political connections, and firefighters remained active players in local political arenas. Ironically, many firefighters and fire chiefs had been appointed or promoted by boss politicians, against whom they later agitated. As fire chiefs negotiated this slippery terrain, the twin ideals of professionalism and brotherhood advocated by the IAFE provided a compass indicating what constituted appropriate organization, work routines, and protocol. Chiefs especially argued that departments should be rationally organized, staffed by well-paid men who trained regularly to maintain physical vitality and to acquire technological proficiency.

Charles Swingley's twenty-year tenure as St. Louis fire chief shows the savvy with which one professional fireman transformed the landscape, institutions, and culture of firefighting during the first decades of the twentieth century. Swingley's skill at winning political fights and his ability to cultivate a relationship with insurance and business interests made his tenure as a reformer possible. As he negotiated a tightrope of competing interests, the career of this consummate professional fireman was emblematic of the tensions between efficient meritorious service and political skill, and reveals the complex interrelationship between labor, capital, municipal governance, and urban growth in turn of the century American cities. In 1898, an especially tumultuous period of politics in St. Louis, Republican Mayor Henry Ziegenhein pressured the Republican-appointed Swingley to increase patronage hires. According to one local newspaper that cited a well-connected politician, Swingley had been told by the administration to dismiss all men who violated departmental rules. This, the paper reasoned, would winnow the large number of Democratic firemen (which it estimated at approximately 85 percent of the department) because, according to the same source, "those who are dismissed under this system will be replaced by Republicans." This action would recast the character of the fire department, presumably making it more favorable to the Republican machine, which threatened to replace Swingley if he did not comply with its wishes.[5]

Although Swingley had been put in a bind, he and other department leaders were not without recourse. And, in the case of Swingley, the mayor and his lieutenants underestimated the chief's political savvy. As his rapid rise through the ranks suggests, Swingley was a seasoned veteran of the city's political maneuvering. When he did not comply with the mayor's demands, he drew upon that reservoir of experience and used his connections to control the situation. Ziegenhein's administration exerted pressure on the chief by using local newspapers to announce Swingley's impending removal as his date of reappointment neared. At the same time, within the department, Assistant Fire Chief Gross emerged as a rival, as did Julius Wurzburger, a political appointee to the job of assistant street commissioner. Swingley's removal seemed so certain that local newspapers anointed Gross the city's next fire chief.[6]

Swingley did not bend to the machine. He turned to his patrons in the business and insurance communities, whom he had cultivated in previous years. In response to his call for help, over one hundred prominent local business leaders advocated for Swingley's candidacy with letters and cards. In addition, the St. Louis Board of Fire Underwriters publicly supported Swingley. Other mercantile interests also expressed their preference directly to the mayor and in city newspapers. Such backing turned the struggle in Swingley's favor, in part because Ziegenhein did not want to alienate the primary constituency of the Republican Party. In behind-the-scenes maneuvering, the Democratic council and Republican mayor reappointed the chief. Swingley's victory underscored the authority wielded by the insurance community and local merchants, an influence that they would exercise repeatedly on his behalf.[7]

Not more than four years after his battle with Ziegenhein, Swingley entered another protracted debate about departmental leadership that again tested his ability to negotiate among the city's mercantile interests, reform concerns, and the rough world of firemen. This time, however, a less favorable political context complicated his effort to control the department. A reformist Democrat, Rolla Wells, had been elected mayor, negating much of the influence of Swingley's Republican connections. Additionally, the year began badly when, following a season of unusually heavy fire loss, over twenty insurance companies abandoned the city, and the remaining companies raised rates 20 percent. After a particularly disastrous blaze, local insurers and business leaders expressed dismay at the increasing losses and met with the mayor (the aggressive reformer and pro-business Democrat Wells). They expressed concern that the city's firemen were "lying down" on Swingley. According to one account, assistant chief John Barry had directed the city's firemen in a manner that increased fire damage. Moreover, Barry apparently had petitioned local business leaders to work against Swingley's reappointment, and had used the fire to discredit the chief further. Whether true or not, such claims frightened business and insurance leaders as much as firefighters. Everyone knew that politics mattered, but such reports raised a significant question. Had the work of the department been compromised in a struggle for control? Was the city being protected adequately; were firemen's lives in jeopardy? The business community demanded action from the men whom they had supported several years earlier.[8]

Swingley quickly responded with characteristic savvy by reaching out to both the business community and to rank-and-file firefighters. Almost immediately he met with the mayor accompanied by a small group of loyal business supporters that included local insurance leaders. The chief demanded two reforms: enlarging the city's firefighting service and increasing the amount of protection in the commercial district. His advocacy of expanded fire defenses quelled the dissatisfaction of

Truck Company No. 2, Philadelphia, 1900. Fire departments continued to add more ladder or "truck" companies in the early twentieth century. Firefighters themselves helped to develop many new ladder apparatus, such as the eighty-five-foot Hayes/La France aerial ladder truck, which the Philadelphia Fire Department acquired at the century's end. Courtesy, Fireman's Hall Museum, Philadelphia

the city's commercial interests. It also won the support of the St. Louis Fire Prevention Bureau, which offered vocal public advocacy for Swingley's proposals. Equally important, the chief defended the city's firefighters as heroic and brave workers and urged the city to increase the department's budget. Firefighters repaid Swingley's loyalty with unequivocal support. A fireman reported, "There is not a man on the whole department who would not face death rather than disobey Chief Swingley's orders."[9]

With the business community and firemen squarely in his corner, Swingley struck at his political opponents using the nearly unchecked power that fire chiefs held prior to the regularization of department rules. Swingley shook up the department's leadership, effectively removing his political competition and quelling any doubts about his authority. In particular, he removed two assistant chiefs, including John Barry, who it had been rumored was scheming to replace him. Barry, however, speculated that his close relationship with Eugene Gross—whom Swingley had forced to resign after Gross sought to be appointed chief in 1898—led to his discharge. Whatever the reasons for his removal, Barry refused to resign. He demanded intervention by the mayor and city council, questioned the broad license given to Swingley, and speculated that his political affiliation caused Swingley's action. Barry's appeal succeeded. Within days of Barry's discharge, Swingley backtracked. Pressed by Democratic Mayor Rolla Wells, Swingley reappointed Barry. And, in an attempt to stifle criticism, he told local newspapers of his support for a nonpartisan trial board. Both actions temporarily defused the crisis and may well have been a compromise between Swingley and Wells. The end to the controversy restored order to the city by quelling the political dispute and fears about the adequacy of the fire department. In addition, Swingley's capricious power as fire department boss had not become a public issue and he retained, perhaps even expanded, his authority over the department. At the same time, Wells reassured property owners and business leaders that the city had settled any outstanding departmental issues. More importantly, Wells ended a distracting political battle that detracted from his other goals. An erstwhile reformer, he and other reformist Democrats moved to confront a larger issue—the power of Democratic Party boss Edward Butler.[10]

The three-way contest between reformers, machine politicians, and the city's commercial interests did not end, however. It reappeared ten months later when Swingley came up for reappointment, and the battle lines remained familiar. Machine leaders (associated with the Democratic Party) led by Butler expressed dismay at the inability or unwillingness of the pro-business and reformist Democrats, such as Mayor Wells, to replace a Republican appointee (Swingley) with someone more amenable to their interests. Wells's own reformist Democrats—who had termed themselves the Jefferson Club—also questioned the mayor's choice of Swingley. The machine supported the assistant chief, John Barry, and the Jefferson Club urged the appointment of former fireman Thomas Finnerty. Meanwhile, the city's "insurance men and wholesale business interests" supported Swingley as they had previously.[11]

As had happened four years earlier, the department's reform-minded fire chief and the commercial interests won the day. Wells went against the interests of the reformist Jefferson Club and nominated Swingley for reappointment, which the *St. Louis Globe-Democrat* labeled "sound business judgment." Nonetheless, the battle over the reappointment raged for weeks. Jefferson Club reformist Democrats even considered expelling the mayor from membership and temporarily joined with the machine to block Swingley's confirmation, which languished long after the mayor's other nominees were confirmed. The council eventually confirmed the chief, although the explanation for the outcome remains locked in the political backroom. Democrats and Republicans, however, took credit for keeping the department out of politics and igniting reform. Republicans asserted that they initially appointed Swingley, who had introduced merit into the department, and Democrats claimed the high ground for reappointing a Republican, thereby avoiding partisanship and underscoring reformers' arguments that politics should not shape urban services, especially fire protection. Swingley triumphed again.[12]

Even as he survived treacherous political waters, Swingley dramatically transformed the St. Louis Fire Department, often in a manner closely modeled on the IAFE's reformist agenda. For example, the IAFE—which elected Swingley as its president in 1898—advocated adding new equipment, especially truck companies, expanding department size and geographic coverage, as well as budgets. On these and other points, Swingley succeeded, especially when the interests of firefighting professionalism overlapped with those of local merchants in St. Louis. During Swingley's tenure as chief and later as director of public safety, the STLFD experienced its greatest sustained growth, more than doubling in size. By other measures, too, Swingley improved the position of the department. He increased the total number of operating engine companies from forty-one to seventy-one, and expanded the numbers of hook-and-ladder companies dramatically. In fact, the ratio of engine companies to hook-and-ladder companies decreased from 3.6 to 2.6, allowing firefighters to be more effective in venting buildings and rescuing people trapped in upper floors. Swingley also initiated practices designed to diminish fire danger in the city's commercial districts, such as having company foremen and assistant foremen inspect buildings, and, as early as 1895, Swingley himself routinely toured the heart of the city's business district to assess the fire danger and the adequacy of fire defenses.[13]

Likewise, Swingley emphasized training and efficiency in a manner consistent with IAFE recommendations, and according to local newspapers he succeeded in improving departmental discipline. Early in the twentieth century, for instance, most fire departments established formal training methods, but their approach varied widely, from the formal training course used by the NYFD to the less rigid, though equally pervasive, drilling methods developed in St. Louis. Beginning in the 1890s, Swingley initiated a program of drills aimed at increasing firefighters' proficiency. Each morning, the city's fire companies spent one hour practicing hitching horses to the company's apparatus. According to the *St. Louis Post Dispatch*, "after two months of practice the time of company hitching had been reduced from [between] eleven to fifteen seconds to [between] six to nine seconds." Thereafter, during the first two decades of the century companies performed regular exercises, learned scaling techniques, practiced raising ladders, and manipulated hose. Such drills provided a means to exercise horses and initiated new company members into the camaraderie of an engine house's firefighting team. Newspaper accounts of training exercises also depicted firemen acquiring and practicing lifesaving skills and underscored the heroic culture of firefighters in the popular imagination. By 1906, such drilling grew increasingly commonplace in the training towers that were erected behind many of the city's engine houses.[14]

Swingley, in addition, supplemented training procedures with administrative routines that made the department's daily operation more rational. A primary administrative change in the department involved the creation of more elaborate disciplinary procedures, which supported the expansion of departmental rules. For instance, around 1895, he supplemented the departmental ledger with a card file that documented promotions, disciplinary action, and transfers. This process would continue long after Swingley's retirement. By 1933, the department instituted an even more complete card file, which meticulously recorded health and pension information in addition to other career data.

Likewise, Swingley improved discipline by establishing a "court" to try firemen who violated department rules and principles. At least two other departmental officials—such as the assistant chiefs or district chiefs—served as judges at the trials, in which firemen faced their accusers. For first or second convictions, the department levied a fine; a third or fourth offense brought suspension without pay. Typically, the chief discharged a man only after several offenses. Of all the rules, only drunkenness appears to have been unforgivable, at least in those cases in which a man was reported for it. In 1895, the St. Louis Post-Dispatch wrote that Swingley refused to reappoint a man discharged for continued drunkenness. More importantly, though, departmental personnel records reveal that Swingley used the entire range of punishment options—including reprimand, suspension, fine, reduction in rank or pay, or, finally, discharge. Disciplinary action became an ordinary part of firefighters' everyday work lives; those appointed after 1886 experienced, on average, more disciplinary actions than previous cohorts. In part, the sudden increase may have been the product of improved record-keeping procedures, or perhaps resulted from the formal enumeration of departmental employment practices or rules of conduct. Whatever the explanation, the very act of recording disciplinary incidents in a personnel file indicates an expansion of management and administrative procedures, and suggests the regularization of firefighting employment more broadly.[15]

The expanded administrative regimen brought St. Louis's firefighters toward greater efficiency by means of a carrot as well as a stick. The IAFE had long advocated rewarding firefighters' daring with generous benefits, and reformers also supported pensions, better pay, and more regular working conditions. The organization believed that it would improve the quality of recruits, encourage older workers to retire, and perhaps increase the quality of work. In fact advocates of firefighting professionalism believed that such changes would foster departments whose members were younger, more physically robust, better educated, and less susceptible to political corruption. In St. Louis, Swingley followed the lead of his predecessor when he championed such causes before the municipal government, advocating higher pay, better pensions, and more generous work rules as well. After the turn of the century, the department's pension fund increased in size and dispensations became more generous. In addition, pay advanced steadily; rank-and-file firefighters received 10 percent increases and officers were given 25 percent raises. And, although it would not be realized until after Swingley's tenure, his proposals for a second shift of firefighters eventually would reduce firemen's hours (but not their pay) and would increase the numerical strength of the department by about 20 percent.[16]

A relatively high degree of professionalism and bureaucratic rationality gained sway in St. Louis early in the twentieth century, despite the lack of a formal civil service system and the relatively unfettered power of fire chiefs. In fact reform had been taking place for twenty years following in the footsteps of Swingley's predecessor Henry Clay Sexton, who had vested hiring, firing, and disciplinary procedures solely in the office of the chief. In 1880, for instance, Sexton argued that the only way to reform firefighting was to give the fire chief unprecedented authority—mirroring the NAFE's arguments about professionalism. He argued that "members of a Fire Department, more than, perhaps, those of any other public service, should be under the immediate and only control, both in matter of discipline and of being retained in their positions, of the Chief of their Department." Just two years later, at the IAFE's annual convention, a special committee advised that the best way to keep politics out of fire departments would be to "appoint Chiefs of fire departments for life, or during good behavior, and the Chief to appoint members of the department under the same conditions, and the Chiefs and members to be . . . [prohibited] from taking part in politics, except for the right to vote." If such a mode of organization did not itself embody reform, it nonetheless served successive chiefs in St. Louis well, especially as they advanced the IAFE's professional agenda.[17]

Swingley adopted his predecessor's and the IAFE's stance that the chief had undisputed authority over the department. Although he often boasted about the department's administrative rules, many procedures remained ambiguous or unwritten. For instance, the chief told local newspapers that he promoted firefighters when they proved "qualifications as an officer." Swingley deemed a man qualified if he had demonstrated "sobriety, efficiency, and ability" while serving as an "acting" officer. But how did one become an acting officer? What precisely were the attributes of efficiency and ability? Did this refer to demonstrated acts of heroism, particular technical skills, leadership qualities, or simply a commitment to Swingley's reformist agenda? Firemen seeking clear guidelines found few written administrative procedures. Published to support the firemen's pension fund, the *History of the St. Louis Fire Department: With Review of Great Fires, and Sidelights upon the Methods of Fire-Fighting from Ancient to Modern Times, from which the Lesson of the Vast Importance of Having Efficient Firemen May Be Drawn* was likely the primary avenue through which the department socialized new recruits. Nonetheless, the *History* did not discuss specific rules; it only suggested broad policies and ideals.[18]

With or without a written code of conduct, Swingley effectively conveyed a distinct message in the *History*. New recruits and seasoned veterans alike received this missive clearly: the department would not tolerate behavior that ran against the ideals of efficiency and sobriety. The *History* did not need to outline regulations to make this point. At every turn, the narrative adopted the rhetoric of progress and reform. For example, it did not offer a subtle interpretation of the transition from volunteer firefighting to a professional department. To the contrary, it adopted the language of professionalism to explain the change, arguing that professional firefighters replaced volunteers precisely because volunteers had been too rough and too undisciplined. It did not matter that many volunteers had directed the new department and forged many of its work values. No, indeed, the *History* was not written to elucidate the past. It was commissioned—as so many fire department histories were—to guide the process of professionalization in American fire departments early in the twentieth century. Indeed, the *History* affirmed the message of the IAFE when it urged firefighters to innovate and to gain better control over themselves, their tools, and the environment. The *History* buttressed the reformers' belief in fire department progress by supporting the message of forward-thinking chiefs like Swingley. It praised the contemporary department for advocating drills, rational practices, and sobriety, so as to encourage discipline and moral courage.[19]

Fire Chiefs, Manly Risks, and Rational Heroism

The apparent contradictions between Swingley the reformer and politician, and Swingley the heroic fireman illuminates the broader tension between firefighters' occupational culture, which often carried a professional and reformist message, and firefighters' work culture, with its rough-hewn manliness. Reformers applauded Swingley for material reasons (protecting capital and urging partisan "disinterest") as well as cultural reasons. Swingley represented competing definitions of manhood prevalent among fin-de-siècle American men and firemen. On one hand, he embodied bold, almost reckless, heroism; on the other, he stood for the increasingly hegemonic values of efficiency, rationality, and technology. Swingley's political struggles were emblematic of the contradictions inherent in the two sets of beliefs. At the turn of the twentieth century, firefighters' heroic manhood be-

Firemen training in Philadelphia, ca. 1925. In 1913, Philadelphia established a formal training school for firemen, thus introducing common procedures and standard terms for tools in a city where firefighting technique had varied from engine house to engine house. Courtesy, Fireman's Hall Museum, Philadelphia came increasingly complex—and difficult to negotiate. Commitment to service, physical virility, and daring-do did not suffice; firemen were increasingly called upon to demonstrate efficiency and sobriety.

The depth and extent of Swingley's support in local business and reform communities undoubtedly grew not just from his reputation as a reformer, but also from performances as a man. As Swingley attempted to discipline the city's firemen, he embodied the manliness characteristic of the sober, professional middle class. In fact, a feature story in a local newspaper distinguished Swingley from his predecessors using language commonly reserved for middle-class business leaders: "Swingley is methodical, cautious, and reserved—not at all the jovial sort we used to know and lionize, men like Clay Sexton and John Lindsay." With administrative responsibilities akin to those of middle-class reformers and urban businessmen, Swingley did not seek "glory"; in other words, he did not affect the rough-and-ready cultural displays of working-class men. The newspaper also reported, "He has an office and stays at it when he's not at a fire." Strikingly, Swingley worked from an "office"—a symbol of middle-class male work, and he executed his duty in a dependable, "emotionally unflappable" fashion. Even so, Swingley's vitality and active manliness protected him from what many popular commentators described as the greatest danger facing office-bound middle-class men: physical weakness. Tempered by self-control and rational leadership, Swingley's interpretation of firefighting manhood struck a chord deep within a middle class concerned about the changing nature of manliness.[20]

If St. Louis's business leaders and middle-class reformers supported Swingley because he represented values of efficiency, rationality, and technology, the city's working-class and ethnic firemen embraced Swingley because of the abandon and ferocity he demonstrated at fires. Like Phelim O'Toole and other heroes from the department's past, Swingley battled fire close-up, saved lives, and performed hard labor to vanquish fires. Time and again during his tenure as chief, local newspapers reported on Swingley's bravery. In 1901, he saved a fireman's life as he worked his way through a fire scene; on several other occasions, he suffered injuries and barely escaped death, as in the Simmons Hardware fire, in which he was dragged unconscious from the acidic fumes six times. One firefighter described his respect for Swingley: "He rose from the ranks himself and he is therefore just the man who knows how to treat those in the ranks now. He has seen thirty-six years of service and during that time has never had any trouble with any one on the department. Another thing, he will never send one of his men to a part of a building where he himself would not go. And as a rule he usually goes first. We all know he is fearless and all the firemen follow his example." Swingley repeatedly demonstrated moxie and battled the elements alongside rank-and-file firefighters, which endeared him to his men.[21]

Like Swingley, most fire chiefs faced the same hazards as rank-and-file firefighters. Furthermore, chiefs certainly knew about the dangers from department records of the hundreds of firefighters treated by the department's physician. In Philadelphia, a department that kept such extraordinary records, Chief James Baxter nearly perished as he evacuated his men and rescued one firemen from the falling walls of a factory blaze; fourteen firefighters died. In Chicago, James Horan and over twenty other firemen died when the walls of a beef warehouse fell. Though their work was usually neither lethal nor heroic, in the normal course of a day firefighters engaged a world made chaotic by environmental danger; simply entering a flaming structure required courage and the ability to eat smoke. Although fire chiefs may not have used ladders, axes, claw tools, and ceiling hooks along with their men, fire engineers entered deep into buildings, directing firemen, searching for victims, ventilating structures, and drenching fires with water. Choking smoke, chemicals, falling walls, and a growing list of hazards in the unstable and densely built urban environments hampered but did not deter firefighters and their leaders.[22]

To a large degree, fire chiefs embodied the contradictions so inherent in firefighting—the tension between rationality and heroism. Although political considerations continued to play important roles in the selection of fire chiefs, fire service lore almost always reported that chiefs received their positions because of merit and mettle. As a result, to the general public and even to their men, fire chiefs exemplified the complex skills that twentieth-century firefighters used to control the problem of fire. For instance, when the PFD rewarded forty-year veteran and fire chief James Baxter by appointing him fire inspector, local newspapers simultaneously lauded Baxter's judgment, experience, bravery, and connection to his men. The *Telegraph* celebrated his ability to maintain control even during "trying excitement"; he was "never reckless" and "always brave," further noting that he had saved "many thousands of dollars for merchants, which would have been lost under a less experienced mind." Baxter showed both prudence and courage; more importantly, according to once source, "he asked none to go where he would not go himself." The chief suffered more than a dozen injuries, but reportedly never left the fire scene until the blaze was under control; "his bravery was a great magnet around which his men did much splendid work."[23]

Although such accounts have the saccharine taste of hyperbole, such veneration had a basis in Baxter's career, with which local firefighters would have been familiar. Baxter had served with the city's volunteer fire companies in the 1860s and later became one of the first appointees to the newly reorganized municipal fire department in Philadelphia. He had been instrumental in shaping the PFD, helping to transform firefighting from an avocation into a vocation. As chief, he propelled it toward the ethos of professional efficiency long championed by the IAFE. Even so, like other firefighters, Baxter remained rooted in the work culture so present in engine houses, seeking to fight fires up close. Just months before his promotion, Baxter narrowly escaped the falling walls of a factory that killed fourteen Philadelphia firemen. Leading the attack from inside the flaming structure, Baxter suffered injury while rescuing a brother fireman as the wall collapsed around them. After he supervised the removal of the firefighters' bodies, he then led the city's efforts to create a relief fund—eventually over $100,000 was raised—for the families of the firefighters. Philadelphia firemen knew that Baxter came from their ranks, and they shared much in common with him. Like so many fire chiefs and fireman, Baxter straddled the line between being a rational protector of urban order and having what it took to enter burning buildings for a living.[24]

As men everywhere in the nation struggled with changing definitions of manliness, the brotherhood of firefighters possessed and exemplified an almost transcendent vision of what it meant to be a fireman and a man. Although their self-assuredness and extraordinary work differentiated them from their fellow urban dwellers, their stories nonetheless spoke to people. And, in many respects, the labor issues that firefighters faced resembled those faced by other American workers. For instance, firefighters confronted rapidly shifting workplaces—terrains changed by technology and industrial development. They found their work lives affected by new management and administrative technologies, and they were persistent in demanding better terms of employment. Also, firefighters dealt with an increasingly complex array of tasks that ran the gambit from the physical to the technical and evaluative. They defined themselves and their craft in the face of the most basic issues confronting Americans society in the early twentieth century. Thus, perhaps even more so than in the nineteenth century, profiles of firefighting courage provided American men with a model of how to negotiate the contradictions between team efficiency and individual expressiveness, between bureaucratic rationality and physical strength, between prudence and aggressive action.[25]

Civil Service Reform and the Philadelphia Fire Department

As in St. Louis, reform of the PFD occurred at the intersection of the interests of firefighting professionalism, local politics, and insurance capital. Unlike St. Louis, however, Philadelphia implemented a civil service system in 1885, long before most cities. The city's regimen of civil service produced a fire department with more formal written rules and bureaucracy earlier than in the STLFD. Even so, specific reforms implemented in Philadelphia mirrored those in St. Louis in spirit, if not in letter. But, because Philadelphia's municipal government codified so much of its reform program, the department produced a host of records about how fire departments were reorganized during the Progressive Era and how the work of firefighting became more regimented.[26]

Before Rudolph Blankenberg's mayoralty transformed municipal employment in 1912, there had been two previous attempts to build a systematic civil service. In 1885, the Pennsylvania State Legislature passed a new charter for Philadelphia, and reform groups there, such as the Philadelphia Civil Service Reform League and the Committee of One Hundred, proposed legislation to curtail the power of the city council, to strengthen the mayor's office, and to implement merit-based hiring practices. Led by the city's business, academic, and church leaders, reformers obtained a law that included a civil service board charged with instituting testing for job applicants. After the city's reformed charter failed to garner significant change, reformers mobilized once again and secured a bill establishing a revamped civil service commission. Although that legislation would be overhauled twice over the next half century, it provided the basic framework for municipal employment through the 1950s. Managed by three people appointed by the mayor (with no more than two being members of the same political party), it administered competitive examinations, arranged for job placements, and published yearly results. The commission spent its first years creating written exams to test general skills; by 1912 it instituted practical testing to measure actual skills.[27]

After 1906, the civil service commission made the process of gaining employment in the city's burgeoning municipal bureaucracy more rational, but it did not diminish the power of individual departments in matters of hiring, firing, and promotion. In order to gain employment, candidates had to pass an examination. Test dates appeared in public newspapers and were mailed to applicants. To become eligible, candidates requested (in person) and filed an application with the civil service commission. They registered their vital information, as well as the position to which they were applying, at least five days before the examination date. The commission called for examinations at the behest of departments interested in hiring new employees. However, final employment decisions resided solely with municipal departments; the civil service commission had "no jurisdiction whatever over appointments."[28]

The new regimen did not make municipal hiring into a meritocracy. To be sure, the system rewarded those who scored highest on written tests, but it did not guarantee employment or promotion early in the twentieth century. Prospective employees, such as firefighter James Gilbert, appear to have understood this reality, which was laid out in publications available to prospective applicants. Those guides noted that everyone who scored over 70 percent passed the examination, but only the top four names were placed on an "eligible list," which was submitted to the respective departments. The list was "certified" by an "appointing power" that hired one of the four candidates. For the next job opening, the name of the applicant with the next highest score was added to the list. Once again, one of the four candidates on a certified list was hired. If a candidate's name appeared on four consecutive certified lists, it was removed and another name was added. As a result, scoring well on an examination did not guarantee employment, but it did increase the likelihood of being hired. In fact, it was possible that only one of the top four scorers on a test would receive an appointment. By the 1920s, the hiring rules were changed substantively to favor those who scored well, but test performance still did not assure promotion or hiring.[29]

As Philadelphia regularized its employment practices, the city elected a galvanizing force behind municipal reform (especially for professionalizing city services): Mayor Rudolph Blankenberg. Blankenberg's administration made the provision of municipal services more efficient in a fashion consistent with Frederick Winslow Taylor's theories of scientific management. It especially advocated removing political considerations from the provision of municipal services. In 1912, the director of public safety noted, "As in other Bureaus of the Department, a war has been waged against employees with political affiliations and obligations." Although the city banned employees from paying or collecting political assessments, it achieved only marginal success in removing political influence from municipal departments. Nonetheless, Blankenberg and his lieutenants had begun the process of reform, and their efforts augured the direction of change in all areas of city governance, including the fire department. In fact, Philadelphia's Fire Department experienced an extraordinary period of reform between 1912 and 1930, especially of bureaucratic procedures, training regimens, and firefighting tactics.[30]

The process of making Philadelphia's fire department more rationally organized intensified after the Committee of Twenty (of the NBFU) assessed the city's fire conditions in 1911. Finding that the department was organized "fairly well," it urged dramatic improvement in other areas. It argued that supervision was "unsatisfactory," promotion and hiring decisions were made in a "political" manner, discipline was "lax," drills and training were "lacking," and "individual efficiency [was] low." A committee of Philadelphia's Board of Trade examined the NBFU's charges and supported the organization's claims, noting that the city should spend

Fire at Ovo Mill Corporation, St. Louis, 1924. Large crowds watch as St. Louis firefighters battle a blaze at the plant of the Ovo Mill Corporation in 1924. Firefighters inundated the fire, which engulfed the entire structure, with water from both automobile and steam-powered fire engines. Courtesy, *St. Louis Post-Dispatch*

more than $2 million to modernize the department. According to the Board of Trade, Philadelphia spent less per capita on its fire department than seven other leading cities. In addition, the NBFU and the Philadelphia Board of Trade offered specific recommendations for the department's improvement, which meshed with the IAFE's agenda of firefighting professionalism. Almost immediately, the city responded, and the city upgraded the fire department within the context of its ongoing civil service and bureaucratic reform. During these first waves of change in Philadelphia, the number of firefighters grew by nearly 30 percent (between 1905 and 1917). The PFD's manpower again increased in 1918 when it implemented a two-platoon system and hired over nine hundred new firefighters.[31]

At the same time, the Blankenberg administration initiated a program to reduce the city's "fire waste"—a move that both involved changes in firefighting and galvanized the support of firefighters, local businesses, municipal leaders, and insurers. In 1912, to shore up fire defenses, Philadelphia established a fire prevention commission, which advocated prevention efforts and improving fire defenses. The commission warned that "American cities are safe from conflagration only by the smallest margin, and sometimes, by reason of extraordinary conditions, that margin disappears entirely and then the best fire department in personnel and equipment may become helpless." Toward this end, the Philadelphia Fire Prevention Commission championed a variety of reform measures, and worked with the fire department to improve fire defenses. For example, along with the department, the commission made building inspections—performed informally—into a formal job responsibility of firefighters. As firemen inspected their neighborhoods, they became agents of safety, making fire prevention recommendations. The fire prevention commission also hoped to improve the department's efficiency by encouraging the purchase of motor-driven fire apparatus, which drew upon the advice of the NBFU and firefighting professionals. The IAFE argued that "automobile fire engines" were more efficient and equally reliable, decreased response time, required fewer men to operate, and were less expensive than horses. Although firefighters were loath to give up their horses, whose service they celebrated, they adopted the new technology, continuing to prioritize efficiency and innovation. The PFD purchased automobile engines beginning in the 1910s, and like other departments it procured the new technology as fast as manufacturers could produce them.[32]

As an unlikely coalition remade fire departments, another new voice entered discussions about firefighting reform early in the twentieth century when firefighters in cities across the nation began to form unions. By 1918, rank-and-file firefighters' influence increased substantially as formerly independent efforts at unionization culminated with the formation of the International Association of Fire Fighters (IAFF), which counted about one quarter of the nation's forty thousand firemen as members. Initially, this act of formalizing the brotherhood of firemen into a labor union met with some consternation at IAFE conventions. However, fire engineers' fears about work stoppages eventually dissipated and they disappeared altogether in the 1930s, after the IAFF amended its constitution to include a no-strike pledge. Although issues surrounding workplace autonomy and control remained, the IAFF also lined up with fire engineers' recommendations about professionalism. Ultimately, then, the IAFF did not develop in opposition to the IAFE, but rather it became an alternative avenue for promoting the interests of firefighters. The organization served to amplify firefighters' voices, and it provided firemen with additional political pull, which became especially critical in a century with increasingly formal negotiating environments and interest group politics.[33]

This move for unionization shaped efforts to reform the Philadelphia Fire Department, as firefighters chartered Local 22 and joined their leadership in pressing for better working conditions. In fact, Philadelphia firefighters initiated the union at a time when Philadelphia Fire Chief Engineer William Murphy had organized a "committee on increase in pay and shorter working hours." Murphy drew upon their numbers when he used the incipient bureaucracy to mobilize firefighters, engine house by engine house, to sign written petitions that were sent to the city council, demanding both higher wages and shorter hours, via the implementation of a "two-platoon" system. This became a popular reform ideal in the 1910s and was promoted by the IAFE and NBFU on the grounds that it improved fire department efficiency. Firefighters favored the double shifts because usually their salaries remained constant but their work hours were reduced significantly, sometimes by as much as half. Political leaders, however, were ambivalent about this change because of the costs involved in financing fire department expansion. The support of Local 22 turned the tide in the drive for a second platoon in Philadelphia. After the city dragged its feet over implementing the changes, firefighters used their new organization to compel the city to comply with the state legislation that had mandated the reform. As a result, by Christmas 1918, Philadelphia firefighters worked shorter hours and received a 10 percent increase in their wages.[34]

Concurrent with the expansion of civil service the department developed new methods of personnel management, revised work rules, and established administrative routines—all of which became powerful tools for change. In 1912 the director of public safety updated the fire department's "entire system of records and forms . . . to meet the demands of an increased and growing force." Through this new regimen of communications, programs of reform had a direct impact on the city's firemen. For instance, by using "General Orders" that were typed, reproduced, and distributed to all fire companies, the chief engineer took an important step toward making life and work practices more standard across engine houses. From 1916 through 1919, Chief Engineer William Murphy issued over four hundred general orders on a range of topics related to company management, departmental organization, and personnel practices. Murphy issued new regulations and provided information to firefighters about the fire service. He and later chiefs created and continued to refine procedures for updating personnel data; chiefs delivered instructions on caring for hose and maintaining and operating new technological apparatus, and they recommended changes in firefighting tactics. Additionally, general orders clarified company responsibilities when alarms came in from fire boxes, thereby diminishing confusion at fire scenes.[35]

By the 1920s, this bureaucratic regimen intensified and a had a more direct impact on firefighters as the PFD implemented systematic reporting methods. The department created more than sixty different reports for recording information about company and firefighter performance and communicating that to headquarters. Company leaders filled out this myriad of daily, monthly, and annual reports on several different forms, sometimes submitted in duplicate or triplicate. By the 1940s, the department even listed which officers should receive which of the growing number of forms. To help firemen keep tabs on the expanding volume of information, the department continuously provided company leaders with instruction on which forms to use in different circumstance, and published a guide to completing them properly. As standard procedures gained momentum, companies and firefighters came under increasing scrutiny from a leadership interested in managing resources more effectively and in developing more coordinated firefighting strategies. The department's administrative procedures gradually reduced the differences in the informal culture of the city's engine houses, and departmental leaders acquired more control over firefighting operations.[36]

Training Firemen's Bodies and Standardizing the Heroic

During the twentieth century, firefighters also made systematic training a foundation of their professionalism. The IAFE and firefighting leaders developed and disseminated standard tactics and work routines designed to create common practices in all fire departments, whether volunteer or paid. In 1913, over two decades after the occupation's first sanctioned guide had been circulated, one-time IAFE president and New York City fire chief John Kenlon published a history of fire department organization and training modeled after the New York Fire Department. In 1920, *Fire & Water Engineering* published *The New York Fire College Course*, and the IAFE held a focused discussion of practice and training at its annual meeting, which emphasized the elements of the New York Fire College Course. Using its guidelines, the IAFE especially urged departments to establish formal training schools to familiarize firefighters with the latest innovations in firefighting, such as motor-guided apparatus, scaling techniques, and breathing equipment. As a result of the IAFE's efforts, city after city established fire schools early in the twentieth century. Not only did such academies become common by the 1930s, but also volunteer firefighters had begun to lose their amateur status and to qualify as professional firefighters, because they too underwent extensive training. A 1917 Department of Commerce report on the conditions of fire protection in the United States underscored the importance of training schools in making firefighters into specialists. It noted that firefighting had become far more complex in dangerous urban centers, with high buildings and electrical wires, and that it had grown increasingly technical. Moreover, the report echoed the message of the IAFE, emphasizing training as the basis for firefighting professionalism. It argued that "the largest degree of cooperation between the fire departments of the cities and educational institutions would be in keeping with the spirit of the age, which calls for both specialization and breadth of training as essential to the highest efficiency."[37]

Like other departments, the PFD took up the IAFE's agenda and sought to make firefighting standard by implementing systematic training. In 1914, the department issued an administrative manual, codified the job responsibilities of firemen, and established a training school. Even the most mundane aspects of firefighting became subject to this discipline. According to both James Gilbert and the director of public safety, this remedied a long-standing problem with the city's fire protection; there had been confusion at fires because many firemen did not understand the meaning of orders given by their superiors, in part because there had been no standard nomenclature for tools and equipment. Reformers believed that standard training made firemen more effective; the director of public safety boasted, "New Men are better equipped for their duties, following their course at the training school, and when they arrive at the fire ground are no longer greenhorns who are more apt to be in the way than to be of assistance."[38]

An ongoing conversation among the service professionals divided and subdivided firefighting work into multiple individual tasks, which were taught at training schools. For instance, as the Los Angeles Fire Department Fire College curriculum reveals, firefighters had separated their work into two thousand separate tasks by 1935. In addition, new strategies and approaches appeared in the occupation's leading journal, *Fire and Water Engineering*, and at the IAFE's annual meeting. Reformers typically adopted these training and organization suggestions quite literally, an approach taken during the reform of the PFD. For instance, in 1913 Philadelphia's director of public safety reported that the fire departments in New York and St. Louis required that firefighters hook into building sprinkler systems as a means to quash blazes in the commercial district. The director noted that although Philadelphia was "filled with sprinkler systems, there has been no such connection on the part of the firemen for years. We unqualifiedly recommend that such an order be issued at once." The expansion of firefighting professionalism, then, developed in the context of a national conversation, as the IAFE set the agenda for training, organization, and administration in departments.[39]

When the Philadelphia Bureau of Fire identified the general organization of, and specific functions performed by, rank-and-file firefighters, it took a major step toward making firefighting practice more efficient and more routine. As the department organized a training school and published *Information for Firefighters*, job responsibilities became codified along much the same lines as the professional mission put forth by the IAFE. In 1920, for instance, the city's compendium of job classifications defined firefighting in terms of extinguishing fire, saving lives, and preserving property, as well as assorted other tasks:

> Placing ladders; operating hose; guiding a fire truck on the streets from the rear seat; operating the hand chemical apparatus; assisting in saving lives; doing salvage work at fires or after fires are extinguished; on occasion acting as driver or chauffeur for a horse-drawn or motor-operated piece of fire apparatus; cleaning windows and floors; polishing bright work; washing wagons; drying hose and drying and cleaning covers; taking care of horses; making adjustments and minor repairs to harness; doing floor duty during an assigned portion of the day or night; connecting and disconnecting hose and turning water on and off; giving instructions to owners and occupants of buildings in the district as to fire hazards and fire prevention and making reports as to bad conditions.

In addition, the rules made it clear that participating in drills and attending regular training sessions also were included in every firefighter's duties. The job description further placed firefighters within a hierarchical chain of command under the direction of the fire chief, the director of public safety, and the mayor. By the late 1920s firefighters reported to superior officers, who were themselves receiving specialized training in the skills of diagnosing the best manner to extinguish any fire.[40]

The PFD organized firefighting into several primary components used throughout the fire service in the 1920s. Philadelphia firemen typically worked in engine companies or truck companies—the two foundations of departmental organization. Hosemen or pipemen working in truck companies established connections to urban water systems and commanded hose streams. Laddermen serving in truck companies raised ladders and ventilated structures, allowing smoke and gas to escape. Far less frequently, firemen worked as members of service companies organized around specialized equipment or tactics, including water tower companies, rescue squads, fireboats, or in other support services, such as the machine shop. The work activities of firefighters—whatever their company—revolved around several primary activities, including: the use of tools to "open up" or "ventilate" a building and to combat special hazards in the landscape, the use of hose to extinguish a fire, and the use of ladders to gain access to upper floors and to perform rescues. Each of these tasks required further knowledge of a number of particular tools and specialized work routines. For example, at the Philadelphia Fire Train-

Fire at Mill Waste Company, Philadelphia, 1945. Philadelphia firefighters scurry away from a falling wall while fighting a five-alarm blaze in 1945. The scene illustrates well one of the most profound dangers faced by firefighters. Courtesy, Fireman's Hall Museum, Philadelphia

ing School firemen learned and then drilled on eleven different "rescue exercises," each involving several different positions or movements. Similarly, the fire department provided examples of "hose line exercises" designed to make firemen proficient using basic equipment such as "stretching in 2½ inch hose lines; taking lines up ladders and the use of ladder straps, and hose rollers; the care and protection of hose on the streets; the removal of burst lengths of hose from various points in a line which has been stretched up and over a building." Just as firefighters acquired increasingly specific instruction in work techniques, so too they received more and more detailed training in the tactics of battling blazes in diverse urban environments; different classifications of buildings and situations increasingly generated different conditions and hazards, requiring varied approaches to fire extinction. To remain familiar with this changing environment, and to mitigate its hazards, the IAFE continued to recommend inspection as an important element of firefighting, a tactic that the PFD embraced.[41]

Fire service experts codified body motions and work strategies using techniques and pedagogical strategies reminiscent of industrial efficiency experts. For example, in 1920 the *New York Fire College Course* and the IAFE described the basic components of firefighting work in abundant diagrams, photographs, and text. About the same time, instructors at Philadelphia's training school divided basic skills, such as hose or ladder work, into specific component techniques. These activities were further broken down into motions and movements that instructors demonstrated to firefighters. For example, firemen learned how to use hook ladders (pompier-style ladders), 18-foot, 20-foot, 25-foot, and 35-foot ladders. Instructors subdivided use of each type into simple procedures, such as raising, extending, and positioning them. In the PFD's training materials approximately a dozen photographs depicted each of the motions needed to become proficient with each type of ladder. After their lessons, firefighters were tested on the technical proficiency and speed with which they performed each task. Between 1922 and 1925, over 250 firefighters—all of the city's recruits—attended the Philadelphia Fire Training School, and by 1940 over 3,000 firemen, most of them rookies, had graduated from the academy.[42]

In addition to making work procedures more regular, training schools also introduced firefighters to the occupation's customs and lore, and they socialized each generation of firefighters entering departments after the 1920s and 1930s. In fact, transforming training into a formal procedure, by taking it out of engine houses and removing it to special schools, had a profound impact on life in engine houses. If academies disciplined the bodies of individual firemen and molded them into a team unit, they also furthered the development of a common occupational identity in other subtle ways. For instance, firefighters learned the importance of efficiency and sobriety. In addition, they also developed their principal loyalty not to the men of their engine houses but to their fellow firemen more generally. In 1922, the PFD taught its forty-second training class this lesson in a somber fashion. The young men were "instructed and exercised in formation for a funeral detail." At the funeral, the training class escorted the fallen fireman along with his engine company. Thus recruits learned what it meant to be part of the brotherhood of firefighting and at the same time came face-to-face with the real dangers of their new occupation.[43]

About this time, beginning in the 1890s and continuing through the 1940s, firefighters wrote the histories of their fire departments, and in the process interpreted the history of firefighting according to the doctrine of efficiency and professionalism. For instance, Philadelphia's manual *Information for Firefighters*, like the *History of the St. Louis Fire Department*, placed the development of firefighting as an occupation within a historical framework that emphasized continual progress toward an increasingly rational organization of firefighting. The volume included a brief departmental history that began with an anecdote about the superiority of steam engines to hand engines. This narrative explicitly placed the firefighting techniques of twentieth-century firefighters along a continuum that began with the advent of steam technology. To distinguish twentieth-century firefighters from their forebears, the story explained that firefighting had once been the purview of ruffians who set fires for the sport of extinguishing them. According to the manual, until the advent of steam technology, bands of undisciplined volunteer firemen threatened the stability of the city. Steam engines and a small cadre of professional firefighters had transformed firefighting service, making it more effective. Through digesting this pithy narrative—with its emphasis on progress and efficiency—firefighters came to perceive themselves as a vanguard leading the way toward the future. Although it is not clear to what degree firefighters internalized such histories, these narratives gradually became part of the broader historical record. Elements of this story cropped up in popular writing about the history of firefighting, and even in the interpretive scholarship on the subject.[44]

Expanded record-keeping and bureaucratic procedures also supported the mission of the training academy. The increasing and voluminous communication helped to facilitate the spread of common firefighting methods and a common culture within the department. Fire department leaders recognized that if they wanted to make firefighting standard, then they would have to make training an ongoing activity in engine houses. Toward this end, the PFD demanded that companies practice various techniques during roll call. Specifically, the department recommended that firefighters review a host of basic skills, such as "knots & hitches," "physical exercise," "hoisting ladders," and "tools & their uses." It also urged discussion of more advanced methods and broader firefighting strategies: "gas mask exercises," "ventilation of buildings," removing people from burning structures, improvising a "cellar pipe" to fight smoky basement fires, and various questions about the use of hose, hydrants, and nozzles for the best water pressure. Finally, the department asked companies to review basic aspects of department culture, including the "funeral drill," "receiving officers," and answering questions for the proficiency exams, which undoubtedly would have included knowledge of departmental history.[45]

Companies also sent a steady flow of reports back to headquarters, which underscored the department's efforts to inculcate a common culture. Indeed, the avalanche of paper and forms made the work of each company leader and each firefighter subject to scrutiny by providing a window into the daily life and work of the company. Much of the flood tended to mundane personnel matters: address changes and updates, resignations, vacation time, sick days, and injury reports (under newly enacted worker's compensation laws). Others documents gave leaders methods for managing insubordinate firefighters, such as recording misconduct, including intoxication. To the degree that such methods of personnel monitoring and record-keeping reminded firefighters of their obligations as workers and firemen, they also formally made firefighters members in the city's and the nation's firefighting brotherhood.[46]

Likewise, the PFD disseminated orders about the best methods for attacking a fire, proper maintenance of equipment, and departmental organization, and it also required that fire companies document their work in the field. By the mid-1920s, fire companies provided detailed accounts of the specific tasks they performed during each call. Each time the fire department received an alarm, a fire company responded and investigated the situation; even if firefighters encountered no blaze, they inspected the site and returned a report to headquarters. Returning these two forms—a "Report of False Alarm" and "Defection of Buildings—advanced significantly efforts to make the built environment less susceptible to fire. Not only did the PFD use its inspections to push for the construction of fire escapes, but Fire Chief Ross Davis used them to create a "fire-spotting" map, to identify hazards and to prevent conflagrations in especially dangerous or congested districts—long a priority of the department's allies in the insurance industry. During the late 1920s, this type of inspection activity even became part of the battle to enforce prohibition.[47]

Communications between headquarters and fire companies especially helped to refine firefighting techniques and to make them standard across the city. After each fire, fire companies returned a comprehensive written description of the call. The account included plenty of mundane descriptive information: the company and platoon, the names of the men who fought the blaze, as well as the commanding officer, the time, date, duration, and location of the call, how the alarm came to the company and at what alarm box it originated, and the number of alarms sounded. In addition, it included information about the built landscape, including: data on the fire hydrant, the pressure at the hydrant at the beginning and at the end of the fire, and a detailed description of the building. Finally, the report asked the company officer to report on the strategies, methods, and equipment used to extinguish the blaze. For instance, in April 1927, Engine Company No. 56 recorded its 101st alarm during the year when it spent an hour extinguishing a fire at a two-story brick building. The company stretched over eight hundred feet of hose with several different types of nozzles to extinguish the cellar fire. Likewise, in April 1939, Truck Company No. 29 answered its 190th call of the year and spent over two hours bringing a fire under control, and then extinguishing it. Judging from the report, the company followed usual procedures; it reported that the commanding officer "ordered to ventilate the entire building, after which [he] ordered to stretch 2½ [inch] water line to the 2nd floor via stairway, after which [he] ordered to attach 1½ [inch] line to 2½ hosepipe to wet down and do general truck work." The company used "4 fire axes, 2 claw tools, 2 6-foot ceiling hooks, 2 12-foot ceiling hooks, 1 8-foot & 20-foot portable truss ladder, and 1 18-foot & 28-foot solid bean portable ladder" in the process of fighting a fire.[48]

The process of recording, writing out, and then submitting such detailed information greatly assisted in the process of training firefighters in a variety of ways. At a most basic level, the forms produced a record of departmental practice that could later be reviewed. Whether or not such reports were carefully scrutinized matters little; the possibility of such assessment forced officers to follow procedures as laid out in manuals, at the fire training school, or by general orders coming from headquarters. This allowed the chief engineer to assess whether their directives were being followed. In 1927, for instance, Davis ordered companies to make better use of basic tools—axes, claw tools, and ceiling hooks, perhaps as a response to reports returned to headquarters. Ross also offered advice on the best methods for stretching hose line to improve water pressure (i.e., by selecting the hydrants closest to a fire), and instructed companies on how to use the "Ross Thawing Device" to keep water flowing from hydrants during winter. Several years later, in the 1930s, the department modified its forms, asking for more detailed information about company actions at fires, especially regarding their use of basic firefighting tools. Additionally, the process of writing reports encouraged company leaders to reflect on their work and the methods used. Ultimately, this reporting system helped to change firefighting; it placed firefighters under a behavioral discipline similar to the values that fire underwriters had begun to encourage in the general population. Firefighters, like the built environment they protected, were becoming standard.[49]

The new administrative and training regimens appear to have had a contradictory effect on five departments. On the one hand, the erstwhile rogue work cultures in fire engine houses came under the framework of a common occupation. On the other hand, by creating guidelines for firefighters' work, dividing it into particular tasks, and teaching it in schools, fire departments threatened to demystify the almost mythical elements of firemen's life-saving profession. Not only had firefighting been transformed into a standard job, with regular and often banal work responsibilities, but even heroism became subject to the department's reporting system. In Philadelphia, for example, officers recorded deeds of "meritorious service" on report number 25, which was completed in duplicate on a standard form. In 1926, Captain Harry Jones reported that ladderman Rockhart of Truck Company No. 2 saved two trapped citizens from a third floor "front room," by "tying rope around chimney, sliding rope to window, entering window, lower[ing] person with rope, and slid[ing] same himself." The account also credited two of Rockhart's fellow firefighters with assisting in the rescue. However, when these sorts of written descriptions adopted the self-effacing style with which firefighters often reported such gallantry, they made heroism seem mundane. Certainly, detailing rescues does not mean that firefighters became less dashing or that the press stopped reporting the incidents. Rather, the reports captured a subtle shift in how firefighters and the press created narratives of firefighting. Accounts of valor grew less dramatic; they emphasized that firefighters' extensive training had enabled them to be heroic. Ironically, as firefighters emphasized and became more proficient at performing rescues, training for and formally recording them seemed to diminish some of the élan of firefighters.[50]

Reconstructing Firefighters' Careers—A Quantitative Analysis

As departments and firefighters reformed and disciplined themselves, careers became more standard and increasingly longer—an effect that manifested itself in the personnel files of the Philadelphia and St. Louis fire departments. Not only did careers increase in length, on average, but firefighters increasingly chose the time and reason for their final exit from service. A growing similarity in career patterns suggests that the message of professionalism profoundly shaped firefighting as an occupation. Although difficult to pinpoint, a combination of factors led to the standardization of firefighting careers. More formal work rules, better wages, more generous pensions, and improved training all contributed to the shifting work patterns. In addition, the IAFE's advocacy of professionalism facilitated the development of a common occupational identity among firefighters, as did the support of underwriters and Progressive Era reforms more broadly.[51]

Generally speaking, several major changes occurred in firefighters' employment in the century following the Civil War. The length of firefighters' careers serves as the starting point for any analysis of the occupation. With each passing decade, firefighters worked for a longer period of time in both St. Louis and Philadelphia. The lengthening of careers suggests that firemen exercised a remarkable degree of

Fire in Race Street apartment, Philadelphia, 1946. Philadelphia firefighters quickly ascend ladders to ventilate the building, locate possible victims, and play water on the fire at this apartment building blaze in 1946. Although firefighting work had been codified into instructional guides and training schools, every fire posed different challenges. Courtesy, Fireman's Hall Museum, Philadelphia

control over the terms by which their employment would end. This is confirmed by the fact that increasingly firefighters could chose to retire or resign from the department, as opposed to being dismissed or working until they died. At the same time, firefighters' career patterns became less variable within particular departments, and patterns between departments converged, though not completely. In addition, the amount of bureaucratic activity experienced by firefighters, such as disciplinary procedures or transfers between houses, increased over time. Taken together, changes in career rhythms suggest that firefighters' choices became more and more structured and thus more similar.

The degree to which firefighters increasingly determined how they would exit the fire department is remarkable. In St. Louis, firefighters appointed in the first decade of the twentieth century left the department by dismissal about 18 percent of the time. Of firemen appointed a decade later, in the 1910s, only 11 percent were fired. And, finally, of cohorts appointed after 1930, less than 8 percent of firefighters left because they were dismissed. Concomitantly, the number of men who resigned from the department increased with each successive cohort. Of the men appointed in the 1940s, few worked until they died (6.5 percent), and most followed the rules and avoided being discharged (4 percent). The vast majority, nearly 90 percent, chose the time and manner of their exit. Similar patterns also became the norm in Philadelphia—especially when compared to the work experiences of firefighters in the nineteenth century. Of the men who joined the PFD after 1910, over two-thirds left of their own volition, and only about 10 percent were discharged. Except for the striking proportion of Philadelphia firefighters who worked until they died—over 20 percent—career patterns of men working in the PFD and the STLFD were converging.[52]

Such overwhelming and consistent data underscores the increasing benefits of firefighting employment, especially improvements in the working conditions of firefighters across the nation. Moreover, when compared to other jobs in the American economy, working as a firefighter carried with it many attractive conditions. For instance, even though firefighters did not earn especially high salaries, working for the municipal government, and in such a critical occupation, buffered them from the vagaries of economic instability, especially layoffs and unemployment resulting from recession, as Alexander Keyssar points out in his study of unemployment. In fact, many workers left better paying positions for spots on urban police and fire departments, and census records confirm that firefighters and other municipal employees suffered less unemployment than most of their contemporaries from the nineteenth into the early twentieth century.[53]

Undoubtedly, firefighters' increasing control over the conditions of their employment, especially the terms on which they exited the fire service, contributed to this remarkable record. Consider, for instance, the Philadelphia labor market, as a point of comparison. Between 1870 and 1936 approximately 38 percent of all workers left their jobs involuntarily in Philadelphia. Of course, this data covers workers in the region's entire labor market, but it sets a context against which to judge firefighters' career patterns. Alternately, the situation of workers in a single industry or corporation might offer a better comparison. Although such data is difficult to come by, workers who labored in Amoskeag's textile mills between 1912 and 1921 involuntarily left their positions over 34 percent of the time, which *in-*

creased over time. Finally, the best comparison might be to contrast the experience of twentieth-century firemen to their nineteenth-century colleagues. In the nineteenth century, firefighters in St. Louis and Philadelphia, on average, were discharged over 30 percent of the time—a figure consistent with the experience of other industrial workers. By contrast, firefighters who entered the STLFD or the PFD after 1910 could expect to be dismissed infrequently, not more than 10 percent of the time. During the twentieth century, then, the careers of firefighters became more similar, even as they began to differ more and more from the work lives of their neighbors.[54]

Administrative reform fostered the creation of this common occupational experience and identity in St. Louis. The STLFD personnel files document the degree to which formal bureaucratic rules replaced informal procedures. Take, for instance, the administrative practice of reducing firemen in rank. Relatively common in the nineteenth century, this practice decreased over time. Examples of "reduction" in rank abound in the personnel files and were occasionally reported in the local media. For instance, in 1899 Chief Swingley reduced several firemen "due to charges brought against the men which, to his mind, impaired their efficiency in the department. . . . Both of these men admitted their inability to do all the work required of their positions." Such reductions, coming as they did on the heels of the controversy surrounding Swingley's reappointment, might have been punitive, or the men did not fit the department's work culture. Either explanation would seem to fit the profile of D. R. Rowe, who was "reduced" by Swingley. Appointed at age thirty-four in 1889, Rowe was only forty-four when Swingley reduced his rank; four years later Rowe resigned from the department.[55]

However, reducing a firefighter's rank may have had a functional purpose; it might have been both a form of humanitarian aid to a brother fireman and a way to upgrade a department's firefighting capabilities. Fire chiefs appear to have used the practice to retain the experience of long-serving firemen without endangering their lives or the lives of their fellow firemen. At the same time, keeping an old-timer on the payroll could have provided a surrogate pension to men with unusual ability, need, or circumstances at a time when such pensions were neither generous nor widely available. Indeed, this practice was consistent with the proposals of the IAFE and other proponents of firefighting professionalism that recommend improving or establishing pensions as a way of improving departments. Perhaps it is not surprising, then, that most reductions appear to have occurred toward the end of firemen's careers and involved demoting a man to the position of "watch." This was the case of the second firefighter reduced by Swingley in 1899. Appointed in 1871 at age twenty-four, William Connors served the department for over twenty-eight years and was fifty-two years old when reduced. However, unlike Rowe, Connors's employment card notes that he was reduced to watchman. As a watchmen Connors would no longer be expected to perform active physical labor but would serve the critical function of operating the company's fire alarm telegraph, which meant staying awake at night, maintaining the equipment, interpreting telegraph messages, and directing the company to fires. After his demotion, Connors served another nineteen years in the department, resigning after nearly fifty years of continuous service.[56]

Seen in this light, the practice of reduction may have benefited fire departments as well as the men affected, especially in the nineteenth century. Reductions appear to have been an informal mechanism through which the department retained experienced firemen. Before fire departments had formal training facilities, keeping old-timers on the force also served socialization purposes, such as acquainting new recruits with the department's culture. Unfortunately, personnel files do not say why some men were reduced at old age and others were not. They do reveal, though, the prevalence of the practice in the nineteenth century. In St. Louis, for instance, approximately one in six firemen could expect a reduction in rank if he was appointed in the 1860s or 1870s, and the likelihood of a reduction increased to just under one in four for those appointed in the 1880s. About one in three of those men appointed in the 1890s could expect to be reduced in rank. At a time of limited disability or retirement pensions, firemen reduced to watch retained their income, comrades, and culture. In turn, their experience and skill continued to benefit and serve the city's fire service.

However, as employment conditions in the STLFD became more standard, the practice of reducing firefighters in rank disappeared. After peaking for firemen appointed in the 1890s, the proportion of firefighters who were reduced decreased with each successive cohort. Few firefighters appointed to the department after 1920 experienced a reduction in rank. Several reforms, concurrently advocated by the IAFE, explain this. First, the advent of more generous pensions, both for retirees and those who had been injured on the job, allowed the department to replace older firemen with younger more robust men without abandoning the old-timers to poverty. Second, the codification and dissemination of firefighting expertise into written documents made the information possessed by experienced firefighters widely available, reducing their critical training role in the department. Similarly, by removing older firefighters, departmental leaders also eliminated the informal cultural lessons taught by them—training that sometimes emphasized the particular work community over common departmental identity. Lastly, the implementation of the disciplinary court replaced reduction as a punitive measure with a range of bureaucratic procedures and punishments.

When departments established work rules and training procedures, they helped to foster a common occupational identity, but they also changed the demographic profile of their fire departments. In St. Louis, for example, the department gradually became younger, and men left the department at an increasingly young age. Although, over time, firemen left the department at more or less the same rate, each successive cohort retired at a younger age. The distribution of firefighters' age at final exit demonstrates this trend strikingly, revealing that approximately 66 percent of all firefighters who entered the department between 1930 and 1950, exited before their sixtieth birthday. Work conditions and occupational identity had once prompted firefighters to toil until they died or were discharged. However, by the 1950s, though firemen continued to dedicate a remarkable portion of their lives to firefighting, they began to temper their devotion to that calling. Work conditions had improved dramatically, and firemen increasingly retired while still relatively young to pursue other opportunities—or a new possibility being afforded to many Americans, retirement.[57]

More striking, perhaps, is the degree to which firefighters' careers in St. Louis grew more patterned over time. The average length of service remained more or less constant for firemen appointed between 1857 and 1950. However, the proportion of firemen serving long (or extremely short) careers decreased dramatically. After peaking at 29 percent among firefighters who began careers during the 1890s, the number who served careers longer than forty years decreased with each successive cohort. The does not mean that firefighters became less enamored of their jobs or benefits. Indeed, a remarkable number of firemen, in all cohorts, continued to work for more than thirty years, and each successive cohort produced more and more twenty-year veterans. This peaked among the cohort that entered during the 1940s; two-thirds of new recruits had careers longer than twenty years. The data suggest that, although firefighters remained devoted to their occupation, they grew increasingly cognizant of the benefits waiting at retirement.

Recruits who joined fire departments after 1945 entered an environment that would have been unrecognizable to firemen of the mid–nineteenth century. Not only would they have learned their skills and regimen at a training academy, but firemen likely would have found that the "bully-boy" culture was less pronounced, though not lost altogether. Working under more stringent and standard conditions circumscribed differences between engine houses and between firefighters, paradoxically helping to foster a common identity but perhaps muting its expressiveness. Indeed, from the inside, many firefighters might have mused that their lives seemed more banal and regulated than heroic. Even so, firefighters would have recognized the benefits of reform—a good salary, an excellent pension, and the respect of their neighbors—and welcomed the choices afforded by them. Nonetheless, firemen's careers and work lives became more standard during the twentieth century.

Rescue Squads and Efficient Heroism

Even as work routines became increasingly standard, firefighters continued to invent new ways of attacking the ever-changing problem of fire. They not only developed more regimented training but demanded that departmental leaders become responsible for developing a much deeper understanding of how firefighting tactics varied in different situations. In 1920, the *New York Fire College Course* provided meticulous instruction about the chemistry of fire, explaining how it reacted to different fuels and conditions. These lessons also translated into particular training efforts for officers, and eventually resulted in the development of a greater complement of firefighting specialists. Likewise, by 1930 Philadelphia instituted a training course for officers that distilled the latest in firefighting tactics. Beyond the mounting number of bureaucratic procedures, officers' training included detailed instruction in the chemistry of fire, special toxic hazards, such as ammonia in cooling units, the city's electrical grid, and strategies for firefighting in many particular situations—such as cellar fires. The officers' training embodied the IAFE's explicit recognition of the multiple fire environments present in an urban setting. In their response to the changing conditions, the IAFE and professional leaders recommended that departments establish squads of specialists trained in the art of "modern" rescue—a twentieth-century variant of the nineteenth-century pompier corps.[58]

Rescue squads allowed firefighters to reconstruct their occupation within the context of the modern city—according to new arrangements of risk brought by technological change. Developed within weeks of a 1915 New York City subway fire that shook that city, the origins of rescue squads more generally represented firefighters' concerns about the increasingly dangerous twentieth-century landscape. And, as early as 1920, the IAFE recommended that departments create rescue squads to explore "buildings, subways and similar locations heavily charged with smoke, dangerous gases and vapors." Departments specifically organized the squads to perform basic firefighting operations in the most extreme conditions, such as reaching "the seat of fire in difficult locations" or "entering and ventilating chemical plants and other establishments, where dangerous vapors or fumes are being generated." To help firefighters confront toxic gases or a "deficiency of oxygen," breathing apparatus became standard equipment on rescue squads, along with an especially prominent emphasis on working in two-person teams. Similarly, rescue squads represented a direct response to new conditions in recovering bodies, resuscitating victims, and rescuing people, especially those entombed by other new dangers, which included train wrecks, trolley wrecks, automobile wrecks, collapses of ever-taller buildings, and elevator accidents. Additionally, they represented a comment on new construction methods, such as the use of concrete. Impervious to axes and trapping superheated and noxious gases, concrete buildings made extinguishing blazes more difficult and dangerous. According to Philadelphia chief Ross Davis, "It is necessary to have a special squad with the newest sort of apparatus to fight the modern fire. Buildings are made stronger and are, therefore, more difficult to tear down when necessity arises." Once again, firefighters re-created heroism by developing a specialized organization to deal with the increasing fire danger. By the 1930s, following the lead of the IAFE and experiments in New York, Chicago, and Philadelphia, departments began to implement this "adjunct to modern firefighting."[59]

Rescue squads signaled an intensification of the role that technology and training played in fire departments. Like New York, Philadelphia in 1926 formed such a crew of elite firemen, whose virtuosity and skill at saving lives represented a shifting definition of firefighting heroism. Three years after its formation, the *Philadelphia Evening Bulletin* effused: "Rescue Squad Takes Extra Risks; Specially Picked and Trained Experts Fight Fire Through Gas and Concrete; Innovation a Success." In emphasizing equipment, tactics, and training, the department constructed the rescue squad according to the tenets of twentieth-century firefighting professionalism. When forming the rescue squad, Chief Davis "hand-selected" the department's best men, evidenced by their excellent performance at training school. He even selected a hero—Lieutenant Alfred Broadbent—to lead the group. In addition, Davis chose firefighters with previous building construction experience—those who had formerly held jobs as welders, plumbers, or builders. Enumerating the squad's many technological gadgets reinforced the impression that the men of the rescue squad represented the highest level of efficiency, skill, and firefighting professionalism. The squad represented a new and more intense connection between firemen and their equipment. Men and machine came together into a single efficient technological instrument. One physician provided a testimonial to the *Philadelphia Evening Bulletin:* "Never have I seen a more efficient team. Each man does his job as if he is a part of the apparatus."[60]

Yet, no matter how closely departments adhered to the dictates of professional reform, they could not eradicate the legacy of the nineteenth-century firefighting culture. The power of the all-male work group remained strong and continued to dictate the tenor of life in an engine house despite the creation of multiple shifts, an increased number of transfers, and a more elaborate bureaucracy. In truth, such bureaucratic rules did not annihilate the local customs of individual engine houses. Despite efforts to keep firefighters busy with calisthenics, drills, and basic chores, firefighters continued to have time to socialize, play games, and engage in sport. In many cases, departmental leaders encouraged hobbies "to keep the men from going stale." In his memoir of African American firefighters, Joseph Marshall affectionately recalled the legacy of Philadelphia Fire Chief Ross Davis. Known as "leather lungs," Davis, a departmental reformer, would sometimes discipline his men according to nineteenth-century leadership methods. According to Marshall, "he would take a 'bad boy' or 'tough guy' down to the cellar, close the door, hang his rank on a nail, and mete out the necessary punishment. This method of 'taking a man to the front' never appeared on the official record—as a fifteen or thirty day fine or suspension would have." Davis even abandoned the mantra of departmental efficiency and stuck by one fireman during his "little drinking problem" because that fireman was "able to stay with him when the going was tough—like taking a 'rich feed' of heavy smoke." Surely the dynamics of working in close quarters fostered the continuation of the "bully-boy" system to some degree, even though formal bureaucratic procedures militated against it.[61]

Likewise, fire companies remained strongly attached to their local communities, although ironically the close ties remained bound to the everyday politics of the department and city. Although firefighters repeatedly received warnings not to work in local elections they appear to have done so anyway, and political considerations remained central to departmental management, as evidenced in the long-running feud between Philadelphia's Fire Chief Ross Davis and Assistant Chief Charles Gill in the 1930s. During the battle, Gill arranged for the transfer of Davis's supporters to engine houses away from their homes—what was known as being "sent to Siberia." Although not uncommon, such shifts threatened firefighters' ability to render effective service, which firemen felt acutely because they often lived in the neighborhoods in which they worked and quite literally protected their friends and families. Indeed, in 1931 during a similar incident, sixty-eight firemen were transferred away from their engine houses in Roxborough and Manayunk, where 90 ninety percent of them lived. Such transfers complicated the personal lives of firefighters and even imperiled the city's fire protection. Reportedly, the replacements did not know the neighborhood and required street guides to respond to calls. By living in the neighborhoods in which they worked and getting out to inspect the community, as was so much a part of their daily work, firefighters developed exceptionally close ties to their neighbors and became incomparable symbols of political power. One example is the "11s," which, like other companies in Philadelphia during the 1920s and 1930s, maintained a close relationship with its neighbors. Local shop owners and small merchants gave the firemen "something in their stockings" at Christmas, and a manufacturer located a few blocks away held an annual turkey dinner for the company. Firefighters, then, remained rooted in urban neighborhoods, performing a difficult balancing act. They negotiated the competing pressures of their work groups and occupation, their communities and departments, and heroism and rationality.[62]

Conclusion: Eating Smoke

By controlling the pace of reform, by taking responsibility for reorganizing their workplaces and job responsibilities, firefighters preserved the integrity of their work cultures and made their occupational identity ever stronger. Although there was sometimes much internal conflict, firefighters subjected themselves to industrial work disciplines—evident in training schools, comprehensive employment manuals, and the drive for efficiency. By leading the pace of change—as they had done for over a century—firefighters retained extraordinary control over their workplaces and skills; they remade the duties and requirements of their work on a continual basis. As they altered the division of the physical and technical characteristics of their work, firefighters also redefined the ideal of manly prowess that underlay their occupational culture. In addition, new civil service rules severed the connection between political leaders and rank-and-file firefighters. Rough competitiveness became somewhat less important than technical skill combined with bravery and team heroics. As they became more sober and efficient, firefighters developed an affinity for the regular career patterns that came with the bureaucratization and rationalization of their workplace. As administrative procedures replaced informal relationships, firefighters achieved careers of unusual longevity, enviable stability, and with excellent salary and pension benefits.

Firefighting became more professional; conditions improved, and the number of firefighters increased—by more than 40 percent during the first two decades of the century. Work conditions improved with the implementation of civil service reforms, second platoons, and pensions. By 1917 over 60 percent of fire departments located in cities with populations of more than thirty thousand people had implemented civil service reform. In 1919, Philadelphia installed a two-platoon system, following more than twenty-five of the nation's cities that had implemented the practice. At midcentury, Philadelphia added a third platoon, and St. Louis moved to three shifts in the 1970s. Pensions also became commonplace in departments protecting large cities by the 1920s, whether they had municipal plans or employee-funded programs. As a result of reform, differences—both between engine houses and fire departments—declined as cities prescribed a spate of common work techniques, bureaucratic rules, training regimens, and firefighting strategies. However, unlike other twentieth-century workplaces, making firefighting standard did not lead to a loss of skill or workplace authority. Reform only strengthened a well-defined occupational identity among firemen, and it diminished the authority of the intransigent and insular all-male work cultures of local engine houses. Of course, this does not mean that idiosyncratic firefighting work cultures did not continue to thrive. Rather, those cultures no longer determined the character of firefighting careers, which now offered stable salaries, good working conditions, and generous benefits.[63]

Although firefighters rhetorically became efficient machines, fighting fires remained an extraordinarily physical job, and the ability to eat smoke continued to determine a man's worth as a fireman, despite the slow inclusion of breathing apparatus into fire department work routines. Even though such devices became a common feature in training school curriculums and were standard rescue squad equipment as early as the 1920s, firefighters rarely used them. For instance, at a fire in the basement of a Philadelphia upholstery company in 1926, firefighters confronted heavy smoke and fumes from burning leather as they rushed in to fight the blaze and to keep casks of lacquer from exploding. Noxious gas overcame more than fifty firefighters that day, including Battalion Chief William Simler, who was wearing a primitive gas mask. After being revived Simler discarded the apparatus because it impeded his efforts to fight the blaze and to recover unconscious firefighters. Likewise, Joseph Marshall recalled of one cellar fire in which "smoke was so thick you could lean against it" and the men could stay only "as long as two or three minutes" before being relieved. Such fires tested men's courage and mettle and became measures of leadership; according to Marshall, Philadelphia's Chief Davis lived up to his nickname "leather lungs." His stamina, allowing him to remain in the building during the entire blaze, indicated Davis's fitness to lead the department. Experienced rank-and-file firefighters emulated their leaders and often did them one better, setting a daily example to new recruits and establishing a tone among all firefighters. As the *Philadelphia Evening Bulletin* reported of one fire captain, "Chief Waters says that the trouble with Captain Gaw is that he has had smoke for a steady diet too long. Gaw won't leave a building in flames. Hosemen in relays always trot at his heels. When Gaw flops over it is the hoseman's duty to carry him out." Even as firefighting grew increasingly bureaucratic, eating smoke remained a marker of manliness, albeit one that appeared increasingly untenable within the occupation's culture of efficiency.[64]

Conclusion

Fighting Fire in Postwar America

By the 1920s, the insurance industry had established broad authority over matters of fire safety, and the Insurance Company of North America introduced a new icon of fire protection—the White Firemen, who in a broad advertising campaign promised to protect American society in a way no firefighter ever could. Instead of reacting to danger, the White Firemen removed fire hazards proactively. As consultant to municipalities, the White Fireman, representing the insurance industry and the power of "paper," made fire protection more efficient and effective by providing expertise. He embodied the industry's new ethos, especially its claims to promoting public safety by sheltering urban centers, businesses, and individual residences from the problem of fire. Through this idealized image, the insurance industry recast itself as heroic; the standards produced by underwriters symbolically replaced firemen. Businessmen, not firemen, saved the day.[1]

Of course, the industry did not really want to replace firemen. After all, firefighters occupied a critical position in the fire-protective infrastructure and fire departments were financed publicly and managed by municipalities. Rather, the advertisements reflected the industry's aggressive prevention agenda, which was based in the standardization of underwriting practices, safety engineering, and the industry's surveillance of the landscape—all of which were developed during the nineteenth century. The loss-prevention engineer inspected property, researched fire hazards, and advised society about safe building practices. According to a feature in the *Underwriters' Review*, the loss-prevention engineer "studies manufacturing processes, and wherever practicable, suggests the substitutions . . . of lower inflammability." His work extended to the most mundane aspects of prevention, as he searched for ways to reduce the cost of insurance and fire loss. His heroism derived from his rationality, expert knowledge, and selfless protection of the economy.[2]

Symbolically, as the red fireman placed his arm around the white firemen, he passed the torch of public fire safety from virile, working-class men to rational middle-class men—from one set of specialists who confronted physical danger to experts that battled risk with financial tools. The first step in this process had been taken in the 1850s when control over firefighting had moved from specialized associations of volunteers into the hands of paid experts. As the heroism of firemen was being etched in the popular imagination during the 1880s, insurers focused on mundane business issues and profited from the problem of fire. However, they met with mixed success, as hundreds of companies went bankrupt in the face of escalating fire danger. Nonetheless, the industry developed an alternative standard of safety as it represented fire danger as an abstraction, using statistics and maps. By the twentieth century, insurers began to realize the fruits of their approach to risk, but only after they became advocates for fire prevention. The industry supported improving public safety by promulgating building codes and behavioral values that defined safety as the responsibility of consumers, who purchased safety for their families and community by obtaining insurance. Like underwriters, firemen promoted the drive for more regimented fire protection. Despite making their work more routine and implementing formal administrative procedures, firefighters maintained their identities as heroes, negotiating the contradiction between the bureaucratic and the exceptional. Although the White Fireman did not signal the demise of the heroic fireman, he did reflect a reorientation toward the bureaucratic, systematic, and rational; the locus of control had shifted from local communities to groups of professional experts. Moreover, just as middle-class manhood had gained hegemony over American society more broadly, businessmen acquired authority over the nation's fire safety.[3]

The advertising campaign also reveals the broader shift in the societal responsibility for fire safety from experts to individuals. The insurance industry pushed a new standard for safety—one embodied in legal building codes and prevention behaviors. The White Fireman advertising campaign proposed a somewhat contradictory message that touted the importance of safety engineers, but also underscored that safety was *everyone's* responsibility. As the campaign phrased it, "He wears citizen's clothes and rides in an ordinary car. He doesn't answer alarms—he prevents them." If the choice of language suggested that anyone could perform such work and that everyone should do so, it nonetheless connected the new expectations about fire safety to the accumulated expertise of underwriters and engineers. The industry had transformed safety into a commodity traded in a consumer society. Consumers, for instance, purchased safety when they bought products tested by the Underwriters' Laboratories, which identified a product as being safe from fire. Specialized experts—be they firemen, fire underwriters, or engineers—created the possibility for safety, but eradicating the hazard became the task of every American. The danger was ameliorated only by living up to the cultural standards of middle-class America, with its idealized gender norms (stay-at-home moms and male breadwinners) and its emphasis on individualism and economically rational decision-making.[4]

Containing fire danger had helped to make urban America a reality and structured its expansion into the suburbs, but the problem remained complex and continued to change in relation to the environment. Indeed, the fire hazard never had a fixed value; it shifted each time Americans remade the physical and cultural spaces in which they lived. Firefighters and underwriters developed a technological infrastructure that helped to bring discipline to the urban fabric—to people and to place. They minimized the problem of fire through institutions, social practices, and cultural beliefs that made city life more regimented and rationally organized, at least in regard to fire danger. In their battle with this hazard, firefighters and underwriters generated two countervailing, though not purely opposing, approaches to the problem that gave structure to the fire-protective infrastructure: a distinction between fire as human danger and fire as quantitative risk, between the physical and the abstract, between safeguarding life and protecting property. The story of fire protection—the shift from smoke to fire to water to paper—reveals how organizational technologies—management techniques, classification schemes, and codes of behavior and law—became paramount. Of course, the physical aspect of this battle remained critical and widely celebrated, but even firefighters came to emphasize the technical and bureaucratic elements of their work by the middle of the twentieth century. Moreover, this story encapsulates the broader history of cities, in which defining danger and safety in terms of risk—in terms of abstract numerical calculations—overwhelmed recognition of their material dangers. And, indeed, this duality between metaphor and physicality expresses well many of the conflicts at the heart of the city-building process.

In the United States, the problem of fire reached a critical point at the turn of the nineteenth century as cities began to emerge in greater numbers on the North American continent. As if obscured by smoke, Americans had a difficult time grasping the full dimensions of the problem as they altered the landscapes. To combat this ill-defined problem, the Philadelphia Hose Company introduced hose in 1803, and transformed firefighting in North America. The company made firefighting more efficient by replacing communal bucket brigades with volunteer fire companies, which came to be defined as organizations of specialists workers, as well as social clubs. Likewise, when the Insurance Company of North America offered its first fire insurance policy in 1794, the firm created a new industry in North America. Distinguishing themselves from a more communal form of corporate organization—that of mutual insurance companies—newly formed stock firms, such as Aetna and INA, began to use crude business tools to apprehend the problem of fire. Insurers and their agents inspected property, created detailed surveys for policies, and classified the risks that they took into categories of danger and safety. Both groups confronted the problem with new forms of social organization in which specialists took charge of representing and confronting fire. In the uncertain and relatively impermanent world of early-nineteenth-century cities, however, volunteer firefighters provided the most important line of defense against a new and more vexing problem that emerged with increasing population and structural density in cities.

As people flooded from country to city and industrial growth soared, firefighters and underwriters altered their strategies and tools for dealing with fire. Although insurers had long treated the hazard as an economic risk, in the 1850s they dramatically altered their approach by applying statistics to their business—by keeping actuarial tables of fire loss and relating those losses to the premiums that they charged for insurance. Not only did insurers compile quantitative data according to categories of danger that they daily used, but they also organized them spatially and began to publish maps of cities, which classified the built environment visually. By tying together several representational techniques—statistics, maps, and categories of danger and safety—underwriters abstracted fire hazard from its physical setting, completely transforming it into an abstract economic risk. Simultaneously, firefighters updated their organizations; they made them more efficient and rationally organized. Firefighters sought closer alliances to municipal governments, created management associations, and defined their service more narrowly. Additionally, despite some initial ambivalence about steam technology, and after repeatedly outperforming steam engines with hand-pumped apparatus, firefighters introduced steam technology into their organizations. After the Civil War, they redefined their service around the new machines, emphasizing especially

White Fireman Advertising Campaign, 1927, Insurance Company of North America. INA's White Fireman advertising campaign symbolized the broad changes in urban fire protection under way during the twentieth century. Firefighters retained significant cultural cachet—as heroes, rescuers, and specimens of physical vigor—but underwriters played a more significant role in determining the character of fire protection. Courtesy, ACE America Corporate Archives the skills required to operate them. In the process, firefighters transformed what had once been an avocation into an occupation of skilled experts.

Getting water onto fires became more difficult for firefighters in the last three decades of the nineteenth century, as the problem of fire grew in relation to the urban landscape—which reached higher into the sky and became more complex with the industrial development resulting from laissez-faire capitalism. To confront this problem, insurers created expansive urban atlases etched with classification schemes that were tied roughly to company actuarial tables and loss statistics. They also established trade associations to enforce discipline within their industry, especially seeking to set rates in accordance with actuarial tables. Yet, the major industry trade association, the National Board of Fire Underwriters, did not prioritize fire-preventive measures. Rather, insurers embraced the market, believing that the capitalist economy would educate people about fire risk and create order in society. Into the 1890s, the fire insurance industry mostly eschewed responsibility for public safety. Meanwhile, firefighters, recognizing the shifting landscape and its increasing dangers, developed new, more assertive work strategies. They channeled the energy they once devoted to pumping apparatus into a struggle that brought them face-to-face with fire on a regular basis. For the first time, firefighters routinely dragged their hoses deep inside mammoth structures and battled blazes "from the inside." Facing choking smoke and falling walls, firefighting became increasingly dangerous, especially because it was battled so aggressively. Out of this intensely personal confrontation with fiery landscapes—in which hazards were constantly changing—firefighters developed the defining rationale for their occupation. They began to emphasize preserving human life as their most pressing obligation, and became staunch advocates of safe construction practices. However, this new focus was best represented by renewed interest in the use of ladders, the quick spread of truck companies, and especially the creation of specialized rescue units—the pompier corps. Prioritization of human life became the defining character of firefighting work. As firefighters performed rescues, they became firemen; they also transformed themselves into icons of manhood, offering the first systematic standard of public fire safety to Americans.

At the dawn of the twentieth century, however, both firefighters and underwriters shifted strategy, and paper provided the means of triumphing over fire and trumped the use of water in making cities materially safer. In 1905, the fire insurance industry revolutionized its battle to manage fire risk by promoting expansive business, legal, and behavioral practices. The resulting code drew upon the latest techniques of business management and marketing, as well as five decades of loss experience and a growing fund of engineering and construction expense. Accompanying the codes was a focused effort to harmonize the industry's daily routines with its new emphasis on stopping fires before they began. Additionally, there was an equally aggressive push to change how ordinary Americans perceived fire hazard and to encourage new standards of behavior that prioritized fire safety. Altogether, this coordinated program of fire prevention eventually became a significant force in reshaping the American landscape. Firefighters, too, embraced prevention, and because they remained popular symbols of safety, they were among its most effective advocates in schools and everyday life. Moreover, fire departments reorganized how they battled fires, adopting modern management strategies. They developed extensive bureaucracies and standard work rules, codifying firefighting techniques into regimented training schools. Even officers and departmental leaders took formal training on the strategy of battling blazes, which was now available in published form as well as in the heads of long-serving firemen. As firefighters developed a more standard approach to fighting fires, they also earned better wages, secured improved benefits, and developed highly patterned careers. Most poignantly, fire departments established rescue squads that melded men and machines together, and they included breathing apparatus as a critical element in their struggle to bring order to an ever-changing urban landscape. Even so, at least through the 1950s, eating smoke remained a crucial marker of firefighting prowess, even as firefighters had remade themselves, somewhat paradoxically, into rational heroes.[5]

Ultimately, firefighters' and fire insurers' battle against the problem of fire—their physical and intellectual labor—created systematic safety for the social, cultural, and physical landscapes of the United States. The impact of their labor can hardly be overstated, since without the ability to control this environmental danger, urbanization would not have occurred as rapidly or as intensely as it did. The significance of this story, however, extends beyond the city; to some degree, the development of systematic fire protection in urban America is the story of the United States writ large. Firefighters' and underwriters' struggle to contain fire both predicted and mirrored broad patterns in American history. During the nineteenth century, for example, the organization of fire protection reflected the impact of industrialization and managerial capitalism even as it illuminated the incomplete and contested manner in which these large social processes shaped the nation. During the twentieth century, the expansion of fire prevention was among the most successful of Progressive Era reforms, and this effort to preserve the material foundation of American society accompanied and supported the rise of a mass consumer society. At the same time, both firefighters' and fire underwriters' efforts were shaped by common understandings about proper male behavior and faith in technological solutions. Strikingly, these beliefs became embedded in the institutions of fire protection as Americans readily disciplined their behavior and society. Firefighters' and underwriters' definitions of safety and risk—their systematic approach to the problem of fire—had helped to give structure to the process of urbanization, and this ordered physical and cultural landscapes, municipal institutions, and everyday urban life.

Firefighting in Postwar America

How did the problem of fire shift in the postwar period as the built environment and American society itself continued to experience dramatic change? Perhaps most obviously, the massive redistribution of the population in the United States following World War II had an impact on the nation's fire protection, as it did on so many other aspects of the American experience. By the 1970s and 1980s, the shifting population would once again alter the location and character of fire danger in the United States, creating a new set of issues for insurers and firefighters. For instance, the suburban housing boom helped to diminish the threat of conflagration in both cities and suburbs, in part because newly constructed dwellings were being built more safely, according to the codes recommended by the insurance industry, architects, and builders. In addition, new suburban homes were often built on separate lots, detached from one another, and the suburbs themselves developed municipal services according to the advice of professional experts—much of it embedded in updates of the legal codes that were created during the Progressive Era. The long-term process set in motion by firefighters and especially fire insurers had begun to pay its most significant and obvious dividends.[6]

In the context of this reconfiguration of danger and the expanding housing market, the Insurance Company of North America introduced a new "bundled" insurance policy that soon became common throughout the industry. Capitalizing on the fact that the fire risk to dwellings was more predictable than ever before, and seeking new ways to sell insurance, INA bundled nine different insurance tools together with its fire insurance policies, including tornado insurance and home-accident insurance, to make a single comprehensive policy. Of course, INA's new products developed in the context of a broader and complex business strategy, including an effort to improve market share in a critical segment of the market. Nonetheless, the introduction of homeowner's insurance in the 1950s suggests the lessening of fire as a singular threat, at least in the minds of insurers and the owners of single-family dwellings. Other threats now entered into the popular consciousness, competing for attention, and the new policies rendered fire risk invisible, indistinguishable from the other hazards that possibly affected homeowners. Homeowner's insurance contributed to the process of removing fire risk from the everyday consciousness, which had begun with the introduction of elaborate codes pertaining to building construction in the early twentieth century.[7]

Curiously, the organization of firefighting remained relatively static, despite the changing nature of the fire hazards. Through the 1970s the organization of fire departments in major cities and suburbs continued to reflect a nineteenth-century approach to fire risk, predicated on stopping conflagrations from getting started. Other innovations, such as rescue squads and a continued emphasis on ladder equipment, underscored the degree to which the lifesaving culture invented by firefighters in the nineteenth century remained a fixture of late-twentieth-century fire protection. However, the redistribution of money away from cities toward the suburbs has led to a new organization of fire departments that is "separate and unequal." Fire extinction resources are moving to wealthier suburban neighborhoods, and overly stressed urban fire departments have less time than ever before to perform building inspections in crumbling urban neighborhoods. In addition, building codes continue to be enforced haphazardly. The net result has been a heightening of danger in older, deteriorating neighborhoods, which often are allocated the least resources.[8]

The shifting nature of fire danger in postwar America also may have reawakened the possibility of conflagration—this time, beyond the suburbs in the "exurban" hinterlands. In 1956, a blaze in Malibu, California, burned 38,000 acres and destroyed 120 buildings. The fire heralded the development of a new fire regime caused by ever-expanding suburbs. Indeed, the lines between rural and urban fire risk have become blurred with intensifying development on the previously rural fringes of cities. Sweeping fires in Malibu (again) in 1993, the hills around Oakland in 1991, and Cerro Grande near Los Alamos, New Mexico, in 2002 reiterated the need to take account of a new type of conflagration and the reorientation of fire danger in the twenty-first century. The Oakland fire destroyed over 3,300 homes and apartments and caused twenty-five deaths. The Cerro Grande blaze destroyed 43,000 acres and 235 homes, although nobody was killed, and over 8,000 threatened homes were spared by dint of good fortune. As the Oakland conflagration demonstrated, in particular, the methods traditionally used by urban and suburban firefighters to subdue blazes are often ill-suited to exurban blazes. Rather than employing techniques commonly used in the battle against rural wildfire—such as setting backfires—firefighters employ the same defensive postures that work so well against urban blazes and in the context of saving lives. They fight exurban conflagrations house to house—a tactic that increases firefighters' risk and is not

Fire at Hamilton Street, Philadelphia, 1954. When fighting a blaze in a garage in 1954, firefighters encountered heavy smoke. Several used breathing apparatus to enter the building, at ground level, as others climbed ladders to ventilate upper stories. Prior to the arrival of Scot Air packs, few firefighters used breathing devices, and many resisted their use. Courtesy, Fireman's Hall Museum

adequate to the task of controlling such conflagrations. Even developers and insurers have come to realize that the dispersed housing patterns so typical of suburban expansion are no longer sufficient impediments to sweeping fires; given the right environmental conditions, suburbs can burn as easily as did nineteenth-century cities. More broadly, the current system of fire protection—divided as it is between urban and rural fire environments—is poorly prepared to deal with the changing conditions. Recognizing this, as well as the intensification of the exurban hazard, after the devastating fire season of 2002, the federal government established a National Fire Plan to deal with the shifting fire environment in a systematic fashion.[9]

Simultaneous with this shift in fire danger has been a change in the demographics of fire departments, which has affected the heroic and relatively autonomous culture of firefighters significantly. Initially in the postwar period, demographic changes may have served to reinforce the singularity of firefighters' work as a battle against nature. Indeed, the influx of returning veterans and the cold war regimen touched fire departments just a significantly as it affected the broader society. By 1960, over 60 percent of the nation's firefighters were veterans of the military, creating greater cohesiveness in firefighters' occupation and binding work groups yet more tightly together. Whereas previous generations of firefighters appear to have earned their training in a variety of skilled blue-collar occupations, firefighters increasingly earned their stripes in the military. Although the proportion of firefighters with such backgrounds gradually declined in ensuing decades, the number of firefighters who had served in the armed forces remained high through the 1980s.[10]

However, a more significant demographic shift set in motion the fracturing of unity among firefighters. As African Americans moved into cities in ever greater numbers following World War II, they gradually gained access to firefighting employment. And, in each subsequent decade after 1960 African Americans comprised a larger portion of the nation's firefighters. This was especially true in large northern cities, where postwar migration and "white flight" created black majorities. Building on the civil rights movement, African American communities demanded greater access to politics, consumption, and employment, including municipal jobs in fire departments. In the 1970s, the women's movement also helped to create new access to work in fire departments. Although relatively few women cracked the ranks of fire departments, they nonetheless gained a narrow foothold in the occupation. African Americans, by contrast, saw more opportunities in fire departments with each passing decade and entered the ranks of firefighters in growing numbers. By the 1990 census, fewer than 2 percent of all firefighters were women, well below the proportion of firefighters who were African American (13 percent).[11]

In the last decades of the twentieth century, questions about race and gender became major issues in fire departments, producing discord that resonated within the occupation. In a calling that defined safety, risk, and heroism in terms of white male prowess, how could a heroic identity be maintained as those people defined as outsiders—women and African Americans—entered the work group? For the first time, the cohesiveness of firefighting would begin to crack, as the generation of firemen who entered departments in the 1960s would face the legacy of race and gender created by their predecessors. The issues being confronted by the Washington, D.C., department in the late 1970s are emblematic of those faced by firefighters in cities throughout the nation. Like many of the nation's fire departments, Washington employed African Americans prior to World War II, but in small numbers and in segregated engine houses. And African Americans seeking to enter the department faced a number of hurdles, including physical examinations in which candidates were disqualified for bogus conditions. Segregated engine houses, in fact, produced two separate departments and cultures, though they operated according to much the same rules. African American firefighters developed similar occupational values and identities to their white counterparts. They created a culture that stressed physical strength, technical proficiency, and heroic professionalism. African American firefighters felt intense competitiveness with other African American fire companies, and especially with white companies in neighboring districts. Black and white firefighters alike appear to have reveled in this competitive culture much as their predecessors did.[12]

Yet, African Americans faced pervasive discrimination in these segregated departments, which shaped their experience of firefighting. In New York City, the Vulcan Society emerged as a separate organization of black firefighters to combat pervasive discrimination in the 1920s. With the help of the Vulcan Society, the NAACP documented and sought to redress the systematic racism of the department. In correspondence with the NYFD's Uniformed Firefighters' Association and its leadership, the NAACP reported that African American firefighters were not assigned to the most prestigious companies, including NYFD rescue squads, and were not allowed to participate on the department's band or baseball team. And, even though the department had integrated work units, it remained segregated in subtle ways. Indeed, the department's integration occurred in a discriminatory fashion that reflected a conscious strategy on the part of departmental leaders, which kept a separate list of black firemen. The NYFD assigned African American firefighters to companies in pairs, but it placed them on different shifts. Further, the African American firefighters were forced to use the same bed and take the same vacation; the arrangement also made it difficult for black firemen to be transferred. Additionally, the NAACP reported that blacks were not made acting officers as warranted by their seniority and they were passed over for special assignments.[13]

By contrast, other departments, including those in Philadelphia and Los Angeles, segregated black and white firefighters into separate engine houses. In Los Angeles, for instance, in 1953, blacks comprised less than 3 percent of the department (but 10 percent of the city's population) and worked in only two of the city's ninety-one engine houses. In an effort to end the discriminatory practices, which limited the number of African Americans who could be hired, the NAACP pressed the case for ending the segregation of the city's engine houses, and although the organization received at least tepid support from the mayor, it met with stiff opposition form Fire Chief John Alderson and most white firemen. White firefighters "unequivocally rejected forced integration" because it was "inimical to morale and efficiency and not in the public interest." For his part, the fire chief appealed to his regional professional association, the Pacific Coast Inter Mountain Association of Fire Chiefs, for assistance. The association attacked the municipal government for "political" meddling, turning the IAFC's long-standing agenda of professionalism into an instrument of discrimination. After a two-year struggle, Alderson resigned and the mayor appointed William L. Miller as chief. Almost immediately, the new chief sent eight white firemen, who had supported integration, to work in black firehouses, and by 1956 African Americans worked in seventeen of the city's ninety-one stations. Although the controversy died, reports of hazing and discrimination continued to surface.[14]

By the 1970s, overtly discriminatory practices became less common, but other more subtle forms of discrimination remained, often inhibiting the ability of firefighters to perform effective service or to receive sufficient training. For instance, in the Washington, D.C., department, officers refused to recognize the experience of black firemen, and many white firefighters greeted African Americans with silence. In a culture that transmitted occupational lore, techniques, and work culture through informal storytelling, such silence could pose serious impediments to a new fireman's ability to develop work skills, not to mention his safety. In his study of D.C. firefighters, Robert McCarl reported, "Some of the white firefighters expressed their hostilities toward blacks by breaking the plates and cups used by black fire fighters, refusing to eat or sleep in the same room with blacks, and even cutting television cords so that blacks couldn't watch television with whites." On the rare occasion when African Americans rose to leadership within an integrated company, they had to carefully maneuver the minefields of race that threatened to break the cohesiveness of firefighters' culture. And, as in so many other American workplaces, African Americans could gain entrée into the culture of an engine house only by achieving more than mere competence—only by demonstrating superlative performance.[15]

As tensions mounted in fire departments following the turbulent 1960s, and as more African Americans entered into the fire service, black and white firemen found less and less common ground—except perhaps in their agreement that women could not measure up to their job. Women have faced as many and perhaps more obstacles to gaining admission into the rough occupational culture of firefighting. Early opposition to admitting women dealt with issues of facilities. This "red herring," according to McCarl, was in fact only a minor barrier to entrance. To many observers the aggressive physicality of the work poses the most serious limitation on women, although others recognize that effective firefighting requires a diversity of skills, not all of them associated with size and strength. Indeed, teamwork, determination, experience, and other intangible qualities often have counted for more than sheer physical power. Many men, however, have continued to question women's physical ability, despite the fact that, in some departments at least, fitness levels are not what they should be. Strikingly, black and white men alike provide females little support and even less training in the informal traditions of firefighting culture—so important to learning the occupation. If women are creating emergent firefighting work traditions, the silence they receive from their male colleagues reveals the deep divides that have developed between firefighters. Women's presence in the work team challenges the tradition of the all-male group, and perhaps the manliness of firemen. As one female firefighter recounted to McCarl, "Our being here takes away from their pride in the job—their egos are deflated having women on the job."[16]

For more than a century, firefighters developed their occupation in the absence of women and African Americans, and the influx of these groups into departments has tested the cohesiveness of firefighters' culture more than any other challenge, including the changing risks of urban and exurban fire environments and even the rationalization of departments early in the twentieth century. The cohesive occupational brotherhood forged in the nineteenth century and long defended by white male firefighters has been shattered. Tensions in firehouses have led to the breakdown of significant communal activities, the distribution of racist and sexist propaganda, and even occasional tensions on the fire ground. Just as the NAFE's ethos of professionalism was used to overcome the parochialism of local communities and work cultures tied to political machines in the nineteenth century, its modern incarnation, the IAFC, is seeking to keep diversity from diminishing the cohesiveness and strength of firefighters' occupational culture. Judging from the acrimony that attends debates about race and gender in fire departments throughout the nation, a shared culture has not completely reemerged in firehouses, which are divided by multiple understandings of the occupation.[17]

Molding efficiency and heroism, and now racial and gender diversity, into a common work culture will challenge future generations of firefighters as much as the changing built environment. And, like previous generations of firefighters, they too will look toward their past to recreate the future. And, more often than not, that history has been preserved as an integral part of the department's life. For example, Thomas Targee has never disappeared from the collective memory of St. Louis firefighters. During the 1990s, a colossal portrait of him—on loan from the Missouri Historical Society—adorned the entrance to the St. Louis Fire Department headquarters, and Phelim O'Toole's portrait and story occupied a wall just a few feet beyond Targee. In 1995, the St. Louis Fire Department also erected a memorial sculpture to its fallen heroes. The statue, a firefighter clad in full lifesaving gear, appears in running pose and clutches a baby to its breast. The memorial emphasized what firefighters continue to stress, the priority of rescuing people trapped by fires or other hazards. Undoubtedly, the magnitude of this task continues to make firefighters into icons of safety. And firefighters remain heroes precisely because they seem so undaunted by the built landscape, penetrating deeper and deeper into burning buildings and fighting fires from closer and closer range. Although firefighters no longer eat as much smoke as they did—instead they use self-contained breathing apparatus to enter fires—their mission remains the same, and continues to be marked by an emphasis on professionalism, teamwork, technical skill, and physical ability. A continued presence of these values—removed from its historic connections to gender, race, and ethnicity—might someday help firefighters to forge a more inclusive definition of professional heroism, and a new basis for occupational unity.

APPENDIX ONE

Firefighting by the Numbers

Quantitative data played an important role in this study of the history of fire protection. Over the course of the project, I examined a variety of quantitative materials. These included several data sources whose procedures for use have been amply discussed by historians, including the Integrated Public Use Microdata Series (IPUMS) and the United States Census. However, in addition to exploring these more standard sources, I also analyzed two databases that I created from the personnel files of the St. Louis and Philadelphia fire departments. Although this data has helped me to better understand how the experiences of firefighters changed over time, making sense of them required me to move into territory not normally traversed by historians. With this appendix, I want to outline, briefly and generally, the procedures and approaches that I followed in order to make the data in the personnel files of these fire departments suitable for analysis.*

* St. Louis Fire Department, Personnel File; Philadelphia Fire Department, Roster, FH; IPUMS, http://www.ipums.umn.edu/usa/; for more discussion, see Mark Tebeau, "'Eating Smoke': Masculinity, Technology, and the Politics of Urbanization, 1850–1950" (Ph.D. diss., Carnegie Mellon University, 1997), 463–79; also, on quantitative methods, more broadly, see for instance, Konrad H. Jarausch and Kenneth A. Hardy, *Quantitative Methods for Historians: A Guide to Research, Data, and Statistics* (Chapel Hill: University of North Carolina Press, 1991); Stephan Thernstrom, "On the Socioeconomic Ranking of Occupations," in *The Other Bostonians* (Cambridge, Mass.: Harvard University

The process began with studying the methods that each department utilized to keep personnel records. The Philadelphia Fire Department kept a membership ledger, divided alphabetically by last name, though within alphabet category, the names were listed by order of the firefighters' year of first entry into the department. The ledger to which I was given access covered the period from the department's inception in 1871 until 1955. Unfortunately, the PFD appears to have stopped using the ledger abruptly in 1955. As a result, any information about firefighters whose careers continued after that date was absent from the ledger. The PFD ledger, then, represents a complete record of the department between 1871 and 1955. However, it does not contain an exhaustive record of all those firemen's careers. The St. Louis Fire Department kept personnel records in three different card files that were used over different but overlapping time periods. The card files to which I was given access included firefighters who began their careers at the department's inception in 1857 through about 1953. The cards were used and updated regularly after 1953. Thus these cards represent a more complete record.

Organizing the data in a fashion that made it amenable to analysis required a long, though relatively simple process that I performed in consultation with quantitative historians, especially John Modell. The first step involved entering data. I first assigned a basic identifying tag (a page, card file, and item number), and then entered the name of the firefighter, the date of entrance, the date of exit, and the reason for his final exit from the department. After entering this information, I "cleaned" the data, looking for duplications, correcting entry errors, and ensuring consistency in each database. This process took many hours, turning into months, in that it involved the painstaking process of checking my computerized database against the original records, again and again. Over this time, I organized, sorted, and reorganized the data. Later, as I finished this process, I performed quality checks as well as some preliminary analyses to learn more about the databases, especially whether systematic errors appeared. Once this process was complete, I had created a database of all firefighters who entered the St. Louis Fire Department between 1857 through 1950, which had 3,437 unique names. I also had developed a database of all the firefighters who entered the Philadelphia Fire Department from 1871 through 1955, which contained 8,035 individuals.

Press, 1975); Theodore Hershberg, Michael Katz, Stuart Blumin, Laurence Glasco, and Clyde Griffen, "Occupation and Ethnicity in Nineteenth-Century Cities," *Historical Methods Newsletter* 7 (1974), 174–216; Charles M. Dollar and Richard J. Jensen, *Historian's Guide to Statistics* (New York: Holt, Rinehart, and Winston, 1971); Stuart Blumin, *The Emergence of the Middle Class: Social Experience in the American City, 1760–1900* (New York: Cambridge University Press, 1989).

Part of the process of exploring the data involved creating a common framework for interpreting the data. It meant examining the different nomenclature used for personnel decisions, which varied slightly over time and between the departments, and then categorizing it. For instance, both organizations used multiple words to describe firefighters' final separation from the department: discharged, dismissed, dropped, resigned, retired, transferred, died, and killed in action. From these mostly self-explanatory categories, I created three categories: resigned or transferred (voluntary separation), dismissed (involuntary separation), and died. Only one category was ambiguous and that was the category of "transfer." Although how transfers were determined is cloudy, sometimes it appears to have been used in cases in which a firefighter was either in the twilight of a long career or physically unable to perform his job fully, but not disabled. However, in every instance, these firemen were transferred to other city departments. I kept this category separate, but for the sake of the tables presented here, I treated it as a resignation. Only 1 percent of all Philadelphia firefighters left the department through transfer. Finally, neither fire department consistently listed firemen killed in action differently from firefighters who died at fires from natural causes—say of a heart attack rather than from falling walls.

The process of developing and examining the data revealed certain limitations, especially in terms of record-keeping procedures, which affected its analysis and interpretation. The St. Louis Fire Department recorded personnel information using cards, and the department used three different file systems between 1857 and 1950. As a result these record-keeping methods overlapped. The earliest group of cards, however, contained more missing information than the later card systems. Therefore, the database of the STLFD did not reflect the experiences of firefighters who entered the department prior to 1870, and especially during the Civil War, as fully as it did those who entered after that year. It also seems likely that the data for this era was skewed slightly toward long-serving firefighters, because it stands to reason that they would more likely be recorded on the personnel cards. However, there is little evidence to back this suspicion or to quantify this effect. Even so, this flaw in the data does not lessen the value of using the entire dataset to compare change over time or as a tool for contrasting the careers of St. Louis firefighters with the careers of firefighters serving in the Philadelphia Fire Department.

By contrast, I discovered exactly the opposite dilemma existed with the records of the Philadelphia Fire Department. The PFD kept its personnel records on an oversized ledger, beginning in 1871 and running through 1955. The ledger contains little or no information on the service of any firefighter after 1955. Very basic information on careers is missing after that date—especially data on the date of exit and reason for a firefighters' separation from the department. As a result, the ends of firefighters' careers are systematically absent in this database—which is evident in records dating to the first decade of the twentieth century. This characteristic became more pronounced over the century. For example, 1.8 percent of firefighters who entered the PFD in the first decade of the twentieth century remained active in 1955; for those who entered in the 1910s the figure was 7.6 percent; for the 1920s, 24 percent; for the 1930s, 80.4 percent; for the 1940s, 91.0 percent; and of those firefighters who entered from 1950 through 1955 91.4 percent were active in 1955. (I calculated the extent of this systematic error by placing all firefighters who were still active in 1955 into a unique category. I then determined what percentage of firefighters in each cohort—1900s, 1910s, 1920s, etc.—were active.) As a result of the missing exit information, the database of the Philadelphia Fire Department systematically omitted the experiences of long-serving firefighters. Rather than compensate for this condition, I chose not to evaluate the records of those long-serving firefighters. Ultimately, the database of the Philadelphia Fire Department contained not 8,035 unique records, but 4,947. Although this flaw made the data more difficult to interpret, its effect was clear, and I organized my analysis and interpretation accordingly. The tables presented in the appendix include the percentage of active firefighters as a category, thus allowing readers to draw their own conclusions.

Once the databases had been organized, I began to analyze them. Generally speaking, this was a straightforward process, and I examined the data in three ways. First, I analyzed them by cohort of entrance. Studying the databases in this fashion provided a longitudinal portrait of firefighters over their careers, from their first entrance to their exit. These firefighters experienced the same shifts in their occupation and departments: the organization of labor, bureaucracy, procedures, and techniques. Though categorizing firefighters into cohorts, delineated by decade, is artificial, these groupings indicate how structural changes affected firefighters actually working in engine houses. Additionally, I categorized the data by decade of a firefighters' final separation from the department. This approach produced a cross-sectional portrait of the folks leaving the departments at different moments, offering a subtly different perspective on the changes in firefighting and fire departments. Finally, because the database of the St. Louis Fire Department held more complete information for all firefighters who entered the department between 1858 and the 1950s, I examined these records with greater attention to administrative details, such as transfers, promotions, disciplinary incidents, and other categories. In order to do this in a parsimonious fashion (and rather than entering all the administrative data for each firefighter) I created a random and representative sample of firefighters who entered the St. Louis Fire Department between the years 1857 and 1950. To be sure that my sample included enough firefighters from the earliest period, I oversampled records of firefighters who began careers before 1880. Later, I assigned a weight to these early records that accounted for the oversampling. Next, I reentered the relevant information from the personnel files for the 732 unique names that my sampling procedure yielded. (Once the sample weight was figured in, the sample size was 628.) Creating this robust sample allowed me to examine more detailed longitudinal questions about firefighter careers in a reasonable amount of time. Ultimately the data helped shape this study in many ways not explicitly mentioned here or in the text. The analyses resulted from hundreds if not thousands of hours of running and rerunning cross-tabs.

As the book, and chapters 6 and 8 in particular, demonstrate, these quantitative records had great influence on this project—all of which began with a relatively simple hypothesis. Before beginning my analyses, I hypothesized that firefighting careers would become longer over time, on average, and that firefighters' final separation from the fire service would increasingly be voluntary. I believed that this would occur as firefighting became a unique occupation, with a distinct work culture, improved working conditions, and more formal administrative procedures. Broadly speaking, the data confirmed my hypothesis, and the text discusses my interpretations in greater depth. Most significantly, the data on firefighters' careers and fire departments paralleled other trends in firefighting, including especially the standardization of work routines and administrative practices. Lastly, this quantitative analysis of firefighters' careers informs our understanding of the changing work experiences of American workers by providing insight into an exceptional group of workers in the American economy—workers whose longevity in the same job, for the same employer, and sometimes in the same work site, distinguished them from other American workers in the period from the Civil War through World War II.

Firefighting Careers in St. Louis and Philadelphia, 1857–1955

Chart 1. Department Staffing, new hires, and exits, by decade, St. Louis Fire Department, 1857–1949

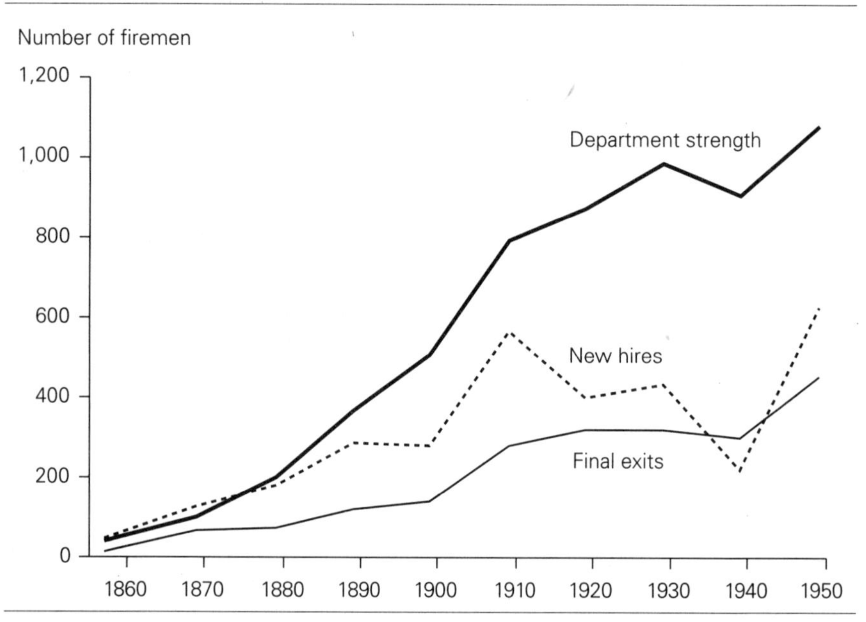

Source: Personnel Files, St. Louis Fire Department; checked against data in St. Louis Firemen's Pension Fund, *"Justifiably Proud": The St. Louis Fire Department* (St. Louis, Mo.: Walsworth, 1978).

Note: STLFD, n = 3,437; the points plotted on the STLFD chart represent data for the following years: 1857, 1858–69, 1870–79, 1880–89, 1890–99, 1900–09, 1910–19, 1920–29, 1930–39, 1940–49.

Chart 2. Department staffing, new hires, and exits, by decade, Philadelphia Fire Department, 1871–1949

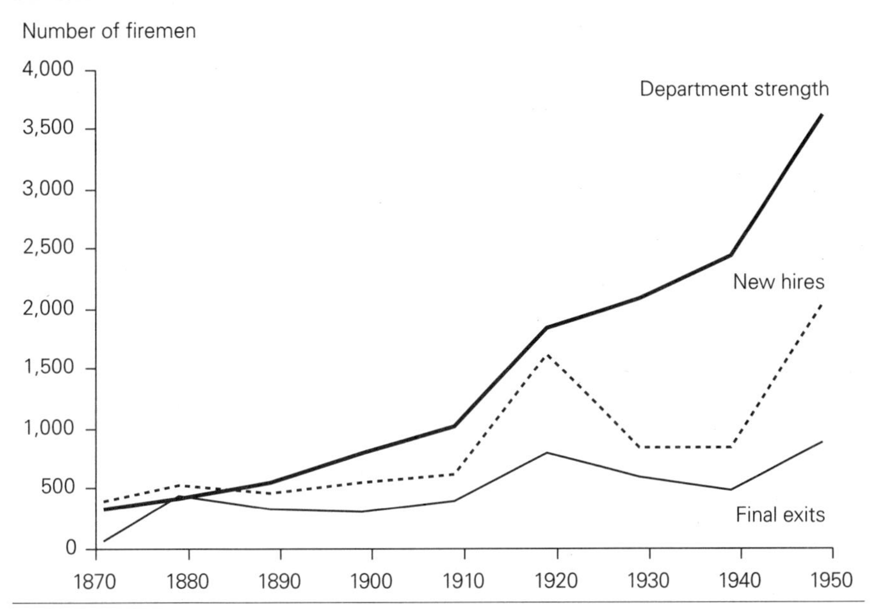

Source: Philadelphia Fire Department Roster, FH; checked against data in Philadelphia Fire Department Historical Corporation, *Hike Out: The History of the Philadelphia Fire Department* (1999).

Note: PFD, n = 8,035; the points plotted on the PFD chart represent data for the following years: 1871, 1872–79, 1880–89, 1890–99, 1900–09, 1910–19, 1920–29, 1930–39, 1940–49.

Table 1. Length of service for firefighters serving in "fully paid" fire departments, 1888–1891 (in percentages)

Years of Service	All reporting depts., NAFE, 1888–89	PFD, 1891	STLFD, 1889
Less than 5	19.6	37.2	39.4
At least 5	80.4	62.8	60.6
At least 10	50.2	46.0	36.2
At least 15	36.2	27.8	26.5
At least 20	8.5	13.1	12.9
At least 25	1.1	n.a.	7.2
At least 30	0.3	n.a.	5.1

Source: Proceedings of the NAFE, 1889, 151–52; Personnel Files, STLFD; PFD Roster, FH.

Note: PFD, n = 580; STLFD, n = 373; NAFE, n = 3,854. I created a cross-section of the STLFD for the year 1889, and of the PFD for 1891; these years were chosen to approximate most closely the data that fire departments across the nation submitted to the NAFE. Because the PFD was not established until 1871, I chose 1891 to make it possible to include data in the category "at least 20 years of service." It would be impossible for any Philadelphia fireman to have "at least 25 years of service" or more; therefore the category is "n.a."

Table 2. Experience of firefighters in the STLFD and PFD, for selected years, 1870–1950 (in average years of departmental service)

	STLFD	PFD
1870	6.4	n.a.
1880	8.7	5.5
1890	9.8	8.7
1900	12.9	11.6
1910	11.7	12.9
1920	21.3	18.4
1930	16.2	13.3
1940	20.3	16.3
1950	14.5	12.5

Source: Personnel Files, STLFD; PFD Roster, FH.
Note: PFD, n = 8,035; STLFD, n = 3,437; I created a cross-section of the membership of each department for various years, evaluating the average length of service in that year.

Table 3. Reasons for firefighters' final exits from STLFD and PFD, by decade of final exit, 1857–1955 (in percentages)

	Resigned or transferred		Died or killed in action		Discharged	
	STLFD	PFD	STLFD	PFD	STLFD	PFD
1857–69	7.1	n.a.	0	n.a.	92.9	n.a.
1870–79	15.4	53.5	23.1	6.5	61.5	40.0
1880–89	21.7	40.2	34.8	13.5	43.5	46.2
1890–99	4.3	47.2	47.8	19.7	47.8	33.1
1900–09	33.9	53.4	32.1	25.1	33.9	21.5
1910–19	38.6	63.3	35.1	20.9	26.3	15.8
1920–29	45.0	52.6	38.3	31.8	16.7	15.6
1930–39	52.3	54.8	33.9	35.5	13.8	9.7
1940–49	72.7	75.1	25.2	22.3	2.0	2.6
1950–55	82.5	77.5	15.5	15.8	1.9	6.7
19th century	12.0	47.9	32.0	12.0	56.0	40.0
20th century	63.5	64.8	25.8	24.2	10.7	10.9
Average	57.3	61.0	26.4	21.4	16.3	17.6

Source: Personnel Files, STLFD; PFD Roster, FH.
Note: PFD, n = 4,947; STLFD, n = 628 (weighted). For the purposes of this table, the PFD dataset contains only the records of those firefighters for whom there is available information on their final exit from the department. Of the 8,035 individual records in the database, this excludes 3,038 firefighters who were still active in 1955—when the PFD stopped using the personnel ledger from which this database is drawn. For further discussion of the PFD database—referenced below in tables 4, 5, and 6—see appendix 1; for discussion of the sampling of the STLFD—referenced below in tables 4, 7, 8, and 9—see appendix 1.

Table 4. Reasons for firefighters' final exits from STLFD and PFD, by decade of entrance, 1857–1949 (in percentages)

	Resigned or transferred		Died or killed in action		Discharged		*PFD firefighters still active in 1955 (in percentage)*
	STLFD	PFD	STLFD	PFD	STLFD	PFD	
1857–69	18.9	n.a.	32.4	n.a.	48.6	n.a.	n.a.
1870–79	29.0	53.8	38.8	14.5	32.3	31.7	n.a.
1880–89	35.1	53.2	36.8	20.1	28.1	26.7	n.a.
1890–99	38.5	58.4	44.3	22.1	17.3	19.5	n.a.
1900–09	54.0	55.9	28.3	27.4	17.7	16.7	1.8
1910–19	61.3	67.6	27.5	23.2	11.3	9.3	7.6
1920–29	63.5	68.2	24.0	21.9	12.5	9.9	24.0
1930–39	67.9	57.3	24.6	30.0	7.5	12.7	80.4
1940–49	89.7	60.9	6.5	20.1	3.7	19.0	91.0
Average	57.3	61.0	26.4	21.4	16.3	17.6	n.a.

Source: Personnel Files, STLFD; PFD Roster, FH.

Note: PFD, n = 8,035; STLFD, n = 628 (weighted). The PFD dataset includes only the records of firefighters for whom there is extant information on final exit. Thus, of the 8,035 records, only 4,947 were used in the calculation of the second, fourth, and sixth columns. The records of the 3,038 firefighters still active in 1955—when the PFD stopped using the personnel ledger from which this database is drawn—were used to determine the percentages in the seventh column, at the far right.

Table 5. Average career length for firefighters employed in the STLFD and PFD, grouped by decade of entrance, 1857–1949

	Average career length (in years)		*PFD firefighters still active in 1955 (in percentage)*
	STLFD	PFD	
1857–69	17.3	n.a.	n.a.
1870–79	19.9	10.6	n.a.
1880–89	22.7	13.9	n.a.
1889–99	21.7	18.6	n.a.
1900–09	21.9	19.6	1.8
1910–19	23.5	19.6	7.6
1920–29	23.9	21.0	24.0
1930–39	21.4	16.6	80.4
1940–49	22.3	9.7	91.0
Average	22.1	14.7	n.a.

Source: Personnel Files, STLFD; PFD Roster, FH.

Note: PFD, n = 8,035; STLFD, n = 3,437. The PFD dataset includes only the records of firefighters for whom there is extant information on final exit. Thus, of the 8,035 records, only 4,947 were used in the calculation of the second column—average career length for PFD firefighters. The records of the 3,038 firefighters still active in 1955—when the PFD stopped using the personnel ledger from which this database is drawn—were used to determine the percentages in the third column.

Table 6. Average career length for firefighters employed in the STLFD and PFD, grouped by decade of exit, 1857–1949 (in years)

	STLFD	PFD
1857–69	2.7	n.a.
1870–79	5.1	2.4
1880–89	8.4	6.1
1889–99	10.7	10.5
1900–09	12.5	13.9
1910–19	15.3	11.4
1920–29	20.5	15.4
1930–39	25.1	22.6
1940–49	25.7	23.6
Average	22.1	14.7

Source: Personnel Files, STLFD; PFD Roster, FH.

Note: PFD, n = 4,947; STLFD, n = 3,437. For the purposes of this table, the PFD dataset contains only the records of those firefighters for whom there is available information on their final exit from the department. Of the 8,035 individual records in the database, this excluded 3,038 firefighters who were still active in 1955—when the department stopped using the personnel ledger from which this database is drawn.

Table 7. Prevalence of disciplinary incidents, by decade of entrance, STLFD, 1857–1949

	0 incidents (in percentage)	1 or more incidents (in percentage)	2 or more incidents (in percentage)	*Average number of incidents per firefighter*
1857–69	97.3	2.7	0	0
1870–79	100	0	0	0
1880–89	89.5	10.5	7.0	0.33
1889–99	90.4	9.6	3.8	0.31
1900–09	74.6	25.4	9.6	0.58
1910–19	82.5	17.5	5.0	0.26
1920–29	80.2	19.8	7.3	0.43
1930–39	74.0	26.0	20.0	0.84
1940–49	88.3	11.7	7.8	0.19
Average	84.1	15.9	7.4	0.37

Source: Personnel Files, STLFD.

Note: STLFD, n = 628 (weighted).

Table 8. Prevalence of transfers, by decade of entrance, STLFD, 1857–1949

	0 transfers (in percentage)	0 or 1 transfers (in percentage)	2 or more transfers (in percentage)	3 or more transfers (in percentage)	4 or more transfers (in percentage)	*Average number of transfers per firefighter*
1857–69	56.8	81.1	43.2	18.9	5.6	0.71
1870–79	65.6	78.1	34.4	21.9	6.3	0.69
1880–89	49.1	63.2	50.9	36.8	26.3	1.37
1890–99	25.0	36.5	75.0	63.5	51.9	2.48
1900–09	25.4	37.7	74.6	62.3	50.0	2.93
1910–19	21.3	31.3	78.8	68.8	63.8	4.24
1920–29	15.6	26.0	84.4	74.0	58.3	3.38
1930–39	14.3	24.5	85.7	75.5	67.3	4.49
1940–49	4.9	19.4	95.1	80.6	68.0	4.25
Average	25.2	37.9	74.8	62.1	50.6	3.08

Source: Personnel Files, STLFD.
Note: STLFD, n = 628 (weighted).

Table 9. Prevalence of reductions, by decade of entrance, STLFD, 1857–1949

	0 reductions (in percentage)	1 or more reduction (in percentage)	*Average number of reductions per firefighter*
1857–69	83.3	16.7	0.17
1870–79	87.5	12.5	0.14
1880–89	71.9	28.1	0.28
1890–99	67.9	32.7	0.45
1900–09	88.6	11.4	0.11
1910–19	88.8	11.3	0.11
1920–29	94.8	5.2	0.11
1930–39	100	0	0
1940–49	100	0	0
Average	88.7	11.3	0

Source: Personnel Files, STLFD.
Note: STLFD, n = 628 (weighted).

Abbreviations

ACE	ACE America Corporate Archives	
	INA	INA Collection
	Aetna	Aetna Collection
BCHS	Bucks County Historical Society	
	FFC	Fire Fighting Collection, Manuscript Collection 285, Spruance Library, Bucks County Historical Society (Material culture available in the museum.)
BHS	Brooklyn Historical Society	
FA	Fire Association of Philadelphia	
FH	Fireman's Hall Museum	
FISOP	Fire Insurance Society of Philadelphia (Material available at the archive, headquarters, Fire Insurance Society of Philadelphia.)	
FJ	*Fireman's Journal* (1877–1886)	
F & W	*Fire and Water* (1886–1902)	
F & WE	*Fire and Water Engineering* (1903–1925)	
HSP	Historical Society of Pennsylvania	
IAFC	International Association of Fire Chiefs (1932–current) (At its 1896 convention the NAFE changed its name to the International Association of Fire Engineers. By 1932, the IAFE had become the International Association of Fire Chiefs. I change the title of the *Proceedings* accordingly.)	
IAFE	International Association of Fire Engineers (1896–1931)	
INA	Insurance Company of North America	
LCP	Library Company of Philadelphia	
MDHS	Maryland Historical Society	
	Manuscript Collection 662, Maryland Historical Society	
MOHS	Missouri Historical Society	
	STLVVFC	St. Louis Veteran Volunteer Firemen's Collection
	Tiffany	Tiffany Collection
	CSS	Charles E. Swingley Scrapbooks, volumes 1–3
NAFE	National Association of Fire Engineers (1873–1895)	
NBFU	National Board of Fire Underwriters	
NFPA	National Fire Protection Association	

NYFD	New York Fire Department
PCA	Philadelphia City Archives
PFDRG	Philadelphia Bureau of Fire/Philadelphia Fire Department, Record Group 74, Philadelphia City Archives
PFD	Philadelphia Fire Department
SAAF	*Safeguarding America Against Fire*
STLCA	St. Louis City Archives
STLFD	St. Louis Fire Department
TUUA	Temple University, Urban Archives
PEBCF	Philadelphia Fire Department, Box 76, *Philadelphia Evening Bulletin*, Clippings Morgue
UL	Underwriters' Laboratories
WDFR	*Whipple Daily Fire Reporter*

(Archived at St. Louis Fire Department Records, St. Louis City Archives.)

Notes

Introduction: The Problem of Fire

1. *Evening Chronicle*, August 10–12, 1877; Record of Fires in the City of St. Louis, 1887, STLFD, STLCA; *Proceedings of the NAFE, 1880*, 76; "The Chicago Training School," *Fireman's Herald* 24 (November 24, 1892); *Proceedings of the NAFE, 1882*, 45; *Fire and Water* 6 (1889), 62, 87; *F & W* 18 (1895), 477; St. Louis Firemen's Fund, *History of the St. Louis Fire Department* (1914), 178; *Proceedings of the NAFE, 1887*, 65; *Proceedings of the NAFE, 1888*, 87.

2. *Whipple's Daily Fire Reporter*, August 10–11, 1887.

3. Risk and safety have received much recent study; see, for instance, Ulrich Beck, *Risk Society: Towards a New Modernity* (London: Sage, 1992); Peter L. Bernstein, *Against the Gods: The Remarkable Story of Risk* (New York: John Wiley and Sons, 1996); Roger Cooter and Bill Luckin, eds., *Accidents in History* (Amsterdam: Rodopi Press, 1997); Mark Aldrich, *Safety First: Technology, Labor, and Business in the Building of American Work Safety, 1870–1939* (Baltimore, Md.: Johns Hopkins University Press, 1997); for insights into studying work and occupation, see Andrew Abbott, *The System of Professions: An Essay on the Division of Expert Labor* (Chicago: University of Chicago Press, 1988); on technology as reflecting broad societal relations, see Bruno Latour, "Technology Is Society Made Durable," in John Law, ed., *A Sociology of Monsters: Essays on Power, Technology, and Domination* (New York: Routledge, 1991), 103–31.

4. For the classic study of fire and its relation to American society, especially in rural settings, see Stephen J. Pyne, *Fire in America: A Cultural History of Wildland and Rural Fire* (Princeton, N.J.: Princeton University Press, 1982).

5. October 3, 1851, Missouri Fire Company No. 5, Record of Fires, 1846–1855, STLVVFC, MOHS; David D. Dana, *The Fireman: The Fire Departments of the United States . . .* (1858), 358–65; Thomas Scharf, *History of St. Louis City and County* (Philadelphia: Everts and Co., 1919), 819 ff.; *Missouri Republican*, May 24, 1849; "Great Fire of 1849," Folder 16, Box 1, STLVVFC; on the danger, see Amy Greenberg, *Cause for Alarm: The Volunteer Fire Department in the Nineteenth-Century City* (Princeton, N.J.: Princeton University Press, 1998), 30–39; Andrea Stulman Dennett and Nina Warnke, "Disaster Spectacles at the Turn of the Century," *Film History* 4, no. 2 (1990), 101–11; Thomas A. Edison, Inc., *Spectacular Scenes during a New York City Fire* (United States: Thomas A. Edison, Inc., 1905); Thomas A. Edison, *Destruction of Standard Oil Company's Plant at Bayonne, N.J., by Fire on July 5th, 1900* (United States: Edison Manufacturing Co., 1900); Thomas A. Edison, James White, producer, *Morning Fire Alarm* (United States: Edison Manufacturing Co., 1896). Evaluating fire loss, either relative to the population or the amount of property exposed is

a dicey proposition. I have chosen to use data that emphasizes the dollar amount of fire loss, per capita and indexed for inflation and the total amount of property exposed; see *Statistical Abstracts of the United States, 1879–1960;* also, see National Fire Protection Association, *Conflagrations in America Since 1900* (1951). For a different approach to evaluating how the threat of fire changed in the nineteenth century—which counts the number of major fires that destroyed "at least fifty houses" and then represents that number per capita, in terms of the populations of select major cities—see L. E. Frost and E. L. Jones, "The Fire Gap and the Greater Durability of Nineteenth-Century Cities," *Planning Perspectives* 4 (1989), 333–47.

6. On fire as "good servant and bad master," see, especially, Margaret Hindle Hazen and Robert M. Hazen, *Keepers of the Flame: The Role of Fire in American Culture, 1775–1925* (Princeton: Princeton University Press, 1992); Johan Goudsblom, *Fire and Civilization* (New York: Penguin, 1992), esp. 176–77; on changing fire environments and fire as metaphor, see Pyne, *Fire in America,* 20–33, 137–42; for a broader cultural view of fire as metaphor, see Robyn Cooper, "The Fireman: Immaculate Manhood," *Journal of Popular Culture* 28, no. 4 (1995), 139–70; Gaston Bachelard, *The Psychoanalysis of Fire,* trans. Alan C. M. Ross (London: Routledge and Kegan Paul, 1964); Mary Douglas, *Natural Symbols* (Harmondsworth: Penguin, 1973); Amy Greenberg, *Cause for Alarm: The Volunteer Fire Department in the Nineteenth-Century City* (Princeton, N.J.: Princeton University Press, 1998); for an overview of city building in America, see Eric Monkkonen, *America Becomes Urban: The Development of Cities and Towns, 1780–1980* (Berkeley: University of California Press, 1988); on the nineteenth-century process of urban development, see William Cronon, *Nature's Metropolis* (New York: W. W. Norton, 1991); also, for overviews of urban and environmental history, literatures to which this study contributes, see Timothy Gilfoyle, "White Cities, Linguistic Turns, and Disneylands: The New Paradigms of Urban History," *Reviews in American History* 26, no. 1 (1998), 175–204; Joel Tarr and Jeffrey K. Stine, "At the Intersection of Histories: Technology and the Environment," *Technology and Culture* 39 (1998), 601–40.

7. On conflagrations, see Christine Meisner Rosen, *The Limits of Power: Great Fires and the Process of City Growth in America* (New York: Cambridge University Press, 1986); Karen Sawislak, *Smoldering City: Chicagoans and the Great Fire, 1871–1874* (Chicago: University of Chicago Press, 1994); on the urban technological infrastructure, more broadly, see Joel Tarr and Gabriel Dupuy, *Technology and the Rise of the Networked City in Europe and America* (Philadelphia: Temple University Press, 1988); Donald Reid, *Paris Sewers and Sewermen: Realities and Representations* (Cambridge, Mass.: Harvard University Press, 1991); on order, see Robert J. Wiebe, *The Search for Order, 1877–1920* (New York: Hill and Wang, 1967); on the chaotic city, see Carl Smith, *Urban Disorder and the Shape of Belief: The Great Chicago Fire, the Haymarket Bomb, and the Model Town of Pullman* (Chicago: University of Chicago Press, 1995); Georg Simmel, "The Metropolis and Mental Life," in Kurt Wolff, trans. and ed., *The Sociology of Georg Simmel* (London: Free Press, 1950), 409–24.

8. On the advantages of studying the mundane, see Mary Douglas, *Purity and Danger: An Analysis of the Concepts of Pollution and Taboo* (New York: Routledge, 1966); Nicholas Dirks, Geoff Eley, and Sherry B. Ortner, eds., *Culture/Power/History: A Reader in Contemporary Social Theory* (Princeton, N.J.: Princeton University Press, 1994); on the life insurance industry, see Viviana Zelizer, *Morals and Markets: The Development of Life Insurance in the United States* (New York: Columbia University Press, 1979); Angel Kwolek-Folland, *En-*

gendering Business: Men and Women in the Corporate Office, 1870–1930 (Baltimore, Md.: Johns Hopkins University, 1994); on early fire insurance, see Clive Trebilcock, *Phoenix Assurance and the Development of British Insurance* (Cambridge: Cambridge University Press, 1985); on information management and classification, see JoAnne Yates, *Control through Communication: The Rise of System in American Management* (Baltimore, Md.: Johns Hopkins University Press, 1989); Geoffrey C. Bowker and Susan Leigh Star, *Sorting Things Out: Classification and Its Consequences* (Cambridge, Mass.: MIT Press, 1999); Theodore Porter, *Trust in Numbers: The Pursuit of Objectivity in Science and Public Life* (Princeton, N.J.: Princeton University Press, 1995); Michel Foucault, *The Order of Things: An Archaeology of the Human Sciences* (New York: Random House, 1970).

9. Bruce Laurie, "Fire Companies and Gangs in Southwark," in Allen F. Davis and Mark Haller, eds., *The Peoples of Philadelphia: A History of Ethnic Groups and Lower-Class Life, 1790–1940* (Philadelphia: Temple University Press, 1973), 71–87; Greenberg, *Cause for Alarm*, esp. 41–108; Cooper, "The Fireman," 139–70; Paul Ditzel, *Fire Engines, Fire Fighters* (New York: Crown, 1976); Rebecca Zurier, *The American Firehouse: An Architectural and Social History* (New York: Abbeville Press, 1982); Miriam Lee Kaprow, "Magical Work: Firefighters in New York," *Human Organization* 50, no. 1 (1991), 97–103; Robert McCarl, *The District of Columbia Fire Fighters' Project: A Case Study in Occupational Folklife* (Washington, D.C.: Smithsonian Institution Press, 1985); Carol Chetkovich, *Real Heat: Gender and Race in the Urban Fire Service* (Brunswick, N.J.: Rutgers University Press, 1997).

10. On this use of the word *discipline*, see Michel Foucault, *Discipline and Punish: The Birth of the Prison* (New York: Vintage, 1990).

11. Of course, other very practical concerns also entered into my decision about which fire departments and insurance companies to study most closely. These included, in no particular order: quality of access to archival materials, financial resources available, the presence of sources sufficient to cover the entire period of this study, and, critically, the ability of those local stories to illuminate broader patterns of change. As I searched for primary sources, archives, and other materials for this study, I had examined the *National Union Catalog* thoroughly, visited over ten cities, called scores of archives, and telephoned nearly a dozen insurance companies.

ONE: Workshops of Democracy: The Invention of Volunteer Firefighting

1. *People's Organ*, May 21, 1849, *St. Louis Star*, April 7, 1931, Folder 16, Box 1, STLVVFC, MOHS; *Missouri Republican*, May 24, 1849.

2. Rebecca Zurier, *The American Firehouse: An Architectural and Social History* (New York: Abbeville Press, 1982), 54–55; Matthew Hastings, *Thomas Targee (Captain, Missouri Fire Company No. 5)*, oil on canvas, 1902, MOHS; *Missouri Republic*, April 11, 1902, Accession Files, Hastings Paintings, MOHS.

3. Alexis de Tocqueville, *Democracy in America*, trans. George Lawrence (New York: Harper and Row, 1966).

4. Thomas C. Cochran, *Frontiers of Change: Early Industrialism in America* (New York: Oxford University Press, 1983).

5. Zurier, *American Firehouse*, esp. 21–31; Andrew M. Neilly, "The Violent Volunteers: A History of the Volunteer Fire Department of Philadelphia, 1736–1871" (Ph.D. diss., University of Pennsylvania, 1959), 8–23; William Joseph Novak, "Salus Populi: The Roots of Reg-

ulation in America, 1787–1873" (Ph.D. diss., Brandeis University, 1992), 163–79; J. A. Fowler, *History of Insurance in Philadelphia for Two Centuries, 1683–1882* (Philadelphia: Review Publishing and Printing, 1888), 290–320; Glen Holt, "Volunteer Firefighting in St. Louis, 1818–1859," *Gateway Heritage* 4, no. 3 (1983–1984), 2–13; David Paul Brown, *Oration on the Centennial Anniversary of the Organization of the Fire Department of Philadelphia* (1839), 8; Carl Bridenbaugh, *Cities in the Wilderness: The First Century of Urban Life in America* (New York: Ronald Press, 1938), 56–57, 208–10.

6. James P. Parke, "A History of the Origin and Establishment of the Philadelphia Hose Company" (ca. 1804), Samuel Hazard Papers, HSP (this handwritten manuscript is not paginated and appears to have been written in 1804); Neilly, "Violent Volunteers," 24.

7. Parke, "Philadelphia Hose Company"; John L. Androit, ed., *Population Abstract of the United States* (McLean, Va.: Androit and Associates, 1980).

8. Walter S. Nichols, *An Illustrated History of the Fire Engine* (New York: C. C. Hine, 1872), 26–28.

9. Parke, "Philadelphia Hose Company."

10. Parke, "Philadelphia Hose Company."

11. Parke, "Philadelphia Hose Company."

12. See, for instance, Missouri Fire Company No. 5, Record of Fires, 1846–1855, vol. 10, ser. 1, STLVVFC, MOHS; A. W. Brayley, *A Complete History of the Boston Fire Department, Including the Fire-Alarm Service and the Protective Department, from 1630 to 1888* (Boston: J. P. Vale, 1889), 234; on the fire threat more broadly, see Margaret Hindle Hazen and Robert M. Hazen, *Keepers of the Flame: The Role of Fire in American Culture, 1775–1925* (Princeton: Princeton University Press, 1992), 3–64.

13. David Dana, *The Fire Departments of the United States . . .* (Boston: E. O. Libby, 1858), 159–66; Thomas Scharf and Westcott, *History of Philadelphia, 1609–1884* (1884), 1,914; Thomas Scharf, *History of St. Louis City and County* (Philadelphia: Everts and Company, 1919), 819 ff.; *Missouri Republican*, May 24, 1849; Recollections of Witnesses, News Clippings, 1849–1939, Folder 16, Box 1, STLVVFC; Official Accounts of the Great Fire, 1849–1902, Folder 17, Box 1, STLVVFC; *Mayor's Messages and Accompanying Documents, City of St. Louis, 1848* (1848), 16.

14. Parke, "Philadelphia Hose Company"; Resolutions Concerning Donations from the Phoenix Assurance Company of London, 1809, FFC; Fire Miscellaneous, List of Donations to Hose Companies, 1807–1824, INA, ACE; Thomas Lynch, *The Volunteer Fire Department of St. Louis, 1819–1850* (1880), 81 ff.

15. Garvan and Wojtowicz, *Catalogue*, 123; on balls see Invitation to the Good Will Library Association Citizens' Dress Ball, 1841, Good Will Fire Company Announcement Regarding Annual Ball, 1868, FFC.

16. March 26, 1850, Firemen's Association, Minutes, 1850–1859, STLVVFC; Neilly, "Violent Volunteers," 48–51; Perseverance Hose Company, Certificate of Exemption from Military Duty for William Thorp, January 9, 1844, FFC; on gender and the obligations of citizenship, see Linda K. Kerber, "A Constitutional Right to Be Treated Like American Ladies: Women and the Obligations of Citizenship," in Linda K. Kerber, Alice Kessler-Harris, and Kathryn Kish Sklar, eds., *U.S. History as Women's History: New Feminist Essays* (Chapel Hill, N.C.: University of North Carolina Press, 1995), 17–36; Iver Bernstein, *The New York City Draft Riots: Their Significance for American Society and Politics in the Age of the Civil War* (New York: Oxford University Press, 1990), 18–22.

17. See, for example, Liberty Fire Company, parade hat, ca. 1850, MOHS; Western Engine Company banner, ca. 1840–1850, Shiffler fire hat, n.d., and John Magee, *The Death of George Shiffler, Kensington, Pennsylvania* (lithograph, 1844), BCHS. The beehive appeared on several hats in the BCHS collection; for possible connections between the symbol and liberalism, see Bernard Mandeville, *Fable of the Bees: Or Private Vices, Publick Benefits* (1728; reprint, Indianapolis, Ind.: Liberty Classics, 1988). On liberalism and republicanism, see Daniel T. Rodgers, "Republicanism: The Career of a Concept," *Journal of American History* 79, no. 1 (1992), 11–38; J. G. A. Pocock, "Virtue and Commerce in the Eighteenth Century," *Journal of Interdisciplinary History* 3 (1972), 119–34; Joyce Appleby, *Capitalism and a New Social Order: The Republican Vision of the 1790s* (New York: New York University Press, 1984); Philip J. Ethington, *The Public City: The Political Construction of Urban Life in San Francisco, 1850–1900* (New York: Cambridge University Press, 1994).

18. Compare, for example, *Act of Incorporation, Philadelphia Hose Company* (n.d.), 3; *Charter and By-Laws of the Pennsylvania Fire Company* (Philadelphia, 1835), 3; *Constitution and By-Laws of the Good Intent Hose Company* (Philadelphia, 1840), 3; *Charter and By-Laws of the United States Fire Company of Philadelphia* (Philadelphia, 1838), 3.

19. For a sense firefighting work and company life in this period, one need only consult the hundreds of extant fire company ledgers located in archives across the county. Although I examined scores of such volumes in various cities, this study draws especially upon logbooks produced by St. Louis and Philadelphia firefighters. For an especially good sense of the dynamics of company life, see, for example, Union Fire Company No. 2, Minutes, 1833–1840, vol. 13, ser. 1, Union Fire Company No. 2, Minutes, 1841–1850, vol. 14, ser. 1, STLVVFC. For an overview of the history of nineteenth-century volunteer firefighting and fire companies, see Bruce Laurie, "Fire Companies and Gangs in Southwark," in Allen F. Davis and Mark Haller, eds., *The Peoples of Philadelphia: A History of Ethnic Groups and Lower-class Life, 1790–1940* (Philadelphia: Temple University Press, 1973), 71–87; Richard Stott, *Workers in the Metropolis: Class, Ethnicity, and Youth in Antebellum New York City* (Ithaca, N.Y.: Cornell University Press, 1990), esp. 212–76; for a reassessment of research on volunteer firefighting, and an interpretation different from the one presented here, see Amy Greenberg, *Cause for Alarm: The Volunteer Fire Department in the Nineteenth-Century City* (Princeton, N.J.: Princeton University Press, 1998).

20. June 12, 1851, January 13, 1852, Missouri Fire Company, Record of Fires, 1846–1855, STLVVFC; Edward Edwards, *History of the Volunteer Fire Department of St. Louis* (St. Louis: Veteran Volunteer Firemen's Historical Society, 1906), 36, 280–82; Robert DeSilver, *De Silver's Philadelphia Directory and Strangers Guide, 1830* (Philadelphia: Robert DeSilver, 1830).

21. See, for example Scharf, *History of St. Louis*, 793–95; Lynch, *Volunteer Fire Department*, 10–11; February 1, 1842, January 9, 1845, April 5, 1849, Union Fire Company, Minutes, 1840–1851, STLVVFC; October 4, 1855, Franklin Fire Company, 1855–1859, STLVVFC; Nichols, *An Illustrated History*, esp. 19–23, 26–34; Eugene S. Ferguson, ed., *Early Reminiscences (1815–1840) of George Escol Sellers* (Washington, D.C.: Smithsonian Institution Press, 1965).

22. *Nile's Register*, October 5, 1839; November 1, 1848, Union Fire Company, Minutes, 1840–1851, STLVVFC; *St. Louis Post-Dispatch*, December 28, 1890, Folder 13, Box 1, STLVVFC. On the workings of different types of engines, see James Braidwood, *On the Construction of Fire-Engines and Apparatus: The Training of Firemen, and the Method of Pro-*

ceeding in Cases of Fire (Edinburgh, 1830); Nichols, *An Illustrated History*, 18–34, esp. 30–34; Anne Daly and Jack J. Robrecht, *Fire Engines: An Illustrated Handbook of Fire Apparatus, with Emphasis on Nineteenth Century American Pieces* (Philadelphia: INA Corporation, 1980); Kenneth Holcomb Dunshee, *Enjine! Enjine!: A Story of Fire Protection* (New York: Home Insurance Company, 1939), 11–31.

23. March 7, 1850, July 15, 1851, Missouri Fire Company, Record of Fires, 1846–1855, STLVVFC; Edwards, *History*, 81–91; Dunshee, *Enjine! Enjine!*, 11–31.

24. February 8, 1848, October 2, 1851, May 1, 1849, Missouri Fire Company, Record of Fires, STLVVFC.

25. September 10, 1851, Missouri Fire Company, Record of Fires, STLVVFC; Grey Eagle Hose Company, Folder 4, Box 1, STLVVFC; August 1, 1849, September 5, 1849, Union Fire Company, Minutes, 1840–1851, STLVVFC; Lynch, *Volunteer Fire Department*, 10–11; Scharf, *History of St. Louis*, 793–95.

26. Stories and images of rescue appear infrequently in the bulk of firefighting materials produced prior to the Civil War; of the hundreds of photographs, prints, drawings, and certificates at BCHS, few depicted rescues. See Humane Engine Company Commemorative Certificate, 1844, FFC; Thomas A. Downing, *A Register of Fires and Alarms of Fires . . .* (Philadelphia, 1836); Edwards, *History*, 22; Neilly, "Violent Volunteers," 42–44.

27. Diligent Fire Company No. 10, handbill, 1852, FFC; ledgers contain many instances in which firefighters were injured or worse; see, for instance, June 12, 1851, December 10, 1851, Missouri Fire Company, Record of Fires, 1846–1855, STLVVFC; April 18, 1841, Union Fire Company, Minutes, 1840–1851, STLVVFC; Edwards, *History*, 266–82; Richard Boyd Calhoun, "From Community to Metropolis: Fire Protection in New York City, 1790–1875" (Ph.D. diss., Columbia University, 1973), esp. 132–35, 176–81; also, see for example a few of the dozens of departmental histories written about fire departments during the past one hundred years: A. W. Brayley, *A Complete History of the Boston Fire Department . . .* (Boston, 1889); J. Albert Cassedy, *The Firemen's Record* (Baltimore, 1891); Augustine Costello, *Our Firemen: A History of the New York Fire Department* (New York, 1887); Thomas O'Connor, *History of the Fire Department of* New Orleans, *from the Earliest Days to the Present Time . . .* (New Orleans, 1895).

28. David Paul Brown, *Oration on the Centennial Anniversary of the Organization of the Fire Department of Philadelphia* (1839), 9; Downing, *Register*, 1–2; membership certificate, Philadelphia Association for the Relief of Disabled Firemen, 1854, BCHS.

29. Ferguson, *Early Reminiscences*, 2–51; Ric Northrup, "Decomposition and Reconstitution: A Theoretical and Historical Study of Philadelphia Artisans, 1785–1820" (Ph.D. diss., University of North Carolina, Chapel Hill, 1989), 184–206.

30. Mechanics Fire Company, parade hat, n.d, CIGNA Museum and Art Collection; Mechanics Fire Company, parade hat, n.d., BCHS; Harry Rubinstein, "With Hammer in Hand: Working-Class Occupational Portraits," in Howard Rock, Paul A. Gilje, and Robert Asher, eds., *American Artisans: Crafting Social Identity, 1750–1850* (Baltimore, Md.: Johns Hopkins University Press, 1995), 176–98; Thomas Schlereth, *Cultural History and Material Culture: Everyday Life, Landscapes, and Museums* (Ann Arbor, Mich.: UMI Press, 1990), 113 ff.; Northrup, *Decomposition*, 163–206; Ferguson, *Early Reminiscences*, 2–53.

31. *The Pony Gazette*, New York Edition, November 18, 1855, MC 662, MDHS; May 23, 1843, October 29, 1845, November 1, 1848, Union Fire Company, Minutes, 1840–1851, STLVVFC; November 27, 1845–July 13, 1846, Phoenix Fire Company No. 7, vol. 2,

ser. 1, 1843–1849, STLVVFC; February 15, 1853, August 15, 1853, Missouri Fire Company No. 5, Minutes, 1851–1858, vol. 8, ser. 1, STLVVFC; July 12, 1848–April 17, 1899, Laclede Fire Company No. 10, Minutes, 1848–1858, STLVVFC.

32. *Fireman's Herald* 1, no. 11 (1882), 1; "On Fire Engines, Hose, and Some Other Apparatus," *Journal of the Franklin Institute* 2 (1827), 281–82; March 7, 1850, Missouri Fire Company, Record of Fires, STLVVFC.

33. *Fireman's Herald* 1, no. 11 (1882), 1; Richard M. Dorson, "Mose the Far-Famed and World Renowned," *American Literature: A Journal of Literary History, Criticism, and Bibliography* 5 (June 1943), 293; Stott, *Workers*, 223–76.

34. See, for examples, Garvan and Wojtowicz, *Catalogue*, 121–27; Mark Tebeau, Exhibition Notes for "We Hazard Ourselves: Firefighting in Nineteenth-Century Philadelphia," BCHS. Compare, for example, membership certificate, Pennsylvania Hose Company, 1835–1836, BCHS; membership certificate, Hibernia Engine Company No. 1, 1861 BCHS; certificate, Good Will Fire Company, 1860–1871, BCHS.

35. May 7, 1851, August 9, 1851, Missouri Fire Company, Record of Fires, STLVVFC; *The Fireman's Herald* 1, no. 11 (1882), 1; Bruce Laurie, *Working People of Philadelphia, 1800–1850* (Philadelphia: Temple University Press, 1980), 59–62; Stott, *Workers*, esp. 249–56; Edwards, *History*, 81–91; Neilly, "Violent Volunteers," 59–70, esp. 62–63; Calhoun, "From Community to Metropolis," esp. 132–35, 176–81.

36. Great Playing by the Diligent Engine No. 10, May 22, 1952 (typescript, n.d.), FFC; *The Old Philadelphia Fire Department, Period of 1850: The Great Engine Contest . . .* (lithograph, 1882), BCHS; Diligent Engine Company, membership certificate, ca. 1850, CIGNA Museum and Art Collection; February 7, 1849, Union Fire Company, Minutes, 1840–1851, STLVVFC; December 6, 1853, December 22, 1854, Missouri Fire Company No. 5, Minutes, 1851–1858, STLVVFC.

37. Lynch, *Volunteer Fire Department*, 22, 35–36; Edwards, *History*, 134–46, 161–69; March 5, 1850, March 7, 1850, Missouri Fire Company, Record of Fires, 1846–1855, STLVVFC; May 23, 1843, Union Fire Company, Minutes, 1841–1850, STLVVFC.

38. *The Fireman's Herald* 1, no. 11 (1882), 1–2; John V. Morris, *Fires and Firefighters* (Boston: Little Brown, 1953), 110–11; "Up and Down on Her Men" is also a title of a Mat Hastings painting in the Missouri Historical Society Collections; also see, for example, November 17, 1846, Missouri Fire Company, Record of Fires, 1846–1855, STLVVFC; August 29, 1844, Union Fire Co. No. 2, Minutes, 1840–1851; November 12, 1862, Minutes, Board of Directors, Fame Hose Company No. 5, 1862–1871, FFC. For another view of sexuality and fire engines, see Greenberg, *Cause for Alarm*, 70–76; on the connections between gender, sexuality, and domination, see for instance, Sherry Ortner, "Is Female to Male as Nature Is to Culture?" in Michelle Zimbalist Rosaldo and Louise Lamphere, eds., *Women, Culture, and Society* (Palo Alto, Calif.: Stanford University Press, 1974), 67–87.

39. *Report of the Committee Appointed . . . To Investigate the Date of Institution of the Hand-In-Hand Fire Company* (1857); *Statement and Testimony Submitted . . . in Opposition to the Claim of the Hand-In-Hand Fire Company as to Seniority* (1857); *Papers of the Philadelphia Fire Department on the Case of the Hibernia Fire Company (1752) vs. the Hand-In-Hand Fire Company (1823)* (1857); October 16, 1854–February 25, 1855, Ledger, Good Will Fire Company, 1837–1863, FFC; Hibernia Fire Engine Company No. 1, Statement of Fire Attendance, 1870, FFC; Frank H. Schell, "Old Volunteer Fire Laddies, the Famous, Fast, Faithful, Fistic,

Fire Fighters of Bygone Days" (unpublished manuscript), Frank H. Schell Papers, HSP, chap. 4, p. 2.

40. Allen Steinberg, *The Transformation of Criminal Justice: Philadelphia, 1800–1880* (Chapel Hill, N.C.: University of North Carolina Press, 1989), 135–37, 145–46; February 2–5, 1852, October 16, 1852, Missouri Fire Company, Minutes, 1851–1858, STLVVFC; May 30, 1845, Union Fire Company, Minutes, 1841–1851, STLVVFC.

41. March 18, 1848, Missouri Fire Company, Record of Fires, STLVVFC; *Fireman's Herald* (1882), 1–2; Neilly, "Violent Volunteers"; Firemen's Association Minutes, 1850–1858, April 13, 1850, Union Fire Company, Minutes, 1841–1851, STLVVFC; Roll Book, Diligent Fire Company, 1812–1820, HSP; Stott, *Workers*, 212–76; Peter Way, "Evil Humors and Ardent Spirits: The Rough Culture of Canal Construction Laborers," *Journal of American History* 79 (March 1993), 1397–1428; May 7, 1851, August 9, 1851, Missouri Fire Company, Record of Fires, STLVVFC; *Fireman's Herald* 1, no. 11 (1882), 1. On the element of working-class manhood in this period, see Gorn, *Manly Art*, 34–128; Stott, *Workers*, 212–76; Peter Way, "Evil Humors and Ardent Spirits: The Rough Culture of Canal Construction Laborers," *Journal of American History* 79 (March 1993), 1397–1428.

42. Scharf, *History of Philadelphia*, 1906.

43. Scharf, *History of Philadelphia*, 1907. An early painting of a fire shows an African American cowering in the foreground, illustrating popular attitudes about allowing African Americans to help restore public order; see *The 1819 Conflagration of the Masonic Hall*, Philadelphia, 1819, FFC.

44. Dana, *Fire Departments*, 159–66, 229–32.

45. For an overview of membership practices, see for instance, August 12, 1852, September 9, 1852, Kensington Fire Company, 1836–1852, FFC; November 13, 1920, Minutes, Harmony Fire Company, 1814–1822, FFC; Assistance Fire Company No. 8, Ballot Box, BCHS; Laurie, *Working Peoples*, 60; December 6–10, 1855, January 6, 1856, November 6, 1856, Franklin Fire Company No. 8, Minutes, 1855–1859, vol. 5, ser. 1, STLVVFC.

46. Edwards, *History*, 163, 169; Inspectors Report on the Missouri Fire Company, April 1, 1854, ser. 1, STLVVFC; Volunteer Fire Department Muster Rolls, St. Louis, 1845–1848, Folder 1, Box 34, Tiffany Collection, MOHS; membership list and frontispiece, Franklin Fire Company, 1855–1859, vol. 3, ser. 1, STLVVFC; Neilly, "Violent Volunteers"; Scharf, *History of Philadelphia*, 1882–2135; Stuart Blumin, *The Emergence of the Middle Class: Social Experience in the American City, 1760–1900* (New York: Cambridge University Press, 1990), 216–17 and 369, n. 84; Laurie, *Working People*, 59–62.

47. Tracking fire companies is difficult because companies frequently formed, disbanded, or relocated. Laurie, "Fire Companies in Southwark," esp. 76–79; *Philadelphia City Directories*, 1800–1870; Philadelphia Fire Association, Minutes of the Board of Control of the Philadelphia Fire Department, 1839–1845, FFC; Scharf, *History of Philadelphia*, 1882–2135; John A. Paxson, *This New Map of the City of Philadelphia for the Use of Firemen and Others* (Philadelphia, 1817); *Annual Report of the Committee of Legacies and Trusts on the Fire and Hose Establishment of the City of Philadelphia* (1838); *Rules and Regulations of the Board of Engineers of the Middle Fire District* (1845); *Charter of the Trustees of the Fire Association of Philadelphia* (1856); *Journal of the Select and Common Councils*, Appendix 124; *Report of the Chief Engineer of the Fire Department, 1856* (1857), 357–443; *Report of the Fire Department of Philadelphia for the year ending November 1, 1863* (1864); Edwards, *History*, 230–32; Mark

Tebeau, "'Eating Smoke': Masculinity, Technology, and the Politics of Urbanization, 1850–1950" (Ph.D. diss., Carnegie Mellon University, 1997), 70–83.

48. Tebeau, "Eating Smoke," 73–83; Tebeau, Exhibition Notes for "We Hazard Ourselves . . . ," BCHS; Paxson, *New Map;* Steinberg, *Transformation,* 137.

49. Blumin, *Middle Class,* 216–17 and 369, ns. 83–84; *Philadelphia City Directories,* 1810–1855; *The Census of the United States,* 1810–1850; Philadelphia Hose Company, *Historical Sketches of the Formation and Founders of the Philadelphia Hose Company . . .* (1854), 61–93; Tebeau, "Eating Smoke," 70–83.

50. Register of the Fire Department, 1868–1869, FFC; Dana, *Fire Departments;* Stott, *Workers;* Zurier, *American Firehouse,* 29–69; Calhoun, "From Community to Metropolis," 142–48, 354–55; Stephen J. Sullivan, "Ethnic and Social Composition of Brooklyn's Volunteer Firemen" (unpublished manuscript, Fire Fighting in Nineteenth-Century Brooklyn, Exhibition Planning Project, Brooklyn Historical Society, 1992), 22–41. For a different perspective, which emphasizes shared firefighting identity over community and company differences, see Greenberg, *Cause for Alarm,* esp. 41–79.

51. See, for instance, John B. Perry, *The Philadelphia Fireman's Songster* (1852); Humane Engine Company Commemorative Certificate, 1844, FFC; Hibernia Fire Engine Company, *The Hibernia Fire Engine Company . . . Visit to the Cities of New York, Boston, Brooklyn, Charlestown, and Newark . . .* (Philadelphia, Pa.: J. B. Chandler, 1859); *Manuscripts in the Fire Fighting Collection of the Spruance Library: A Finding Aid* (1996), BCHS. On the study of material culture, ritual, and public display, see for instance Susan G. Davis, *Parades and Power: Street Theatre in Nineteenth-Century Philadelphia* (Philadelphia, Pa.: Temple University Press, 1986), esp. 2–3, 110–16, 143–47; Nicholas B. Dirks, Geoff Eley, and Sherry B. Ortner, eds., *Culture/Power/History: A Reader in Contemporary Social Theory* (Princeton, N.J.: Princeton University Press, 1994); Barbara Melosh, *Engendering Culture: Manhood and Womanhood in New Deal Public Art and Theater* (Washington, D.C.: Smithsonian Institution Press, 1991); Steven Lubar and W. David Kingery, eds., *History from Things: Essays on Material Culture* (Washington, D.C.: Smithsonian Institution Press, 1993).

52. See, for instance, August 14, 1857, Franklin Fire Company, Minutes, 1855–1859, STLVVFC; February 2, 1847, March 2, 1847, February 26, 1849, Union Fire Company, Minutes, 1840–1851, STLVVFC; April 16, 1852, Laclede Fire Company, Minutes, 1848–1858, STLVVFC; on the diversity of symbols and costumes, see M. J. McCosker, *The Historical Collection of the Insurance Company of North America* (Philadelphia: Insurance Company of North America, 1945), 70–74, 90–103; Garvan and Wojtowicz, *Catalogue,* 128–39.

53. Garvan and Wojtowicz, *Catalogue,* 128–39; Scharf, *History of Philadelphia,* 1910–12; Western Engine Company banner, ca. 1840–1850, BCHS; Liberty Fire Company, parade hat, ca. 1850, MOHS.

54. Neptune Hose banner, n.d., CIGNA Museum and Art Collection; Scharf, *History of Philadelphia,* 1,911–12; Lynch, *Volunteer Fire Department,* 12–60; October 4, 1848, Union Fire Company, Minutes, 1840–1851, STLVVFC.

55. Poem delivered by Charles Kumle, May 8, 1844, Volunteer Firemen's Benevolent Association Records, 1845–1907, STLVVFC; see also, ceremonial speaking horn presented to the United States Hose Company, 1865, BCHS; see also, for example, invitation to the Good Will Library Association Citizens' Dress Ball, April 30, 1841, BCHS; menu for dinner sent to Pennsylvania Engine Company No. 12 of San Francisco by Confidence Engine Company No. 1, February 8, 1859, BCHS.

56. *The Hibernia Fire Engine Company . . .* ; January 1, 1848 to November 1, 1848, and October 3, 1849 to July 13, 1850, Union Fire Company, Minutes, 1840–1851, STLVVFC; James Boyd Jones Jr., "Mose the Bowery B'hoy and the Nashville Fire Department, 1849–1860," *Tennessee Historical Quarterly* 40, no. 2 (1981), 174; *The Pony Gazette*, New York Edition, November 18, 1855, MC 662, MDHS. On trumpets, see for instance, Sotheby's, *Fine Americana, Sale 6731* (1995): "A Rare American Silver Presentation Fireman's Trumpet, Conrad Bard and Son, Philadelphia, 1845"; United States Hose Company, presentation trumpet awarded to Jacob Tripler, 1871, commemorative certificate, firemen's parade, October 5, 1857, FFC; Garvan and Wojtowicz, *Catalogue*, 112, 118–20.

57. Sam Bass Warner, *The Private City: Philadelphia in Three Periods of Its Growth* (Philadelphia, Pa.: University of Pennsylvania Press, 1968); Harry C. Silcox, "William McMullen, Nineteenth-Century Political Boss," *Pennsylvania Magazine of History and Biography* 110, no. 3 (July 1986), 389–412; Neilly, "Violent Volunteers," 76–82; Laurie, *Working People*, 151; Steinberg, *Transformation*, 134–67; on parades, see for example, Firemen's Parade program, 1843, BCHS; Firemen's Parade program, 1857, BCHS; announcement, Grand Parade, 1865, BCHS; Scharf, *History of Philadelphia*, 1,910.

58. Trumpet presented to Samuel Miller, Good Will Fire Company, 1838, BCHS; Neilly, "Violent Volunteers," 82–84; Nicholas Wainright, ed., *A Philadelphia Perspective: The Diary of Sidney George Fisher Covering the Years 1834–1871* (Philadelphia, Pa.: Historical Society of Pennsylvania, 1967), 174–75, 122.

59. Bernstein, *Draft Riots*, 18–22.

60. Alexander Blackburn, *Philadelphia Fire Marshal Almanac and Underwriters' Advertiser for the Year 1860* (1860), 192; Neilly, "Violent Volunteers," 30–33; *Report of the Fire Department of Philadelphia*, 9–11; Fire Association of Philadelphia, *The Fire Association of Philadelphia: A Short Account of the Origin and Development of Fire Insurance in Philadelphia* (Philadelphia, 1917), 16–29; Resolution Concerning Donation from Phoenix Assurance Company of London, 1809, FFC; Fire Miscellaneous, List of Donations to Hose Companies, 1807–1824, RG 5/14, INA, ACE. Also see *Charter of the Trustees of the Fire Association, 1826* (1826); *Charter of the Trustees of the Fire Association of Philadelphia* (1842); *Charter of the Trustees of the Fire Association, 1856* (1856).

61. Philadelphia Hose Company, *Historical Sketches*, 41–43.

62. August 29, 1839, Minutes, Board of Control of the Philadelphia Fire Department, 1839–1845, BCHS; September, 1839, Articles of Constitution, Board of Control, FFC.

63. "Fire Companies. Ordinance of May 21, 1840," *A Digest of the Ordinances of the Corporation of the City of Philadelphia and of the Acts of Assembly* (1841), 118–19; Neilly, "Violent Volunteers," 51–52; October 25, 1839, June 9, 1840, March 27, 1840, November 22, 1839, Minutes, Board of Control, 1839–1845, FFC; *Legacies and Trusts* (1838); Tebeau, "Eating Smoke," figs. pp. 74, 76.

64. September, 1839 to November 22, 1839, Minutes, Board of Control, 1839–1845, FFC; *United States Gazette*, June 14, 1843; Neilly, "Violent Volunteers," 74.

65. Steinberg, *Transformation*, 43, 49, 37–118.

66. *Constitution and By-Laws of the Firemen's Convention for the Incorporated District of the Northern Liberties and Kensington* (1842); *Annual Report of the Committee of Legacies and Trusts on the Fire and Hose Establishment of the City of Philadelphia* (1838); Minutes, Board of Control, 1839–1845, FFC.

67. *Statistics of Philadelphia Comprehending a Concise View of All the Public Institutions and*

the Fire Engine and Hose Companies of the City and County of Philadelphia, on the First of January, 1842 (1842); *Statistics of Philadelphia Comprehending a Concise View of All the Public Institutions and the Fire Engine and Hose Companies of the City and County of Philadelphia, on the First of January, 1842* (1842); also, for example, see December 15, 1849, Union Fire Company, Minutes, 1840–1851, STLVVFC; February 9, 1851, Membership Roll, Phoenix Hose Company, 1845–1854, FFC; August, 3, 1838, Minutes, Board of Directors, Resolution Hose Company, 1837–1842, FFC; Membership Roll, Fairmount Engine Company, 1854, BCHS.

68. Philadelphia Hose Company, *Historical Sketches*, 61–93; *Statistics of Philadelphia (1842).*

TWO: The Business of Safety: The American Fire Insurance Industry, 1800–1850

1. The notion that underwriters "took risks" emerged from my encounter with the archives, but an example of the pervasiveness of its use can be found in Henry R. Gall and William George Jordan, *One Hundred Years of Fire Insurance: Being a History of the Aetna Insurance Company, Hartford, Connecticut, 1819–1919* (Hartford: Aetna Insurance Company, 1919); for the history of risk, see Peter L. Bernstein, *Against the Gods: The Remarkable Story of Risk* (New York : John Wiley and Sons, 1996); on early information management strategies, see JoAnne Yates, *Control through Communication: The Rise of System in American Management* (Baltimore, Md.: Johns Hopkins University Press, 1989); on the use of quantitative, objective evidence in everyday life, see for example, Theodore M. Porter, *Trust in Numbers: The Pursuit of Objectivity in Science and Public Life* (Princeton: Princeton University Press, 1995); on early-nineteenth-century business and finance, see Naomi R. Lamoreaux, *Insider Lending: Banks, Personal Connections, and Economic Development in Industrial New England* (New York: Cambridge University Press, 1994); Alfred Chandler, *The Visible Hand: The Managerial Revolution in American Business* (Cambridge, Mass.: Harvard University Press, Belknap, 1977).

2. James A. Waterworth, *My Memories of the St. Louis Board of Fire Underwriters: Its Members and Its Work* (St. Louis, Mo.: Skaer Printing Company, 1926), 14–16.

3. Philadelphia Contributionship, *Deed of Settlement* (1752); Elizabeth Gray Kogen Spera, "Building for Business: The Impact of Commerce on the City Plan and Architecture of the City of Philadelphia, 1750–1800" (Ph.D. diss., University of Pennsylvania, 1980), 181–82; J. A. Fowler, *History of Insurance in Philadelphia for Two Centuries, 1683–1882* (Philadelphia, Review Publishing and Printing, 1888), 297–98.

4. *Pennsylvania Gazette*, April 10, 1774, quoted in Fowler, *History*, 306; Mutual Assurance Company, *Deed of Settlement* (1801); Spera, "Building for Business," 184–86 and 205, n. 5.

5. Sharon Salinger, "Spaces, Inside and Outside, in Eighteenth-Century Philadelphia," *Journal of Interdisciplinary History* 26, no. 1 (Summer 1995), 7; Fowler, *History*, 290–307.

6. Marquis James, *Biography of a Business, 1792–1942* (New York: Bobbs-Merrill, 1942), 18–23; Spera, "Building for Business," 180–81; Fowler, *History*, 310 ff.; also, on INA and early underwriting see William H. A. Carr, *Perils Named and Unnamed: The Dramatic Story of the Insurance Company of North America* (New York: McGraw-Hill, 1967); Nicholas B. Wainwright, *A Philadelphia Story* (Philadelphia, 1952); Clive Trebilcock, *Phoenix Assurance*

and the Development of British Insurance: Volume 1, 1782–1870 (Cambridge, Mass.: Cambridge University Press, 1985).

7. Fowler, *History*, 303 ff., 310–20; more broadly, on the practice of business in this period, see Chandler, *Visible Hand*, chap. 1; Lamoreaux, *Insider Lending*, esp. 1–83; Allan Pred, *Making Histories and Constructing Human Geographies* (Boulder, Colo.: Westview Press, 1990), 51–73.

8. James, *Biography*, 40, 20–27; Fowler, *History*, 308–9.

9. Salinger, "Spaces," 3; Spera, "Building for Business," 184–86; Fowler, *History*, 357–58.

10. December 1, 1807, Board of Directors Minutes, INA Collection, ACE (hereafter the collection will be cited only as INA); Fowler, *History*, 16–17; James, *Biography*, 103–4.

11. Fowler, *History*, 342–47.

12. Fowler, *History*, 309–17, 322, 327, 336, 346–47.

13. For instance, see policies 10445 (1821), 10653 (1822), 19279 (1838), 27354 (1847), Sample Policies, 1820–1848, RG 5, INA; Fowler, *History*, 317–19, 310–30, 336–37.

14. Quoted in Fowler, *History*, 336–38.

15. Fowler, *History*, 309–10.

16. William J. Novak, *The People's Welfare: Law and Regulation in Nineteenth-Century America* (Chapel Hill, N.C.: University of North Carolina Press, 1996), 51–83; Fowler, *History*, 300–313.

17. Spera, "Building for Business," 279–317; Fowler, *History*, 303–5; Salinger, "Spaces," table 1, p. 8. It appears that about 18 percent of Philadelphia houses were insured in 1769.

18. *Poulson's Daily Advertiser*, April 13, 1819, quoted in Fowler, *History*, 333, 348.

19. Fowler, *History*, 323, 348, 353–55; Warner, *Private City*, 49–51; Margaret Hindle Hazen and Robert M. Hazen, *Keepers of the Flame: The Role of Fire in American Culture, 1775–1925* (Princeton, N.J.: Princeton University Press, 1992), 65–109, 195–203; Aetna, *Rates and Rules: Instructions for the Use of Agents* (Hartford: Aetna Insurance Company, 1857).

20. Fowler, *History*, 345–47; Henry R. Gall and William George Jordan, *One Hundred Years of Fire Insurance: Being a History of the Aetna Insurance Company, Hartford, Connecticut, 1819–1919* (Hartford: Aetna Insurance Company, 1919), 62.

21. Gall and Jordan, *One Hundred Years*, 57–70.

22. Quoted in Gall and Jordan, *One Hundred Years*, 62, 52–54, 66–70.

23. Quoted in Gall and Jordan, *One Hundred Years*, 62, 57–58, 68–70.

24. September 2, 1834, September 4, 1834, September 9, 1834, Letterbook 2, RG 1/3, Aetna Collection, ACE (hereafter cited only as letterbook with volume number; Aetna Collection will be cited only as Aetna); July 1, 1844, July 3, 1844, Letterbook 9.

25. July 4, 1837, Letterbook 3; quoted in Gall, *One Hundred Years*, 83; Simeon L. Loomis, *To Seth Burton: Being a Consideration of Agency Practice in Writing Fire Insurance in 1848* (republished 1943), Box 4, RG 4/2, Aetna; Aetna, *Directions for Agents* (1857), 6; on the middle class in this period, see Karen Halttunen, *Confidence Men and Painted Women: A Study of Middle-Class Culture in America, 1830–1870* (New Haven: Yale University Press, 1982), 1–32; Mary P. Ryan, *Cradle of the Middle Class: The Family in Oneida County, New York, 1790–1865* (New York: Cambridge University Press, 1981); Stuart M. Blumin, *The Emergence of the Middle Class: Social Experience in the American City, 1760–1900* (New York: Cambridge University Press, 1989); Mark C. Carnes and Clyde Griffen, eds., *Meanings for Manhood: Constructions of Masculinity in Victorian America* (Chicago: University of Chicago Press,

1990); E. Anthony Rotundo, *American Manhood: Transformations in Masculinity from the Revolution to the Modern Era* (New York: Basic Books, 1993), 18–30. For an overview of the concept of moral hazard, albeit in British insurance, see Robin Pearson, "Moral Hazard and the Assessment of Insurance Risk in Eighteenth- and Early-Nineteenth-Century Britain," *Business History Review* 76 (Spring 2002), 1–35.

26. Loomis, *Agency Practice*, 16.

27. Aetna, *Instructions to Agents* (1825), section 3; July 1–3, 1829, August 4, 1830, Letterbook 1; on changes in practice compare to Aetna, *Directions* (1857).

28. Gall and Jordan, *One Hundred Years*, 40–41, 79.

29. June 27, 1829, July 12, 1829, August 10, 1830, Letterbook 1; September 9, 1834, Letterbook 2; compare changes in section 7 of Aetna, *Instructions* (1825); Aetna, *Instructions* (1830).

30. On classification systems as bureaucratic technologies, see Geoffrey C. Bowker and Susan Leigh Star, *Sorting Things Out: Classification and Its Consequences* (Cambridge, Mass.: MIT Press, 1999), 1–32, 135 ff., 319–28.

31. Loomis, *Agency Practice*, 7; Aetna, *Instructions* (1825), sec. 12.

32. Aetna, *Instructions* (1825), sec. 12.

33. September 2, 1834, Letterbook 2.

34. Aetna, *Instructions* (1830), sec. 7; usually the company asked agents to draw diagrams on the back of policies: for instance, January 9, 1838, Letterbook 3.

35. Gall and Jordan, *One Hundred Years*, 72–73.

36. August 21–22, 1843, Diary of Business Trip to Montreal, Calling on Agents, En Route, 14 August 1843 to 31 August, 1843, Box 4, RG 4/2, Aetna (hereafter, Morgan Diary).

37. August 21, 1843, Morgan Diary.

38. August 21, 1843, Morgan Diary.

39. August 21, 1843, Morgan Diary.

40. Quoted in Gall and Jordan, *One Hundred Years*, 82–83; Aetna, *Instructions* (1825), sec. 10.

41. Quoted in Gall and Jordan, *One Hundred Years*, 74.

42. August 10–12, 1830, Letterbook 1.

43. August 10–12, 1830, Letterbook 1; Gall and Jordan, *One Hundred Years*, 75; Novak, *People's Welfare*, 54–79; David Dana, *The Fireman: The Fire Departments of the United States . . .* (1858), esp. 267–367; Margaret Sloane Patterson, "Nicolino Calyo and His Paintings of the Great Fire of New York, December 16th and 17th, 1835," *American Art Journal* 14, no. 2 (Spring 1982), 4–22.

44. For some examples of cross-fertilization of firefighters and insurers compare, Fowler, *History*, 296–99, 406; Philadelphia Hose Company, *Historical Sketches of the Formation and Founders of the Philadelphia Hose Company . . .* (1854), 61–93; for images on policies see Sample Policies, 1820–1848, INA.

45. Gall and Jordan, *One Hundred Years*, 37–39.

46. The author identified six requests for support extant in INA's letterbooks during the first decade of the nineteenth century; on four occasions INA granted support. On two occasions there was no record of donation. Additionally, see Letter from Fire Hose Association of Philadelphia, n.d., Correspondence to INA, Fire Insurance Misc., RG 5/1, INA.

47. March 30, 1813, August 7, 1812, Correspondence to INA, INA.

48. Letter from Fire Hose Association of Philadelphia, n.d., Correspondence to INA, INA; Neilly, "Violent Volunteers," 30.

49. July 1, 1806, Letter from Fire Hose Association of Philadelphia, n.d., Correspondence to INA, INA.

50. Correspondence between Volunteer Fire Companies and INA, 1805–1813, List of Amounts Paid to Hose Companies by INA, 1807–1824, Fire Insurance Misc., RG 5/1, INA; February 24, 1835, February 28, 1838, October 16, 1838, February 20, 1840, Board of Directors Minutes, vol. 4, 1825–1845, INA; March 26, 1850, Fire Association Minutes, 1849–1859, STLVVFC.

51. November 11, 1838, October 4, 1840, Board of Directors Minutes, vol. 4, 1825–1845, INA; Gall and Jordan, *One Hundred Years*, 48; August 22, 1843, Morgan Diary.

52. July 14, 1829, Letterbook 1; July 6, 1858, Letterbook 34; quoted in Gall and Jordan, *One Hundred Years*, 48; July 28, 1851, Letterbook 20.

53. Gall and Jordan, *One Hundred Years*, 83.

54. Loomis, *Agency Practice*, 4.

55. Gall and Jordan, *One Hundred Years*, 65–70.

56. Loomis, *Agency Practice*, 4, 11; Loomis, *Circular Letter of Advice and Instructions to Agents* (April 1850), Box 4, RG 4/2, Aetna.

THREE: Statistics, Maps, and Morals: Making Fire Risk Objective, 1850–1875

1. *"Mr. Binney's Address," Centennial Meeting of the Philadelphia Contributionship for the Insurance of Houses from Loss by Fire, Second Monday of April, 1852* (1852; reprinted 1876, 1885); J. A. Fowler, *History of Insurance in Philadelphia for Two Centuries, 1683–1882* (Philadelphia: Review Publishing and Printing, 1888), 394–95; Henry R. Gall and William George Jordan, *One Hundred Years of Fire Insurance: Being a History of the Aetna Insurance Company, Hartford, Connecticut, 1819–1919* (Hartford: Aetna Insurance Company, 1919), 82 ff.

2. H. C. Watson, *Jerry Pratt's Progress; or Adventures in the Hose House* (Philadelphia: Brown Job Printing Office, 1855), 66; *Report of a Committee, Appointed at a Meeting Held on Friday Evening, December 3rd, 1852, To Consider the Propriety of Organising a Paid Fire Department* (1853), 11 (hereafter, *Report . . . [on] Organising a Paid Fire Department).*

3. On the relation between capitalism and cities, see David Harvey, *The Urbanization of Capital: Studies in the History and Theory of Capitalist Accumulation* (Baltimore, Md.: Johns Hopkins University Press, 1985); William Cronon, *Nature's Metropolis: Chicago and the Great West* (New York: Norton, 1991); for the notion of discipline, see Michel Foucault, *Discipline and Punish: The Birth of the Prison* (New York: Vintage, 1990).

4. *Mr. Binney's Address*, 56–57.

5. *Mr. Binney's Address*, 52–53.

6. *Mr. Binney's Address*, 52–53, 57.

7. Aetna, *Instructions for the Use of Agents* (1857), 34 ff.; Classification of Fire Risk, vol. 1, 1852–1872, item 19, RG 5/1, Aetna Collection, ACE (hereafter the collection will be cited only as Aetna).

8. Classification of Fire Risk, vol. 1.

9. Aetna, *Instructions* (1857), 25–26, 36–50; Aetna, *Directions for Agents* (1857), 18 ff.

10. Classification of Fire Risk, vol. 1; Classification of Fire Risk, vol. 2, 1872–1888, item 19, RG 5/1, Aetna; Aetna, *Instructions* (1857), 25.

11. Aetna, *Directions* (1857), 4–5; J. B. Bennett, "A Central Insurance Agency," circular, 1854, RG 4/2, Aetna.

12. Aetna, *Instructions* (1857), 25; J. B. Bennett, *Aetna Guide to Fire Insurance: For the Representatives of the Aetna Insurance Company of Hartford, Conn.; Cincinnati Branch* (Cincinnati, Ohio: Robert Clarke and Co., 1867), 200 ff.

13. Bennett, *Guide*, 200–202; also see other contemporary manuals: C. C. Hine, *Fire Insurance: A Book of Forms, Containing Forms of Policies, Endorsements, Certificates, and Other Valuable Matter, For the Use of Agents and Others* (New York: Sanford, Harroun, 1865); C. C. Hine, *Fire Insurance: A Book of Instructions for the Use of the Agents of the United States* (New York: The Insurance Monitor, 1870).

14. Aetna, *Directions* (1857), 13; Classification of Fire Risk, vol. 1; Classification of Fire Risk, vol. 2.

15. Classification of Fire Risk, vol. 1, Aetna.

16. Aetna, *Instructions* (1857), 21 ff.; Classification of Fire Risk, vol. 1; Classifications of Fire Risk, vol. 2.

17. Aetna, *Instructions* (1857), 4–5. The first reference that I found to the term *moral hazard* in any guidebook is in the "Aetna Bible" of 1866; see Bennett, *Guide*. The concept itself appeared sometime in the mid–nineteenth century; for an overview and discussion of the concept of moral hazard, albeit in the context of British insurance, see Robin Pearson, "Moral Hazard and the Assessment of the Insurance Risk in Eighteenth- and Early-Nineteenth-Century Britain," *Business History Review* 76 (Spring 2002), 1–35.

18. On the middle class in this period see Mary P. Ryan, *Cradle of the Middle Class: The Family in Oneida County, New York, 1790–1865* (New York: Cambridge University Press, 1981); Stuart M. Blumin, *The Emergence of the Middle Class: Social Experience in the American City, 1760–1900* (New York: Cambridge University Press, 1989), esp. 1–16; Karen Halttunen, *Confidence Men and Painted Women: A Study of Middle-Class Culture in America, 1830–1870* (New Haven: Yale University Press, 1982), 1–55; Mark C. Carnes and Clyde Griffen, eds., *Meanings for Manhood: Constructions of Masculinity in Victorian America* (Chicago: University of Chicago Press, 1990); E. Anthony Rotundo, *American Manhood: Transformations in Masculinity from the Revolution to the Modern Era* (New York: Basic Books, 1993), 22–25, 104–85.

19. Aetna, *Instructions* (1857), 5.

20. Michael Kimmel, *Manhood in America* (New York: Free Press, 1996), 84–85, 103–5, 81–117.

21. Bennett, *Guide*, 140 ff.

22. Mr. Binney's Address, 57–58.

23. See, for instance, Map of Patriot, Ohio, 1846 (corrected 1883), RG 4/1, Aetna. In 1790, the Phoenix Assurance Company commissioned the earliest urban insurance map, of Charleston, and the company also mapped other risks outside of Britain between 1805 and 1837. However, there was not a systematic effort by insurance firms in the United States—and probably globally—to produce or commission atlases of urban fire risk until the efforts discussed in the text. See Library of Congress, *Fire Insurance Maps in the Library of Congress: Plans of North American Cities and Towns Produced by the Sanborn Map Company* (Washington, D.C.: Library of Congress, 1981), introduction.

24. On the early maps, see Walter W. Ristow, "United States Fire Insurance and Underwriters Maps, 1852–1968," *Quarterly Journal of the Library of Congress* 25, no. 3 (July 1968), 194–219; Sanborn Map Company, *Description and Utilization of The Sanborn Map* (New York: Sanborn Map Company, 1953), 3; November 2, 1852, Board of Directors Minutes, vol. 5, 1845–1860, INA; cf. Ernest Hexamer and William Locher, *Maps of the City of Philadelphia, Volume 3, Comprising the 7th and 8th Wards* (1858); Western Bascome and John A. Parr, *Insurance Map of St. Louis, Mo.* (1859).

25. Helena Wright, "Insurance Mapping and Industrial Archeology," *IA: The Journal of the Society for Industrial Archeology* 9, no. 1 (1993), 1–19; cf. Hexamer and Locher, *Maps of . . . Philadelphia* (1858); A. Whipple, *A. Whipple and Company's Insurance Map of St. Louis, Missouri* (1876); William Perris, *Fire Insurance Map of New York* (1852).

26. Wright, "Mapping," 10 ff.; A. Whipple, *A. Whipple and Company's Insurance Maps; 2nd Series, Special Risks* (Rev. ed., 1872); Ernest Hexamer and Son, *General Surveys of Philadelphia, 29 Volumes, 1866–1895* (1866); Joyce A. Post, *A Consolidated Name Index to the Hexamer General Surveys* (Free Library of Philadelphia, 1974); J. B. Post and Joyce Post, "Indexing the Hexamer General Surveys," *Special Libraries* (March 1977), 103–8; for a complete list of Hexamer maps, see Ernest Hexamer & Son, *Alphabetical Index to Hexamer's Insurance Maps of Philadelphia and Suburban Towns* (1913).

27. Fowler, *History*, 426; Hexamer and Locher, *Maps of . . . Philadelphia* (1858); Ernest Hexamer, *Insurance Maps of the City of Philadelphia, Volume 4* (1873; pasteovers to 1901); Perris, *Fire Insurance Map* (1852); Bascome and Parr, *Insurance Map* (1859); Mark Tebeau, "Re-Imagining the Urban Landscape: Fire Risk and Insurance in Nineteenth-Century St. Louis," in Andrew Hurley, ed., *Common Fields: An Environmental History of St. Louis* (St. Louis: Missouri Historical Society Press, 1997), 126–45.

28. Cf. Hexamer and Locher, *Maps of . . . Philadelphia* (1858); Hexamer, *Insurance Maps* (1873); Ernest Hexamer & Son, *Insurance Maps of the City of Philadelphia, Volume III* (1908, revisions to 1915).

29. C. T. Aubin, *St. Louis Fire Insurance Maps, Surveyed and Drawn for the St. Louis Board of Fire Underwriters* (St. Louis, 1874).

30. See, for instance, A. Whipple, *Insurance Map* (1876), front/back covers, frontispiece, examined at MOHS. Pasteovers are evident only on paper copies. Between 1870 and 1895 Whipple's Insurance Protective Agency published *Whipple's Daily Fire Reporter*, which revealed much about how mapping companies operated; see chapter 5 for further discussion.

31. See, for instance, A. Whipple, *A. Whipple and Company's Insurance Map of St. Louis, Missouri* (1870), 8, 11, 20, examined at MOHS; Francis Cruger Moore, *Fires: Their Causes, Prevention, and Extinction, Combining Also a Guide to Agents . . .* (1877), 137–39.

32. Visit to Quebec, 1856, Book 1, Diaries and Notebooks of Special Agent A. A. Williams Covering Visits to Quebec, New England, and Mid-Atlantic States Agencies, 1855–1857, Item 21, Box 5, RG 5/1, Aetna (hereafter, Williams, Visit to Quebec).

33. Williams, Visit to Quebec; the Goad Company began to map Canada shortly after Perris and Hexamer began mapping the United States; Robert Hayward, *Fire Insurance Plans in the National Map Collection* (Ottawa: Public Archives of Canada, 1977); G. T. Bloomfield, "Canadian Fire Insurance Plans and Industrial Archeology," *IA: The Society for Industrial Archeology* 8 (1982), 67–80.

34. Williams, Visit to Quebec; Aetna, *Directions* (1857); see also pages 6–9 of Visit to New England, 1856, Book 5 of Diaries and Notebooks of Special Agent A. A. Williams Cov-

ering Visits to Quebec, New England, and Mid-Atlantic States Agencies, 1855–1857, Item 21, Box 5, RG 5/1, Aetna (hereafter Williams, Visit to New England); Aetna, *Instructions* (1857), appended diagram.

35. Williams, Visit to New England, 2; Williams, Visit to Quebec; Aetna, *Directions* (1857).

36. Williams, Visit to Quebec; Visit to New England.

37. *Mr. Binney's Address*, 34–38, 55–58.

38. *Insurance Monitor* 19, no. 7 (July 1866), 254; E. R. Hardy, *The Making of the Fire Insurance Rate* (1926), 43; F. C. Oviatt, "Historical Study of Fire Insurance in the United States," *Annals of the American Academy of Political and Social Science* 26 (1905), 347; Marc Schneiberg, "Political and Institutional Conditions for Governance by Association: Private Order and Price Controls in American Fire Insurance," *Politics and Society* 27, no. 1 (1999), 67–104; Robert Riegel, *Fire Underwriters Associations in the United States* (New York: Chronicle Company, 1916).

39. *Insurance Monitor* 20, no. 4 (1872), 464; Fowler, *History*, 396.

40. *At a Meeting of the Fire Insurance Companies and Agencies in the City and County of Philadelphia, April 7, 1852* (1852); Fowler, *History*, 416, 396 ff.

41. Fowler, *History*, 396, 397–400.

42. Fowler, *History*, 396 ff.

43. Fowler, *History*, 414–15, 422, 426.

44. Fowler, *History*, 421, 430, 448.

45. *Mr. Binney's Address*, 58, 63–67.

46. Fowler, *History*, 369 ff. In New York City a coalition of insurers and property owners organized a trial of a steam fire engine in 1841; see Rebecca Zurier, *The American Firehouse: An Architectural and Social History* (New York: Abbeville Press, 1982), 71; Paul Ditzel, *Fire Engines, Fire Fighters* (New York: Crown, 1976), 90–93.

47. Although the committee did not list its full membership, its recommendations were signed by several leaders, including local underwriter Frederick Fraley, who was a member of the Philadelphia Hose Company; Philadelphia Hose Company, *Historical Sketches;* Fowler, *History*, 406; *Report . . . [on] Organising a Paid Fire Department; Report of the Committee Appointed to Devise a Plan for the Better Organization of the Fire Department* (1853)—hereafter, *Report . . . [on] Better Organization; Letters of the Judges of the Court of Quarter Sessions, and The Marshal of Police, and Report of the Board of Trade* (1853)—hereafter, *Judges' Letters.* For comparison, see Geoffrey Gigleirano, "History of the Cincinnati Fire Department," typescript, Cincinnati Historical Society, 1980, 34–39; Geoffrey Gigleirano, "A Creature of Law: Cincinnati's Paid Fire Department," *Cincinnati Historical Society Bulletin* 40, no. 2 (1982), 78–99.

48. *Report . . . [on] Organising a Paid Fire Department*, 11; *Report . . . [on] Better Organization*, 4.

49. *Report . . . [on] Organising a Paid Fire Department*, 5; *Report . . . [on] Better Organization.*

50. *Judges' Letters*, 13–17; *Report . . . [on] Better Organization; Report . . . [on] Organising a Paid Fire Department;* Fowler, *History*, 406 ff.

51. *Report . . . [on] Organising a Paid Fire Department*, 9–10.

52. *Report . . . [on] Organising a Paid Fire Department*, 12; Blumin, *Middle Class*, 179–91, 248–97.

53. See, for instance, Henry G. Harrison and William N. Weightman, *The Fireman, Number 3; Near a Fire: Say! Just hold this while I fetch another section, will you (Likely?)* (1858); Anthony N. B. Garvan and Carol A. Wojtowicz, *Catalogue of the Green Tree Collection* (Philadelphia: Mutual Assurance Company, 1977), 146–49; *Report . . . [on] Organising a Paid Fire Department,* 12.

54. Watson, *Jerry Pratt's Progress,* 24, 5–8; *Harper's* quoted in Blumin, *Middle Class,* 183. See chapter 4 for discussion of the class-based nature of this critique, especially its links to the popular stage representation of firemen, Mose; also, see Richard M. Dorson, "Mose the Far-Famed and World Renowned," *American Literature: A Journal of Literary History, Criticism, and Bibliography* 15 (June 1943); Richard Stott, *Workers in the Metropolis: Class, Ethnicity, and Youth in Antebellum New York City* (Ithaca, N.Y.: Cornell University Press, 1990), 223–26, 249–76.

55. Watson, *Jerry Pratt's Progress;* George Rogers Taylor, "'Philadelphia in Slices' by George G. Foster," *Pennsylvania Magazine of History and Biography* 93, no. 1 (January 1969), 36–37.

56. Watson, *Jerry Pratt's Progress,* 66.

57. *Report of the Special Committee of the Select and Common Council in Relation to the Fire Alarm and Police Telegraph, Presented 12 October 1854* (1854); *Report of the Sub-Committee to Put the Steam Fire Engine "Young America" in Service* (1854).

58. *Report . . . [on] Organising a Paid Fire Department,* 11–12.

59. *Report . . . [on] Better Organization,* 2.

60. *Report . . . [on] Better Organization,* 4–6; on consolidation, see Eli K. Price, *The Consolidation of Philadelphia* (1873); Henry Leffman, *The Consolidation of Philadelphia* (1908); Michael P. McCarthy, "The Philadelphia Consolidation Act of 1854: A Reappraisal," *Pennsylvania Magazine of History and Biography* 110 (1986).

61. *Report . . . [on] Organising a Paid Fire Department,* 9–13; *Report . . . [on] Better Organization,* 2–4.

62. *Report . . . [on] Better Organization,* 6; *Pennsylvanian,* January 8, 1853, *Public Ledger,* January 8, 1853, quoted in *Report . . . [on] Better Organization,* 3; Zurier, *American Firehouse,* 36; *Report . . . [on] Organising a Paid Fire Department,* 14.

63. *Public Ledger,* January 8, 1853, quoted in *Report . . . [on] Better Organization,* 3.

64. *Public Ledger,* January 8, 1853, quoted in *Report . . . [on] Better Organization,* 3; *Report of the Paid Fire Department to Common Council* (1854). The *Ledger* exaggerated the number of active firemen; in 1868, only three thousand men served actively, but there were as many as five thousand honorary members; see Register of the Fire Department 1868–1869, FFC.

65. *Report . . . [on] Organising a Paid Fire Department,* 16, 11–12.

66. On the political power of firemen, see for example Andrew M. Neilly, "The Violent Volunteers: A History of the Volunteer Fire Department of Philadelphia, 1736–1871" (Ph.D. diss., University of Pennsylvania, 1959), 94–100; *A Philadelphia Perspective: The Diary of Sidney George Fisher Covering the Years, 1834–1871,* ed. Nicholas Wainwright (Philadelphia: HSP, 1967), 174–75.

67. *Report . . . [on] Organising a Paid Fire Department,* 16–17; *Report . . . [on] Better Organization,* 2–4.

68. *Report . . . Fire Alarm and Police Telegraph* (1854), 254–59; *Address to the Municipal Authorities, Fire Department, Insurance Companies, Merchants, Property Holders, and the Insured of the City of Philadelphia, Against Loss by Fire* (1854); Joel Tarr, "The Municipal Telegraph

Network: Origins of the Fire and Police Alarm Systems in the United States," *Flux* 9 (July–September 1992), 5–18.

69. J. Albert Cassedy, *The Firemen's Record* (Philadelphia, n.d.), 99–103; *Report of the Committee on a Paid Fire Department, Made to Common Council, May 5th, 1859* (1859); *Report . . . "Young America" in Service.*

70. *Report . . . [on] Organising a Paid Fire Department,* 11.

FOUR: Muscle and Steam: Establishing Municipal Fire Departments, 1850–1875

1. May 29, 1855, Phoenix Hose Company Ledger, FFC, BCHS.

2. Allen Steinberg, *The Transformation of Criminal Justice: Philadelphia, 1800–1880* (Chapel Hill, N.C.: University of North Carolina Press, 1989), 134–40, 145–47, 163–67; *The Mysteries and Miseries of Philadelphia, As Exhibited by a Late Presentment of the Grand Jury, and by a Sketch of the Condition of the Most Degraded Classes in the City* (1853); *Report of the Committee on Police on the Circumstances Attending and Connected with the Destruction of the Pennsylvania Hall, and Other Consequent Disturbances of the Peace* (1838); Michael Feldberg, *The Philadelphia Riots of 1844: A Study of Ethnic Conflict* (Westport, Conn.: Greenwood Press, 1975). For an evaluation of fighting among firefighters that finds a similar overstatement of its prevalence but which offers a different interpretation, see Amy Greenberg, *Cause for Alarm: The Volunteer Fire Department in the Nineteenth-Century City* (Princeton, N.J.: Princeton University Press, 1998), 80–108. More broadly, rough physicality was an integral part of daily life in the nineteenth century, including among firefighters; see for instance Peter Way, *Common Labour: Workers and the Digging of North American Canals, 1780–1860* (New York: Cambridge University Press, 1993); Elliot Gorn, *The Manly Art: Bare-Knuckle Prize-fighting in America* (Ithaca, N.Y.: Cornell University Press, 1990); Richard Stott, *Workers in the Metropolis* (Ithaca, N.Y.: Cornell University Press, 1990), esp. 257–62; Bruce Laurie, "Fire Companies and Gangs in Southwark: The 1840s," in Allen F. Davis and Mark Haller, eds., *The Peoples of Philadelphia: A History of Ethnic Groups and Lower Class Life, 1790–1840* (Philadelphia: Temple University Press, 1973), 71–87.

3. George Rogers Taylor, "'Philadelphia in Slices' by George G. Foster," *Pennsylvania Magazine of History and Biography* 93, no. 1 (January 1969), 36–37.

4. Steinberg, *Transformation,* 145 ff.; J. Thomas Scharf and Thompson Westcott, *History of Philadelphia, 1609–1884* (Philadelphia: L. H. Everts and Company, 1884), 1,914; *The Life and Adventures of Charles A. Chester, the Notorious Leader of the Philadelphia "Killers"* (Philadelphia, 1850); *The Almighty Dollar; or, the Brilliant Exploits of a Killer* (Philadelphia, 1847); John B. Perry, *The Philadelphia Firemen's Songster* (Philadelphia, n.d.); Harry Silcox, "William McMullen, Nineteenth-Century Political Boss," *Pennsylvania Magazine of History and Biography* 110, no. 3 (July 1986), 389–412; H. C. Watson, *Jerry Pratt's Progress; or Adventures in the Hose House* (Philadelphia: Brown Job Printing Office, 1855); August 29, 1839, Minutes, Board of Control, 1839–1845, FFC; In re The Northern Liberty Hose Company, 1 Harris 193, HSP.

5. Benjamin Baker and Frank Chanfrau, *A Glance at New York in 1848* (New York, ca. 1848); Richard M. Dorson, "Mose the Far-Famed and World Renowned," *American Literature: A Journal of Literary History, Criticism, and Bibliography* 15 (June 1943), 293; Stott, *Workers,* 223–26, 249–76; Richard Butsch, "Bowery B'hoys and Matinee Ladies: The Re-

Gendering of Nineteenth-Century American Theater Audiences," *American Quarterly* 46, no. 3 (September 1994), 374–403; James Boyd Jones, Jr., "Mose the Bowery B'hoy and the Nashville Fire Department, 1849–1860," *Tennessee Historical Quarterly* 40, no. 2 (1981), 170–81.

6. *A Philadelphia Perspective: The Diary of Sidney George Fisher Covering the Years 1834–1871*, edited by Nicholas Wainwright (Philadelphia: HSP, 1967), 122, 174–75.

7. *Diary of Sidney George Fisher*, 122.

8. Richard Vaux, *Address Delivered before the Philadelphia Hose Company on the 47th Anniversary, December 16, 1850* (1851), 18–19; Andrew H. Neilly, "The Violent Volunteers: A History of the Volunteer Fire Department of Philadelphia, 1736–1871" (Ph.D. diss., University of Pennsylvania, 1959); Scharf and Westcott, *History of Philadelphia*, 1883–1921; on unstable business conditions, see *The Western Underwriter: A Monthly Magazine Devoted to the Interests of Fire, Marine, and Life Insurance* 1, no. 3 (February 1857), 62 ff.

9. Nokes, *The Paid Fire Department* (Philadelphia, 1853), HSP; Anthony N. B. Garvan and Carol A. Wojtowicz, *Catalogue of the Green Tree Collection* (Philadelphia: Mutual Assurance Company, 1977), 143; on the political and social context of the 1850s, see for instance, Eric Foner, *Free Soil, Free Labor, Free Men: The Ideology of the Republican Party before the Civil War* (1970; reprint, New York: Oxford University Press, 1995); David Roediger, *Wages of Whiteness: Race and the Making of the American Working Class* (New York: Verso Press, 1991), 85–87; Alexander Saxton, *The Rise and Fall of the White Republic: Class Politics and Mass Culture in Nineteenth-Century America* (New York: Verso Press, 1990).

10. For an excellent discussion of working-class speech patterns, including their manifestation in fire companies, see Stott, *Workers*, 257–62. On parades, see for example, Firemen's Parade Program, 1843, BCHS; Announcement, Grand Parade, 1857, BCHS; Firemen's Parade Program, 1857, BCHS; Scharf, *History of Philadelphia*, 1910.

11. On changes in the capacities and organization of federal, state, and local government in the nineteenth century, see Stephen Skowronek, *Building a New American State: The Expansion of National Administrative Capacities, 1877–1920* (New York: Cambridge University Press, 1982), 3–35; Eric Monkkonen, *America Becomes Urban: The Development of U.S. Cities and Towns, 1780–1980* (Berkeley: University of California Press, 1988), 89–130; on consolidation, see Steinberg, *Transformation*, 162–67; Howard F. Gillette Jr., "The Emergence of the Modern Metropolis: Philadelphia in the Age of Consolidation," in Howard F. Gillette and William W. Cutler III, eds., *The Divided Metropolis: The Social and Spatial Dimensions of Philadelphia, 1800–1975* (Westport, Conn.: Greenwood Press, 1980); Eli K. Price, *The History of Consolidation in the City of Philadelphia* (1873).

12. Minutes, Board of Directors of the Philadelphia Fire Department, 1853–1855, HSP; Minutes, Board of Directors, Philadelphia Fire Department, 1855–1864, HSP; Minutes, Board of Directors, Philadelphia Fire Department, 1864–1870, HSP; Roll Book, Board of Directors of the Philadelphia Fire Department, 1868–1870, HSP; *Constitution and By-Laws of the Board of Engineers of the Middle Fire District, Reported to the Select and Common Councils, April 11, 1844* (1844); *Rules and Regulations of the Board of Engineers of the Middle Fire District, Revised 1845* (1845).

13. October 24, 1853, January 23, 1854, Minutes, Board of Directors, 1853–1855, HSP.

14. January 23, 1854, Minutes, Board of Directors, 1853–1855, HSP.

15. August 9, 1853, December 26, 1853, Minutes, Board of Directors, 1853–1855, HSP; *Report of a Committee, Appointed at a Meeting Held on Friday Evening, December 3rd, 1852, to*

Consider the Propriety of Organising a Paid Fire Department (1853); *Report of the Committee Appointed to Devise a Plan for the Better Organization of the Fire Department* (1853); *Letters of the Judges of the Court of Quarter Sessions, and the Marshal of Police, and Report of the Board of Trade* (1853).

16. April 10, 1854, Minutes, Board of Directors, 1853–1855, HSP.

17. February 6, 1854, Minutes, Board of Directors, 1853–1855, HSP.

18. April 10, 1854, May 22, 1854, January to May, 1854, Minutes, Board of Directors, 1853–1855, HSP.

19. April 10, 1854, Minutes, Board of Directors, 1853–1855, HSP; Steinberg, *Transformation*, 146.

20. March 26, 1854, September 9, 1854, November 29, 1854, April 23, 1855, Minutes, Board of Directors, 1853–1855, HSP.

21. March–May 1855, esp. April 23, 1855, August 27, 1855, Minutes, Board of Directors, 1855–1864, HSP; September 24, 1855, November, 26, 1855, Minutes . . . 1853–1855, HSP. The newly consolidated municipality recognized the Board of Directors, which was elected by fire companies, as representing firefighters' interests and helping to administer day-to-day operations of the department. However, formal legal authority, including especially financial matters, resided with the Committee of Trusts and Fire Department, which was appointed by the municipal government. See Scharf, *History of Philadelphia*, 1909; Minutes, Board of Directors, 1853–1855, HSP; Minutes, Board of Directors, 1855–1864, HSP; Minutes, Board of Directors, 1864–1870, HSP; Roll Book, Board of Directors of the Philadelphia Fire Department, 1868–1870, HSP; Minutes of the Committee on Trusts and Fire Department, 1854–1871, FFC.

22. Rebecca Zurier, *The American Firehouse: An Architectural and Social History* (New York: Abbeville Press, 1982), 71; Paul Ditzel, *Fire Engines, Fire Fighters* (New York: Crown, 1976), 90–93; Geoffrey Gigleirano, "History of the Cincinnati Fire Department," unpublished manuscript, Cincinnati Historical Society, 1980, 34 ff., 111 ns. 57 and 61; Geoffrey Gigleirano, "A Creature of Law: Cincinnati's Paid Fire Department," *Cincinnati Historical Society Bulletin* 40, no. 2 (1982), 78–99; Firemen's Protective Association, *History of the Cincinnati Fire Department* (1895; reprint, Cincinnati, Ohio: Heskamp Publishing, 1988), 103–30; Alexander Latta and E. Latta, *The Origin and Introduction of the Steam Fire Engine . . .* (1860); William T. King, *History of the American Steam Fire Engine* (1896; reprint, Chicago: Owen Davies, 1960), 13–19.

23. *Report of the Joint Special Committee of the Select and Common Councils on Steam Fire Engines* (1854), 3–4.

24. May 29, 1855, Phoenix Ledger, FFC; *Report of the Joint Special Committee;* Diligent Engine Company, Membership Certificate, ca. 1850, CIGNA Museum and Art Collection; Harry Rubenstein, "With Hammer in Hand: Working-Class Occupational Portraits," in Howard Rock, Paul A. Gilje, and Robert Asher, eds., *American Artisans: Crafting Social Identity, 1750–1850* (Baltimore, Md.: Johns Hopkins University Press, 1995), 183; Mechanics Fire Company, parade hat, n.d., CIGNA; Mechanics Fire Company, parade hat, n.d, BCHS.

25. May 29, 1855, Phoenix Ledger, FFC; *Report of the Sub-Committee to Put the Steam Fire Engine "Young America" in Service* (1856); Philadelphia Hose Company, *Historical Sketches of the Formation and Founders of the Philadelphia Hose Company . . .* (1854).

26. May 29, 1855, Phoenix Ledger, FFC.

27. May 29, 1855, Phoenix Ledger, FFC.

28. *Report . . . "Young America"*, 5; August 23, 1855, to July 11, 1857, Minutes, Committee on Trusts, 1854–1871; J. A. Fowler, *History of Insurance in Philadelphia for Two Centuries, 1683–1882* (Philadelphia: Review Publishing and Printing, 1888), 406.

29. September 25, 1855, November 15, 1855, letter (loose), September 27, 1855, Minutes of the Committee on Trusts and Fire Department, 1854 to 1871, FFC; *Report . . . "Young America,"* 12–13.

30. Alexander W. Blackburn and William A. Strait, *The Philadelphia Fire Marshal Almanac and Underwriters' Advertiser, for the Year 1860* (1860), 180.

31. Blackburn and Strait, *Fire Marshal*, 174–75, 180–84, 191; Philadelphia Hose Company, *Historical Sketches of the Formation and Founders of the Philadelphia Hose Company . . .* (1854); King, *History*, 34–40.

32. Blackburn and Strait, *Fire Marshal*, 185–91; November 15, 1855, July 12, 1858, Minutes, Committee on Trusts, 1854–1871, FFC; *Report of the Visiting Committee of the Board of Delegates of the Fire Association of Philadelphia, December 3, 1860* (1860), 7–8; King, *History*, 34–41, 48–49, 65–67, 75–76, 113.

33. Magee, *A Paid Fire Department As It Is Likely to Be under the Contract System* (n.d.), HSP.

34. Cf. Magee, *A Paid Fire Department*; G. G. Heiss, *Philadelphia Hose Company Steam Fire Engine* (Philadelphia, ca. 1858–1860), CIGNA Museum and Art Collection.

35. *Statistics of Philadelphia Comprehending a Concise View of All the Public Institutions and the Fire Engine and Hose Companies of the City and County of Philadelphia, on the First of January, 1842* (1842); *Journal of the Select and Common Councils, Appendix #124, Report of the Chief Engineer of the Fire Department, 1856*, 357–443; *Report of the Fire Department of Philadelphia for the year ending November 1, 1863* (1864); *Report of the Visiting Committee . . . 1860.*

36. The telegraph especially reinforced the administrative innovations, such as ongoing efforts to restrict fire companies to battling blazes in well-defined fire districts; see for example, Route of the Good Will Steam Fire Engine Company, in Going to and Returning from Fires, ca. 1859–1871, FFC; January 23, 1854, Minutes, Board of Directors, 1853–1855, HSP; June 12, 1866, Minutes of the Committee on Trusts, 1854–1871, FFC.

37. July 12, 1859, Minutes of Committee on Trusts, 1854–1871, FFC; on the common palette of membership certificates, cf. Delaware Fire Company member certificate, 1870, FFC; Anthony N. B. Garvan and Carol A Wojtowicz, *Catalogue of the Green Tree Collection* (Philadelphia: Mutual Assurance Company, 1977), 108–10.

38. January 16, 1869, Minutes of the Committee on Trusts, 1854–1871, FFC; Third Firemen's Footrace, March 12, 1855, FFC; Hibernia Engine Company No. 1, Statement Showing the Results and 3 Months Attendance at Fine-able Fires, December 2, 1869, March 2, 1870, FFC; Announcement of the Grand Parade of the Philadelphia Fire Department, October 16, 1865, FFC; Announcement of the Grand Parade of the Philadelphia Fire Department, October 5, 1857, FFC; Firemen's Parade Program, March 27, 1843, FFC; Firemen's Parade Program, October 5, 1857, FFC. On the changing conventions of parades, see Susan Davis, *Parades and Power: Street Theatre in Nineteenth-Century Philadelphia* (Philadelphia: Temple University Press, 1986), esp. 159–66.

39. Fire companies could thwart the wishes of the Chief Engineer, see April 24, 1860, Minutes of the Committee on Trusts, 1854–1871, FFC.

40. March 26, 1850, Fire Association Minutes, 1849–1859, STLVVFC (hereafter, FA Minutes); James Neal Primm, *Lion of the Valley* (2d ed.; Boulder, Colo.: Pruett Publishing, 1990), 174–75.

41. March 26, 1850, FA Minutes; Primm, *Lion*, 146; Thomas Lynch, *The Volunteer Fire Department of St. Louis, 1819–1850* (1880), 98–99.

42. March 26, 1850, FA Minutes.

43. March 26, 1850, FA Minutes; complaints using similar language were lodged in other cities about the same time; see Jones, "Mose," 178.

44. March 26, 1850, FA Minutes; on the occupations of FA leaders, see *St. Louis City Directory, 1850*.

45. March 26, 1850, FA Minutes.

46. March 26, 1850, FA Minutes.

47. July 2, 1850, July 12, 1850, January 28, 1851, February 7, 1851, February 26, 1851, March 6, 1851, FA Minutes; April 4, 1848, February 20, 1849, June 12, 1849, August 7, 1849, August 21, 1849, vol. 5 (1845–1860), Board of Directors Minutes, INA, ACE.

48. January 28, 1851, February 7, 1851, February 24, 1851, February 27, 1851, March 6, 1851, March 28, 1855, FA Minutes; on the economically privileged members, I refer to Branson, Carroll, Powers, and Miller.

49. March 6, 1851, June 13, 1853, FA Minutes.

50. Union Fire Company, Minutes, 1833–1840, STLVVFC; Union Fire Company, Minutes, 1841–1850, STLVVFC; also see, for example, Edward Edwards, *History of the Volunteer Fire Department of St. Louis* (1906), 75–91, 134–46.

51. Fire Underwriters of St. Louis, *Report of the Fire Departments of Cincinnati and St. Louis and the Use of Steam Fire Engines* (St. Louis: George Knapp and Company, 1858), 7; King, *History*, 13–18;

52. For analyses of steam fire engines and departmental change in which the role of steam engines is overdetermined, see Annelise Graebner Anderson, "The Development of Municipal Fire Departments in the United States," *Journal of Libertarian Studies* 3, no. 3 (1979), 331–59; *Cincinnati Daily Commercial*, January 18, 1853; Gigleirano, "History," 37–41, 111 n. 57; Gigleirano, "A Creature of Law," 78–99; Amy Greenberg, *Cause for Alarm: The Volunteer Fire Department in the Nineteenth-Century City* (Princeton, N.J.: Princeton University Press, 1998), 125–51.

53. This figure is conservative. Comparing names on the St. Louis Fire Department personnel files and the department's 1857 payroll with limited extant membership records from volunteer fire companies revealed that at least twenty-three of sixty-five men had previously volunteered; cf. STLFD, Personnel File; November and December 1857, Payroll of the Paid Fire Department of St. Louis, Tiffany, MOHS; *Proceedings of the NAFE, 1891*, 22.

54. St. Louis Firemen's Fund, *History of the St. Louis Fire Department: . . . from which the Lesson of the Vast Importance of Having Efficient Firemen May be Drawn* (St. Louis: Central Publishing Company, 1914), 156, 168–69 (hereafter, *History STLFD*); Edwards, *History*, 146–52, 186–99; Payroll for the Washington Fire Company to 1 November 1857, Payment of Ordinance Requisitions of the St. Louis Fire Department, 1857, Tiffany.

55. Payroll of the Franklin Fire Company to 1 November 1857, Payroll of the Mound Fire Company to 1 November 1857, Payment . . . 1857, Tiffany; November–December 1857, January–March, 1858, Payroll . . . St. Louis, Tiffany; Franklin Fire Company, 1855, Fire Company Membership Lists, Tiffany; Edwards, *History*, 193–94.

56. November–December 1857, January–March, 1858, Payroll . . . St. Louis; *History STLFD*, 170.

57. November–December 1857, January–March, 1858, Payroll . . . St. Louis.

58. April 3, 1857, April 30, 1857, June 1, 1857, and July 1, 1857, Payment of Fire Department Bills, Tiffany; November–December 1857, January–March, 1858, Payroll . . . St. Louis; about Mawdsley, see Edwards, *History*, 193–94; *History STLFD*, 159; *Report of the Fire Departments of Cincinnati and St. Louis . . .* (1858), 7; *Mayor's Message with Accompanying Documents Submitted to City Council of the City of St. Louis . . . May, 1865*, "Report of the Fire Department," 65–69.

59. *Second Annual Convention of the Association of Fire Engineers* (1875), 24–25; Stephen Roper, *Handbook of Modern Steam Fire Engines* (Philadelphia: Claxton, Remsen, and Haffelfinger, 1876), 110–12; expense reports indicate that responsibility for repairing the city's steam engines was given to local machine shops; see *Mayor's Message . . .*, "Report of the Fire Department," 65–69; King, *History*, 52–62, 148–50.

60. Philadelphia Department of Public Safety, *Information for Firefighters* (1915), 63; PFD Departmental Orders, 1919 (typescript), pages 19–40, Box A3630, PFD Records, PCA.

61. *History STLFD*, 171; Edwards, *History*, 277–79.

62. Zurier, *Firehouse*, 101–10; W. Fred Conway, *Chemical Fire Engines* (New Albany, Ind.: Fire Buff House Publishers, 1987); the *Proceedings of the NAFE* regularly listed the latest inventions by firefighters.

63. For instance, Monkkonen, *America Becomes Urban*, 93–103; Stanley K. Schultz, *Constructing Urban Culture: American Cities and City Planning, 1800–1920* (Philadelphia: Temple University Press, 1989), 111–49; Martin Melosi, *The Sanitary City: Urban Infrastructure in America from Colonial Times to Present* (Baltimore, Md.: Johns Hopkins University Press, 2000), 58–116.

64. Neilly, "Violent Volunteers," 183–96; Albert Cassedy, *The Firemen's Record* (1883), 99–103; *Report of the Committee on a Paid Fire Department, Made to Common Council, May 5th, 1859* (1859); *Report . . . Steam Fire Engine "Young America."*

65. November–December 1857, Payroll . . . St. Louis, Tiffany; Gigleirano, "Creature of Law," 78–99; *History of the Cincinnati Fire Department* (1895); Richard Boyd Calhoun, "From Community to Metropolis: Fire Protection in New York City, 1790–1875" (Ph.D. diss., Columbia University, 1973), 288–344; *Amoskeag Steam Fire Engines and Hose Carriages* (Philadelphia: Edward Stearn and Co., 1896), esp. 34–54; *Ordinance Establishing a Paid Fire Department, 1871;* page 1, Fire Department of Philadelphia, History, PFD, *Bulletin* Clippings, TUUA; Anderson, "Development," 349–50.

FIVE: Disciplining the City: Everyday Practice and Mapping Risk, 1875–1900

1. *Proceedings of the IAFE, 1894*, 97–101.

2. J. B. Bennett, *Aetna Guide to Fire Insurance: For the Representatives of the Aetna Insurance Company of Hartford, Conn.; Cincinnati Branch* (Cincinnati: Robert Clarke and Company, 1867), 193–95, 246; compare, for instance, changes in the manuals that Aetna distributed to agents during the nineteenth century, which include guides published by the firm in 1825, 1830, 1857, 1859, 1867, and 1876, Items 1–9, Section 1, RG 5/3, Aetna; Charles C. Dominge and Walter O. Lincoln, *Fire Insurance Inspection and Underwriting: An Encyclopedic Handbook . . .* (1919; reprint, Chicago: Spectator Company, 1929), 715.

3. J. A. Fowler, *History of Insurance in Philadelphia for Two Centuries, 1683–1882* (1888),

396 ff.; on the complex history of government fire regulation see William J. Novak, *The People's Welfare: Law and Regulation in Nineteenth-Century America* (Chapel Hill: University of North Carolina Press, 1996), 51–82; on the NBFU see Harry Chase Brearley, *Fifty Years of a Civilizing Force: The History of the National Board of Fire Underwriters* (New York: Frederick A. Stokes Co., 1916); also see, variously, *Proceedings of the NBFU, 1866–1900.*

4. *Report of the Proceedings of the National Convention of Fire Underwriters* (1866), 14.

5. *An Interview with Mr. Henry A. Oakley, President of the National Board of Fire Underwriters* (1873), 12; *At a Meeting of the Fire Insurance Companies in the City and County of Philadelphia* (1852); Loss Books Collected by Platt, 1870–1899, RG 12, INA.

6. See, for example, *Annual Report of the Committee of Adjustments & Statistics of the National Board of Fire Underwriters, 1873–1874* (1874), cover, 112; *Proceedings of the National Board of Fire Underwriters, 1867*, 4; National Board of Fire Underwriters, *Circular No. 1, issued April 19, 1872;* National Board of Fire Underwriters, *Circular No. 71, issued June 9, 1873;* NBFU; *Circular No. 81, issued June 30, 1873; Annual Report of the Committee on Statistics of the National Board of Fire Underwriters, 1877–1878.*

7. For an overview, see Lester W. Zartman and William H. Price, eds., *Yale Readings in Insurance: Property Insurance, Marine and Fire* (1910; reprint, New Haven: Yale University Press, 1926), esp. 121–39; Roger Kenney, *Fundamentals of Fire and Casualty Insurance Strength* (Dedham, Mass.: Kenney Insurance Studies, 1967).

8. *Proceedings of a Convention of Fire Insurance Companies Interested in Agency Business Held in the City of New York, November 1, 1865* (1865); *Report . . . National Convention of Fire Underwriters* (1866); Bissell, "Organization of Companies," in Zartman and Price, *Yale Readings*, 121–39; *The Standard: Fire Insurance Tables* (1894).

9. Bennett, *Guide*, 207–9, 245–46; J. B. Bennett, A Central Insurance Agency, Circular, 1854, RG 4/2, Aetna; *Interview with Mr. Henry A. Oakley*, 4.

10. National Board of Fire Underwriters, *Pioneers of Progress, 1866–1941* (New York: H. Wolff, 1941), 120–21; Brearley, *Fifty Years*, 18–19, 25; *Proceedings of the NBFU, 1867*, 4; *Proceedings of the Executive Committee of the NBFU, 1869*, May 11 and July 7–8, 1869, 6–13.

11. Quoted in Brearley, *Fifty Years*, 25–26; NBFU, *Pioneers*, 120–21; *Proceedings of the Executive Committee NBFU*, December 8 and 9, 1869, 6–8.

12. Brearley, *Fifty Years*, 32, 35; *Proceedings of the NBFU, Executive Committee, 1872;* NBFU, *Pioneers*, 120, Fowler, *History*, 512–14, 517, James A. Waterworth, *My Memories of the St. Louis Board of Fire Underwriters: Its Members and Its Work* (St. Louis: Skaer Printing Company, 1926), 52.

13. NBFU, *Pioneers*, 121; Brearley, *Fifty Years*, 37, 51–52; Waterworth, *Memories*, 52–53.

14. Brearley, *Fifty Years*, 38–39, 47–49; *Proceedings of the NBFU, 1873;* Jon C. Teaford, *The Unheralded Triumph: City Government in America, 1870–1900* (Baltimore, Md.: Johns Hopkins University Press, 1984), 200–203; Letty Anderson, "Fire and Disease: The Development of Water Supply Systems in New England, 1870–1900," in Joel Tarr and Gabriel Dupuy, eds., *Technology and the Rise of the Networked City in Europe and America* (Philadelphia: Temple University Press, 1988), 142, 137–56; Martin Melosi, *The Sanitary City: Urban Infrastructure in America from Colonial Times to Present* (Baltimore, Md.: Johns Hopkins University Press, 2000), 117, 103–48; more broadly, on nineteenth-century conflagrations, see Christine Meisner Rosen, *The Limits of Power: Great Fires and the Process of City Growth in America* (New York: Cambridge University Press, 1986), 95–108.

15. Brearley, *Fifty Years*, 54–55, 47–49; Waterworth, *Memories*, 54–55, 57; NBFU, *Pi-*

oneers, 122–23; *Bulletin of the Executive Committee of the NBFU*, vol. 3, no. 1 (January and February, 1876), 1–4; *Proceedings of the NBFU, Executive Committee, 1876; Proceedings of the NBFU, Executive Committee, 1877; Proceedings of the NBFU, 1888*, 5.

16. Waterworth, *Memories*, 50–51.

17. Waterworth, *Memories*, 50–51; Fowler, *History*, 510–11.

18. Waterworth, *Memories*, 51–53; Western Bascome and John A. Parr, *Insurance Map of St. Louis, Mo.* (St. Louis, 1859).

19. Angel Kwolek-Folland, *Engendering Business: Men and Women in the Corporate Office, 1870–1930* (Baltimore, Md.: Johns Hopkins University Press, 1994), 74.

20. *The Underwriter, Fire and Marine: The Men Who Take the Risks, Historical and Biographical* (Chicago: Fidelity Publishing, 1896); F. C. Oviatt, "History of Fire Insurance in the United States," in Zartman and Price, eds., *Yale Readings*, 87; Henry R. Gall and William George Jordan, *One Hundred Years of Fire Insurance: Being a History of the Aetna Insurance Company, Hartford, Connecticut, 1819–1919* (Hartford, Conn.: Aetna Insurance Company, 1919), 98, 165–69.

21. C. C. Hine, *Fire Insurance: A Book of Instructions for the Use of the Agents of the United States* (New York: Insurance Monitor, 1870); C. C. Hine, *Fire Insurance, Policy Forms, and Policy Writing* (New York: Insurance Monitor, 1882), 4.

22. *The Underwriter, Fire and Marine;* Hine, *Instructions*, frontispiece; Francis Cruger Moore, *Fires: Their Causes, Prevention, and Extinction; Combining Also a Guide to Agents Respecting Insurance against Loss by Fire . . .* (1877; reprint, New York, 1881); Francis Cruger Moore, *Fire Insurance and How to Build: Combining Also a Guide to Insurance Agents Respecting Fire Prevention and Extinction, Special Features of Manufacturing Risks, Writing of Policies, Adjustment of Losses, Etc., Etc.* (New York: Baker and Taylor, 1903).

23. Hine, *Instructions*, 6, 26 ff., 77 ff., 98–102.

24. List of Agents, 1855, RG 5/1, Aetna; Marc Schneiberg, "Private Order and Inter-firm Governance in the American Fire Insurance Industry, 1820–1950" (Ph.D. diss., University of Wisconsin, 1994), 119–25, 181, 321; Kwolek-Folland, *Engendering Business*, 78; *Western Underwriter*, September 11, 1902, 6; New York State, "Hearings Before the Joint Committee of the Senate and Assembly, 1909–1911," in *New York Assembly Documents, Number 30* (Albany, N.Y.: J. B. Lyon, 1911), 1444.

25. Moore, *Fires*, 11 ff.; Kwolek-Folland, *Engendering Business*, 70–93.

26. Moore, *Fires*, 18–24; moral hazard achieved systematic use sometime after the middle of the nineteenth century; on the concept, see Robin Pearson, "Moral Hazard and the Assessment of the Insurance Risk in Eighteenth- and Early-Nineteenth-Century Britain," *Business History Review* 76 (Spring 2002), 1–35.

27. Moore, *Fires*, 5–6, 14–15.

28. *Proceedings of the NBFU, 1896*, 19, 64.

29. On manhood in this period, see E. Anthony Rotundo, *American Manhood: Transformations in Masculinity from the Revolution to the Modern Era* (New York: Basic Books, 1993), 222–83; Michael Kimmel, *Manhood in America* (New York: Free Press, 1996), 81–190; Peter Filene, *Him/Her/Self: Sex Roles in Modern America* (Baltimore, Md.: Johns Hopkins University Press, 1986), 86 ff.

30. Moore, *Fires*, 138–41; also on mapping in fire insurance see chapter 3 above; Helena Wright, "Insurance Mapping and Industrial Archaeology," *IA: The Journal of the Society for Industrial Archaeology* 9, no. 1 (1993), 1–19; R. P. Getty, "Insurance Surveying and

Mapmaking," *Cassier's Magazine: An Engineering Monthly* 39, no. 1 (November 1910), 19–25; Walter W. Ristow, "United States Fire Insurance and Underwriters Maps, 1852–1968," *Quarterly Journal of the Library of Congress* 25, no. 3 (July 1968), 194–219; Sanborn Map Company, *Description and Utilization of the Sanborn Map* (New York: Sanborn Map Company, 1953).

31. Compare, for instance, Ernest Hexamer, *Maps of the City of Philadelphia Surveyed by Ernest Hexamer and William Locher, Civil Engineers and Surveyors, 1858; Volume 3, Comprising the 7th and 8th Wards* (1858); Ernest Hexamer, *Insurance Maps of the City of Philadelphia, Volume 9* (1873; pasteovers to 1901); Ernest Hexamer and Son, *Insurance Maps of the City of Philadelphia, Volume III* (1908, revisions to 1915).

32. January 19, 1883, September 2, 1882, *Whipple Daily Fire Reporter* (hereafter *WDFR*).

33. December 13, 1882, January 24, 1882, May 30, 1882, *WDFR.*

34. January 11, 1882, August 3, 1882, January 19, 1883, *WDFR.*

35. There is little scholarly literature that deals with the extent to which municipal building laws were enforced, but such enforcement appears to have been lax; however, on the politics of regulation and property ownership, see Novak, *People's Welfare*, esp. 51–52; Robin L. Einhorn, *Property Rules: Political Economy in Chicago, 1833–1872* (Chicago: University of Chicago Press, 1991).

36. On the concept of warranty, see for example Fontaine Talbott Fox Jr., *A Treatise on Warranty in Fire Insurance Contracts* (Chicago: Callaghan and Co., 1883).

37. George Richards, "Fundamental in the Law of Insurance and Why Adopted," in The Insurance Society of New York, *The Fire Insurance Contract: Its History and Interpretation* (New York: Rough Notes, 1922), 73–80.

38. Dominge and Lincoln, *Insurance Inspection*, 633.

39. January 19, 1883, May 18, 1882, *WDFR.*

40. January 24, 1882, *WDFR.*

41. July 21, 1882, *WDFR.*

42. July 21, 1882, *WDFR.*

43. September 4, 1882, *WDFR;* on the salvage corps in Philadelphia, sponsored by the insurance industry, *Report of the Philadelphia Fire Insurance Patrol, for the Year 1872.*

44. Dominge and Lincoln, *Insurance Inspection*, 555–58; Helena Wright, "Insurance Mapping," 2; Ristow, "Underwriters' Maps," 202; Sanborn Map Company, *Surveyor's Manual for the Exclusive Use and Guidance of Employees* (1905); *The Sanborn History*, 2; L. L. Buchanan, "The Publication of Fire Insurance Maps," *Fire Insurance Club of Chicago*, June 11, 1912.

45. Recently, historian Sara Wermeil has argued that the stock insurance industry changed direction as a result of pressures from mutual insurers; although this was clearly a factor, this chapter adopts a view more in keeping with Marc Schneiberg's argument, which emphasizes the political and institutional conditions reshaping industry practice; compare Sara E. Wermeil, *The Fireproof Building: Technology and Public Safety in the Nineteenth-Century American City* (Baltimore, Md.: Johns Hopkins University Press, 2000), esp. 37–103; Marc Schneiberg, "Political and Institutional Conditions for Governance by Association: Private Order and Price Controls in American Fire Insurance," *Politics and Society* 27, no. 1 (March 1999), 67–104.

46. Schneiberg, "Private Order," 305, 301 ff.; Schneiberg, "Political and Institutional Conditions," 67–104; Robert Riegel, *Fire Underwriters Associations in the United States* (New York: Chronicle Company, 1916); Brearley, *Fifty Years.*

47. H. R. Hayden, *Hayden's Annual Cyclopedia of Insurance in the United States, 1901–1902* (Hartford, Conn.: Insurance Journal Company, 1902), 168–69; Schneiberg, "Private Order," 306; Schneiberg, "Political and Institutional Conditions," 67–104.

48. Waterworth, *Memories*, 157–70; STLFD, Personnel Files; Teaford, *Unheralded Triumph*, 200 ff.

49. *Annual Cyclopedia . . . 1901–1902*, 168–69; *Philadelphia Fire Underwriters Tariff Association, 1884; Philadelphia Fire Underwriters' Association, Ninth Annual Report, November 18, 1892* (1892); *Tariff Adopted by the Association of Fire Underwriters of Philadelphia*, edition of April 1876.

50. *PFUA Report, 1892*, 4; Waterworth, *Memories;* Francis Cruger Moore, *Standard Universal Schedule for Rating Mercantile Risks* (New York, 1902).

51. *Weekly Underwriter* 48, no. 20 (May 20, 1893), 354; *The Annual Cyclopedia of Insurance in the United States, 1893–1894* (Hartford, Conn.: H. R. Hayden, 1894), 546–48, 549; *Annual Cyclopedia, 1901–1902*, 590.

52. *Weekly Underwriter* 48, no. 9 (March 4, 1893), 155, 164–68; *Weekly Underwriter* 48, no. 13 (April 1, 1893), 244–46; *Weekly Underwriter* 48, no. 10 (March 11, 1893), 178–79, 184–86; *PFUA Report, 1892*, 2 ff.

53. *PFUA Report, 1892*, 4; Waterworth, *Memories*, 92–93.

54. Brearley, *Fifty Years*, 78–83.

55. *Proceedings of the NBFU, 1891*, 26–27, 147–51; *Proceedings of the NBFU, 1893*, 90 ff.; *Proceedings of the NBFU, 1890*, 46–48.

56. *Proceedings of the NBFU, 1893*, 83 ff.; *Proceedings of the NBFU, 1900*, 70 ff.; *Proceedings of the NBFU*, 1904, 79 ff.

57. *Proceedings of the NBFU, 1896*, 59 ff.; cf. Brearley, *Fifty Years*, 195–96.

58. *Annual Cyclopedia of Insurance, 1896–1897*, 391; *Weekly Underwriter* 58, no. 24 (1898), 414; *Proceedings of the NBFU, 1900*, 72–73, 82–83; also, on the collaboration, see *Proceedings of the NFPA, 1897–1902*.

59. See, for instance, *Proceedings of the NBFU, 1896*, 19, 64; Albert W. Whitney, *Report of the Co-Insurance Committee to the Board of Fire Underwriters of the Pacific on Percentage of Co-Insurance and the Relative Rates Chargeable Therefore, also on the Cost of Conflagration Hazard of Large Cities* (1905). Betsy W. Bahr, "New England Mill Engineering: Rationalization and Reform in Textile Mill Design, 1790–1920" (Ph.D. diss., University of Delaware, 1988); Wermeil, *Fireproof Building* (2000).

60. Fowler, *History*, 566–67; Joseph A. Schumpeter, *Capitalism, Socialism, and Democracy* (1942; reprint, New York: Harper, 1975), 82–85; Alan Trachtenberg, *The Incorporation of America: Culture and Society in the Gilded Age* (New York: Hill and Wang, 1982); Kimmel, *Manhood*, 75–190; Rotundo, *Manhood*, 224–44; Peter Stearns, *Be A Man!: Males in Modern Society* (New York: Holmes and Meier, 1990), 127–39; Gail Bederman, *Manliness and Civilization: A Cultural History of Gender and Race in the United States, 1880–1917* (Chicago: University of Chicago Press, 1995), 5–20;.

SIX: Becoming Heroes: A New Standard for Urban Fire Safety, 1875–1900

1. St. Louis Firemen's Fund, *History of the St. Louis Fire Department: With Review of Great Fires, and Sidelights upon the Methods of Fire-Fighting from Ancient to Modern Times, from which the Lesson of the Vast Importance of Having Efficient Firemen May Be Drawn*

(St. Louis: Central Publishing Company, 1914), 57–59 (hereafter, *History STLFD);* Arlen Dykstra, "St. Louis Mourns the Untimely Death of Phelim O'Toole," *Missouri Historical Review* (October 1974), 36; *Fire and Water* 30 (1901), 361; *Proceedings of the NAFE, 1885,* 23–25.

2. Donald M. O'Brien, *"A Century of Progress through Service": The Centennial History of the International Association of Fire Chiefs* (International Association of Fire Chiefs, 1973). This project distinguishes between occupational and work cultures; see Barbara Melosh, *"The Physician's Hand": Work Culture and Conflict in American Nursing* (Philadelphia: Temple University Press, 1982); Susan Porter Benson, *Counter Cultures: Saleswomen, Managers, and Customers in American Department Stores, 1890–1940* (Urbana: University of Illinois Press, 1986). Its approach to occupation is informed by the idea of jurisdiction in setting occupational and professional boundaries; see Andrew Abbott, *The System of Professions: An Essay on the Division of Expert Labor* (Chicago: University of Chicago Press, 1988). For an introduction to the historical literature on professionalism, see Robert H. Wiebe, *The Search for Order, 1877–1920* (New York: Hill and Wang, 1967); Burton J. Bledstein, *The Culture of Professionalism: The Middle Class and the Development of Higher Education in America* (New York: W. W. Norton, 1978).

3. Joseph A. Schumpeter, *Capitalism, Socialism, and Democracy* (1942; reprint, New York: Harper, 1975), 82–85.

4. For instance, see *Proceedings of the NAFE, 1873,* 8–12; *Proceedings of the NAFE, 1874,* 26–27; *Proceedings of the NAFE, 1896,* 112–25; *Fireman's Journal,* 44; *F & W* 1 (1886), 3.

5. For the clearest statement on the growing height of cities, especially city centers, see Robert Fogelson, *Downtown: Its Rise and Fall, 1880–1950* (New Haven, Conn.: Yale University Press, 2001), esp. 112–82; see also Vincent Scully, *American Architecture and Urbanism* (2d ed.; New York: Henry Holt, 1988), 103–10, 144–54; Leland M. Roth, *A Concise History of American Architecture* (New York: Harper and Row, 1979), 172–227; for a view of Philadelphia, see Richard Webster, *Philadelphia Preserved: Catalog of the Historic American Building Survey* (Philadelphia: Temple University Press, 1976), esp. 104–49; on the legacy of the St. Louis built environment, see http://stlcin.missouri.org/history/histstruct.cfm; Angel Kwollek-Folland, *Engendering Business: Men and Women in the Corporate Office, 1870–1930* (Baltimore, Md.: Johns Hopkins University Press, 1994), 96–100; John Burchard and Albert Brush Brown, *The Architecture of America: A Social and Cultural History* (Boston: Little, Brown, 1961), 152; *Proceedings of the IAFE, 1896,* 49–53; *Proceedings of the IAFE, 1897,* 54–67; for a definition and discussion of fireproof construction, see Sara E. Wermeil, *The Fireproof Building: Technology and Public Safety in the Nineteenth-Century American City* (Baltimore, Md.: Johns Hopkins University Press, 2000), esp. 1–10.

6. *F & W* 34 (1903), 79; *Proceedings of the NAFE, 1874,* 10; *Proceedings of the NAFE, 1884,* 18–20; *F & W* 1 (1886), 15, 144; *Proceedings of the NAFE, 1889,* 59–68; *F & W* 1 (1886), 144; *F & W* 12 (1892), 247; *F & W* 20 (1896), 414–17, 424–27; on "fire environments," see Stephen Pyne, *Fire in America: A Cultural History of Wildland and Rural Fire* (Princeton, N.J.: Princeton University Press, 1982), xviii.

7. See, for instance, *Proceedings of the NAFE, 1874,* 10–13; 16–17, 21–25; *Proceedings of the NAFE, 1875,* 32–33; *Proceedings of the NAFE, 1876,* 11–15; *Proceedings of the NAFE, 1880,* 38–56; *Proceedings of the NAFE, 1882,* 16–39; *Proceedings of the NAFE, 1884,* 18–32; *Proceedings of the NAFE, 1887,* 29–46; *Proceedings of the NAFE, 1888,* 124–31; *Proceedings of the NAFE, 1889,* 59–67; *Proceedings of the NAFE, 1876,* 11–15. When firefighters battled blazes,

they "produced space," giving cities, danger, and their occupation meaning; they created their identity in direct relationship to cities. On the relationship between physical and mental space, see Henri Lefebvre, *The Production of Space*, trans. Donald Nicholson-Smith (1974; reprint, Cambridge, Mass.: Blackwell Publishers, 1994).

8. *Proceedings of the NAFE, 1874*, 10–12, 16–17, 21–25; *Proceedings of the NAFE, 1875*, 32–33; *Proceedings of the NAFE, 1876*, 11–15; *Proceedings of the NAFE*, 1883, 15–19, 24–25, 27–28, 33 ff.; *Proceedings of the NAFE, 1889*, 59; *F & W* 1 (1886), 144; *F & W* 14 (1893), 286–88.

9. *Proceedings of the NAFE, 1883*, 15–19, 24–28, 33 ff.; *Proceedings of the NAFE, 1887*, 20–26; *Proceedings of the NAFE, 1889*, 35–50; *F & W* 1 (1886), 103; *F & W* 6 (1889), 241, 278; *F & W* 20 (1896), 427; *F & W* 14 (1893), 118; *F & W* 16 (1894), 356; *F & W* 30 (1901), 348; *F & W* 34 (1903), 74, 153–56, 168.

10. *Proceedings of the NAFE, 1887*, 20; *Proceedings of the NAFE, 1883*, 33–34; *Proceedings of the NAFE, 1888*, 124–31; *Proceedings of the NAFE, 1889*, 59–68; *Proceedings of the NAFE, 1892*, 44–56; *Proceedings of the NAFE, 1894:* 82–91; *F & W* 1 (1886), 144; *F & W* 20 (1896), 424–27.

11. *Proceedings of the NAFE, 1883*, 17–19; *Proceedings of the NAFE, 1892*, 44–56; *Proceedings of the NAFE, 1894*, 82–91; *F & W* 34 (1903), 79.

12. *Proceedings of the NAFE, 1874*, 19–20; *Proceedings of the NAFE, 1875*, 33–35; *Proceedings of the NAFE, 1877*, 14–19; *Proceedings of the NAFE, 1880*, 25 ff.; *Proceedings of the NAFE, 1882*, 46–47, 49–50; *Proceedings of the NAFE, 1883*, 33–35, 36 ff; *Proceedings of the NAFE, 1894*, 148–49; *F & W* 12 (1892), 42–43, 52, 72–73, 82–83. On hydraulics and so-called "Siamese" couplings, see *F & W* 1 (1886), 15; *F & W* 19 (1896), 299; Philadelphia Training School for Fire Service, n.d., Box A3624, PFD Records, PCA; *F & W* 1 (1886), 1; *F & W* 17 (1895), 203–5.

13. *F & W* 14 (1893), 122–23; *F & W* 2 (1887), 60–61.

14. *Proceedings of the NAFE, 1877*, 10; *Proceedings of the NAFE, 1884*, 18–33; *Proceedings of the NAFE, 1876*, 11–15; *Proceedings of the NAFE, 1882*, 52–53; *F & W* 34 (1903), 79; *F & W* 14 (1893), 120.

15. *Proceedings of the NAFE, 1894*, 96–101; Bahr, *New England Mill Engineering;* more broadly, on the difficulties of rebuilding after conflagration, see Christine Meisner Rosen, *The Limits of Power: Great Fires and the Process of City Growth in America* (New York: Cambridge University Press, 1986), 89–324; Wermeil, *Fireproof Building*, 73–125.

16. *Proceedings of the NAFE, 1894*, 97–101; *Proceedings of the NAFE, 1891*, 26–27, 146–53; *Proceedings of the National Board of Fire Underwriters, 1893*, 102–3; *Proceedings of the NBFU, 1894*, 127–30; Harry Chase Brearley, *Fifty Years of a Civilizing Force: The History of the National Board of Fire Underwriters* (New York: Frederick A. Stokes Company, 1916), 79 ff.; *Building Code . . . 1905*.

17. *F & W* 2 (1887), 4–5; *Proceedings of the NAFE, 1892*, 75–76; *Proceedings of the NAFE, 1883*, 13 ff.; *Proceedings of the NAFE, 1887*, 29–35; *F & W* 2 (1887), 4–5.

18. *Proceedings of the NAFE, 1877*, 8–9, 20–26; *Proceedings of the NAFE, 1882*, 44; *Proceedings of the NAFE, 1889*, 61, 35–58; *Proceedings of the NAFE, 1884*, 18 ff.; *Proceedings of the NAFE, 1892*, 44–47, 50–53; *F & W* 14 (1893), 119–21; *F & W* 20 (1896), 424–28; *F & W* 33 (1903), 79.

19. *F & W* 1 (1886), 1; *Proceedings of the NAFE, 1885*, 23–25; *Proceedings of the NAFE, 1887*, 18–20; *Proceedings of the NAFE, 1892*, 44 ff.; *Proceedings of the NAFE, 1898*, 10–11;

December 20, 1897, May 19, 1901, Volume 1, Charles Swingley Scrapbook, MOHS; *F & W* 13 (1893), 21, 113–15; *F & W* 19 (1896), 201, 227, 328.

20. Jacob A. Riis, "Heroes Who Fight Fire," *Century Magazine* 60, no. 4 (February 1898), 492; William C. Lewis, *A Manual for Volunteer and Paid Fire Organizations* (New York: Fred J. Miller, 1877), 11–12; see also, *F & W* 1 (1886), 14; *F & W* 18 (1895), 520; *F & W* 20 (1896), 414; *F & W* 34 (1903), 79; *Proceedings of the NAFE, 1880*, 76; *Proceedings of the NAFE, 1882*, 45; *Proceedings of the IAFE, 1897*, 112–125; also compare, for example, Lewis, *A Manual;* Andrew Isaac Meserve, *The Fireman's Hand-Book and Drill Manual* (Chicago: Stromberg, Allen, and Co., 1889); Charles T. Hill, *Fighting a Fire* (1894; reprint, New York: Century Company, 1898).

21. *F & W* 18 (1895), 520; Riis, "Heroes," 490; see also, Miriam Lee Kaprow, "The Last Best Work: Firefighters in the Fire Department of New York," *Anthropology of Work Review* 19, no. 2 (Winter 1999), 5–10.

22. *Proceedings of the NAFE, 1880*, 12, 14; Riis, "Heroes," 486; *Proceedings of the NAFE, 1884*, 110–11.

23. Quotation in Arlen Dykstra, "St. Louis Mourns," 36; *Proceedings of the NAFE, 1885*, 23–25.

24. *History STLFD*, 57–59; *Proceedings of the NAFE, 1898*, 17.

25. *Proceedings of the NAFE, 1874*, 18–23; *Proceedings of the NAFE, 1883*, 31–34; Lewis, *A Manual*, 8; on steam engines, see Stephen Roper, *Handbook of Modern Steam Fire Engines* (Philadelphia: Claxton, Remsen, and Haffelfinger, 1876); Engineer's Association of the Bureau of Fire, *Circular to the President and Members of the Select and Common Council* (1896); *Classification of Positions in the Classified Service of the City of Philadelphia, With Schedule of Compensation* (Philadelphia, 1920). On technology in the fire service, see for example, *Proceedings of the NAFE, 1880*, 77–78; Rebecca Zurier, *The American Firehouse: An Architectural and Social History* (New York: Abbeville Press, 1982), 100–11; *F & W* 1 (1886), 4–5, 32; *F & W* 2 (1887), 207–208; *F & W* 13 (1893), 220–23; *F & W* 30 (1901), 296, 300–305.

26. *Proceedings of the NAFE, 1880*, 77–78; Lewis, *A Manual*, 8; Hill, *Fire*, 36–62; for a pictorial depiction of basic ladder skills, see PCA, PFDRG, Box A3624: Philadelphia Training School for Fire Service, n.d. (ca. 1918–1920); St. Louis Firemen's Pension Fund, *"Justifiably Proud": The St. Louis Fire Department* (St. Louis, Mo.: Walsworth, 1978), 82–101; Jack Robrecht Notes, Book 2, "Apparatus Lists by Year," CIGNA Museum and Art Collection; United States Government, Department of Commerce, *Statistics of Fire Departments of Cities Having a Population over 30,000, 1917* (Washington, D.C.: Government Printing Office, 1918), table 10; *F & W* 6 (1889), 86.

27. *Proceedings of the NAFE, 1880*, 76; "The Chicago Training School," *Fireman's Herald* 24 (November 24, 1892); *Proceedings of the NAFE, 1882*, 45; Hill, *Fire* (1898), 39, 43; *F & W* 6 (1889), 62, 87; *F & W* 18 (1895), 477; *St. Louis Fire Department*, 178; STLFD, Personnel Records; *Proceedings of the NAFE, 1887*, 65; *Proceedings of the NAFE, 1888*, 87.

28. *F & W* 34 (1903), 35; *F & W* 6 (1889), 62, 87; "The Chicago Training School," *Fireman's Herald* 24 (November 24, 1892), William Hyde and Howard Conrad, *Encyclopedia of the History of St. Louis*, vol. 2 (St. Louis: Southern History Company, 1899), 766; *St. Louis Fire Department*, 178; Personnel Records, STLFD; *Proceedings of the NAFE, 1887*, 65; *Proceedings of the NAFE, 1888*, 87.

29. *F & W* 13 (1893), 220–23; *F & W* 14 (1894), 32–33; *Proceedings of the NAFE, 1896;* Edwin O. Sachs, *A Record of the International Fire Exhibition, Earl's Court London, 1903* (London: British Fire Protection Committee, 1903).

30. Lewis, *Manual,* 11–12, 13–22.

31. *Proceedings of the NAFE, 1887,* 66; *Proceedings of the NAFE, 1882,* 45–49; Meserve, *Hand-Book;* Hill, *Fire;* John Kenlon, *Fires and Fire-Fighters: A History of Modern Fire-Fighting with a Review of Its Development from Earliest Times* (New York: George H. Doran Company, 1913).

32. *F & W* 14 (1893), 119–21; *F & W* 19 (1896), 7; *F & W* 20 (1896), 577.

33. *Proceedings of the NAFE, 1887,* 65, 48–66; Riis, "Heroes," 486; *Proceedings of the NAFE, 1880,* 10–14, 60–65; *Proceedings of the NAFE, 1888,* 86–89; *F & W* 34 (1903), 35.

34. See, for example, Andrea Stulman Dennett and Nina Warnke, "Disaster Spectacles at the Turn of the Century," *Film History* 4, no. 2 (1990), 101–11; "The Man with the Ladder and the Hose"—Song Slides (ca. 1904), CIGNA Museum and Art Collection; Fires, Box 179, Lester S. Levy Collection of Sheet Music, Milton S. Eisenhower Library, Johns Hopkins University; June 17, 1911, Volume 2, CSS; Robin Cooper, "The Fireman: Immaculate Manhood," *Journal of Popular Culture* 28, no. 4 (1995), 139–71. Perusing local newspapers, or even the pages of national weeklies such as *Harper's Weekly* and *Frank Leslie's Illustrated,* will reveal hundreds of stories and images of firefighting in the decades following the Civil War.

35. Riis, "Heroes," 489.

36. Meserve, *The Fireman's Hand-Book* (1889), 69; Riis, "Heroes," 488–89.

37. Cooper, "The Fireman," 143; Riis, "Heroes," 489; Miriam Kaprow, "Magical Work," 99–100.

38. Gilbert TS, 143; Riis, "Heroes," 489.

39. James McBride Dabbs, *Heroes, Rogues, and Lovers: Testosterone and Behavior* (New York: McGraw-Hill, 2000), 177; interestingly, firefighters' testosterone levels increase temporarily on the way to a fire, diminishing on return to the engine house. On the cultural elements of firefighting culture, see also Kaprow, "Last, Best Work," 14–19.

40. *Proceedings of the NAFE, 1885,* 17–18; see also Michael Kimmel, *Manhood in America* (New York: Free Press, 1996), 79–190; E. Anthony Rotundo, *American Manhood: Transformations in Masculinity from the Revolution to the Modern Era* (New York: Basic Books, 1993), 222–83.

41. *Proceedings of the NAFE, 1880,* 11–12; see also Elliot Gorn, *The Manly Art: Bare-Knuckle Prizefighting in America* (Ithaca, N.Y.: Cornell University Press, 1990), 179–206; Gail Bederman, *Manliness and Civilization: A Cultural History of Gender and Race in the United States, 1880–1917* (Chicago: University of Chicago Press, 1995), 170–215.

42. *Proceedings of the NAFE, 1880,* 14.

43. The background for this discussion relies on an unpublished essay graciously shared with the author: Karin C. C. Dalton, "Currier & Ives's Darktown Comics: Ridicule and Race," presented at *Democratic Vistas: The Prints of Currier & Ives,* Symposium, Museum of the City of New York, May 2, 1992; compare, for instance, Thomas Worth, *The Darktown Fire Brigade—to the Rescue!* (Currier & Ives, lithograph, 1884), Museum of the City of New York; Louis Mauer, *The American Fireman: Prompt to the Rescue* (Currier & Ives, lithograph, 1858), Library of Congress.

44. Thomas Worth, *Darktown . . . Rescue!;* Thomas Worth, *The Darktown Fire Brigade—Saved!* (Currier & Ives, Lithograph, 1884), Chicago Historical Society; *Fire and Water* 10 (August 8, 1891), 55–57; Marianne Torgovnick, *Gone Primitive: Savage Intellects, Modern Lives* (Chicago: University of Chicago Press, 1990), 159–74, 209–23.

45. *Proceedings of the NAFE, 1885*, 71, 67–73.

46. O'Brien, *Centennial History*, 51–58; *Proceedings of the NAFE, 1920;* Personnel Files, STLFD. "Nativity" was ascertained via self-report, which may have led to an underreport of immigrant status.

47. Personnel Files, STLFD; PFD Roster, FH; *Proceedings of the NAFE, 1889*, 151–52. Note that the NAFE's data was not broken out more finely and provided only a cross-section of several departments.

48. For comparative data on other work environments, see Walter Licht, *Getting Work, 1840–1950* (Cambridge, Mass.: Harvard University Press, 1992), esp. 34, 229; Theodore Hershberg, ed., *Philadelphia: Work, Space, Family, and Group Experience in the Nineteenth Century* (New York: Oxford University Press, 1981), 37–120. Nineteenth-century employment was volatile, and firefighters appear to have had better working conditions than most workers; see Alexander Keyssar, *Out of Work: The First Century of Unemployment in Massachusetts* (New York: Cambridge University Press, 1986), misc. 308–41.

49. James Bryson Gilbert, "An Effort to Portray . . . the Manner[of] the People of the Class to Which I Belonged Lived Their Lives," typescript, 141–42, HSP (hereafter, Gilbert TS); "The Chicago Training School," *Fireman's Herald* 24 (November 24, 1892); *Philadelphia Mayor's Message, 1914*, 36; *F & W* 12 (1892), 65, 192; *F & W* 19 (1896), 311; *F & W* 19 (1899), 75–76.

50. *F & W* 2 (1887), 303; *F & W* 20 (1896), 328; Engineer's Association of the Bureau of Fire, *Circular to the President and Members of the Select and Common Council* (1896); Michael R. Haines, "Poverty, Economic Stress, and the Family in a Late-Nineteenth-Century American City: Whites in Philadelphia, 1880," in Hershberg, ed., *Philadelphia: Work, Space, Family, and Group Experience*, 245–46, table 1.

51. STLFD, Personnel Files; "Why St. Louis Has So Few" (n.d., p. 107), Volume 2, CSS.

52. STLFD, Personnel Files; Gilbert TS, 146.

53. Gilbert TS, 139.

54. Gilbert TS, 153–56.

55. Gilbert TS, 164–65.

56. Pp. 227–28, Truck Company F, January 1, 1887, to September, 30, 1889, HSP; also, cf., Truck Company F, Daily Log Book, 1884–1886; Truck Company F, Daily Log Book, 1887–1889, Truck Company F, Daily Log Book, 1895–1896, Truck Company F, June 4, 1896, June 19, 1897; Truck Company F, June 19, 1897, to January 26, 1898; Engine Company No. 4, Daily Record, 1871–1880, HSP; Greely S. Curtis, *Investigation of the New York Fire Department* (New York: Merchants' Association of New York, 1908), 41–47.

57. *Proceedings of the NAFE, 1876*, 16; *Proceedings of the NAFE, 1880*, 15–19; *Proceedings of the NAFE, 1882*, 50–52; *Proceedings of the NAFE, 1885*, 67–77; *Proceedings of the NAFE, 1894*, 139–46.

58. Jon C. Teaford, *The Unheralded Triumph: City Government in America, 1870–1900* (Baltimore, Md.: Johns Hopkins University Press, 1984), 162–66; *St. Louis Fire Department*, 179; John Lindsay, Personnel Files, STLFD; James Neal Primm, *Lion of the Valley: St. Louis, Missouri*, 2d ed. (Boulder, Colo.: Pruett Publishing, 1981), 372; *F & W* 30 (1901), 290.

59. Page 1, Volume 1, CSS; Charles Swingley, Eugene J. Gross, STLFD Personnel File.

60. Gilbert TS, 148–55.

61. *St. Louis Post Dispatch*, December 22, 1895, Volume 1, CSS; Gilbert TS, 145–55; Licht, *Getting Work*, 36–39, table 2.6.

62. Gilbert TS, 145; Russell F. Weigley, *Philadelphia: A 300-Year History* (New York: W. W. Norton, 1982), 546.

63. A sense of the boss politician's rough manliness suffuses the vast literature on the subject; see Arthur Mann, *LaGuardia Comes to Power* (1965), 86 ff.; Zane L. Miller, *Boss Cox's Cincinnati* (1968); William L. Riordon, *Plunkitt of Tammany Hall* (1963); Seymour J. Mandlebaum, *Boss Tweed's New York* (1965).

64. Riis, "Heroes," 483–97; Eric Monkkonen, *America Becomes Urban: The Development of U.S. Cities and Towns, 1780–1980* (Berkeley: University of California Press, 1988); Stanley K. Schultz, *Constructing Urban Culture: American Cities and City Planning, 1800–1920* (Philadelphia: Temple University Press, 1989); Teaford, *Unheralded Triumph;* Martin Melosi, *The Sanitary City: Urban Infrastructure in America from Colonial Times to Present* (Baltimore, Md.: Johns Hopkins University Press, 2000).

65. On the police, see for instance, Samuel Walker, *Popular Justice: A History of American Criminal Justice* (New York: Oxford University Press, 1980); Eric H. Monkkonen, *Police in Urban America, 1860–1920* (New York: Cambridge University Press, 1981); on sanitation and electrical systems, see Melosi, *Sanitary City*, 74 ff.; Thomas P. Hughes, *Networks of Power: Electrification in Western Society, 1880–1930* (Baltimore, Md.: Johns Hopkins University Press, 1983).

66. *Proceedings of the NAFE, 1888*, 125; *F & W* 33 (1903), 232; O'Brien, *Centennial History*, 41.

SEVEN: Consuming Safety: Fire Prevention and Fire Risk in the Twentieth Century

1. Harry Chase Brearley, *Fifty Years of a Civilizing Force: The History of the National Board of Fire Underwriters* (New York: Frederick A. Stokes Company, 1916), 94; *Proceedings of the National Board of Fire Underwriters, 1917*, 158 ff.; Margaret Hindle Hazen and Robert M. Hazen, *Keepers of the Flame: The Role of Fire in American Culture, 1775–1925* (Princeton, N.J.: Princeton University Press, 1992), 114–15; *SAAF* 5, no. 9 (September 1922), 1 ff.; National Fire Protection Association, *Suggestions for Guidance in Planning the Observation of Fire Prevention Week* (Boston: NFPA, 1926); October 3, 1922, Fire Prevention Week—1951 and Prior, PEBFC, TUUA; on fire-resistant building practices, Sara E. Wermeil, *The Fireproof Building: Technology and Public Safety in the Nineteenth-Century American City* (Baltimore: Johns Hopkins University Press, 2000), esp. 37–103.

2. Quoted in Marc Schneiberg, "Private Order and Inter-firm Governance in the American Fire Insurance Industry, 1820–1950" (Ph.D. diss., University of Wisconsin, 1994), 130.

3. Albert W. Whitney, *Report of the Co-Insurance Committee to the Board of Fire Underwriters of the Pacific on Percentage of Co-Insurance and the Relative Rates Chargeable Therefore, also on the Cost of Conflagration Hazard of Large Cities* (1905).

4. Brearley, *Fifty Years*, 94; *Proceedings of the NBFU, 1899*, 26 ff.

5. See, for instance, *Proceedings of the NBFU, 1896; Proceedings of the NBFU, 1900; Weekly Underwriter* 58, no. 24 (1898), 414.

6. Olivier Zunz, *Making America Corporate, 1870–1920* (Chicago: University of Chicago Press, 1990), 90–101.

7. *Building Code Recommended by the National Board of Fire Underwriters . . . Edition 1905* (New York: James Kempster Printing, 1905).

8. *Building Code . . . 1905; Proceedings of the NBFU, 1894,* 127–30; Jon C. Teaford, *The Unheralded Triumph: City Government in America, 1870–1900* (Baltimore, Md.: Johns Hopkins University Press, 1984), 202–5; Martin Melosi, *The Sanitary City: Urban Infrastructure in America from Colonial Times to Present* (Baltimore, Md.: Johns Hopkins University Press, 2000), 200 ff.

9. See, for instance, *Proceedings of the NBFU, 1894,* 82–83, 87 ff.; *Proceedings of the NBFU, 1895,* 104–5; *Proceedings of the NBFU, 1893,* 109 ff.; *Proceedings of the NBFU, 1900,* 83–86; *Proceedings of the NBFU, 1906,* 68–69; *Proceedings of the NBFU, 1912,* 25–26; *Proceedings of the NBFU, 1925,* 89–92; Harry Chase Brearley, *A Symbol of Safety: An Interpretive Study of a Notable Institution Organized for Service—Not Profit* (New York: Doubleday, Page, and Co., 1923), 1–29; Brearley, *Fifty Years,* 178–96.

10. See, for instance, *Annual Cyclopedia of Insurance, 1896–1897,* 391; *Weekly Underwriter 58 (1898),* 414; *Proceedings of the NFPA, 1897; Proceedings of the NFPA, 1898; Proceedings of the NFPA, 1899; Proceedings of the National Fire Protection Association, 1900; Proceedings of the National Fire Protection Association, 1901; Proceedings of the NFPA, 1902; Proceedings of the NBFU, 1900,* 72–73, 82–83; *Proceedings of the NBFU, 1906,* 66; Brearley, *Fifty Years,* 165–77.

11. *Building Code . . . 1905,* 263.

12. National Board of Fire Underwriters, *Analysis of Reports of the National Board of Fire Underwriters—Committee of Twenty: Being an Explanation of the Reasons for the Inclusion of Various Classes of Information* (1906), 8; *Building Code . . . 1905,* 214–25, 263.

13. *Building Code . . . 1905,* 163, 82–89, 117, 122–72.

14. *Building Code . . . 1905,* 187, 129 ff., 186–90.

15. Classifications of Fire Risk, Volume 2, 1852–1872, RG 5/1, Aetna; *Building Code . . . 1905;* also, see for example, Joel Tarr and Mark Tebeau, "Managing Danger in the Home Environment, 1900–1940," *Journal of Social History* 29 (Summer 1996), 797–816.

16. *Building Code . . . 1905,* 1–4, 17 ff.

17. *Building Code . . . 1905,* 252–53.

18. *Proceedings of the NBFU, 1907,* 18–19, 97; see also *Annual Report of the Committee of Twenty of the Executive Committee of the NBFU,* 20–22.

19. *Proceedings of the NBFU, 1906,* 74–75; Committee on Public Information of the Woman's League for Good Government, *Facts about Philadelphia* (1919), 65–69; Powell Evans, "Fire Waste and Its Prevention," *City Club Bulletin* 4, no. 1 (January 10, 1911), 3–19.

20. *Building Code . . . 1905,* 212; NBFU, *Building Codes and Their Aims* (New York, n.d. [ca. 1940]), 21–22; *St. Louis Globe-Democrat,* January 18, 1902, Swingley Volume 2, MOHS.

21. *Proceedings of the NBFU, 1906,* 73; NBFU, *Building Codes,* 2–3; NBFU, *A Model Building Code* (5th ed.; New York, 1931); *Proceedings of the NBFU, 1914,* 33–34.

22. *Second Annual Report of the Committee of Twenty to the Executive Committee of the National Board of Fire Underwriters, April 1906,* 3; *Proceedings of the NBFU, 1906,* 93–95.

23. *Proceedings of the NBFU, 1906,* 85.

24. NBFU, *Analysis of Reports,* 4–8.

25. NBFU, *Analysis of Reports,* 8–9.

26. NBFU, *Analysis of Reports,* 9–12, 29–30; *Report Committee of Twenty, 1906,* 11 ff.

27. NBFU, *Analysis of Reports,* 8, 17–22.

28. *Report Committee of Twenty, 1905,* esp. 23–26; *Proceedings of the NBFU, 1906,* 81–97; *Report Committee of Twenty, 1906,* esp. 41–48.

29. *Report Committee of Twenty, 1906,* 8–10, 26–27; NBFU, *Analysis of Reports,* 14–15.

30. *FISOP Bulletin* 7, no. 4 (October 1911), 1; *FISOP Bulletin* 7, no. 2–3 (July–August 1911), 2; Trade League of Philadelphia, *Report of the Insurance Committee of the Trade League of Philadelphia, with Suggestions* (Philadelphia: John McFetridge, 1900); pages 1–4, 1914, PFD, Miscellaneous, 1920, and Prior, *Bulletin* Clippings, TUUA; June 1911, PFD, High Pressure, *Bulletin* Clippings; page 1, 1914, PFD, Equipment Description, *Bulletin* Clippings; *City Club Bulletin* 4, no. 1 (1911), 2–19; *FISOP Bulletin* 7, no. 12 (1912); *FISOP Bulletin* 8, no. 3 (1912).

31. Charles C. Dominge and Walter O. Lincoln, *Fire Insurance Inspection and Underwriting: An Encyclopedic Handbook . . .* (1919; reprint, Chicago: Spectator Company, 1929), 807–11; Orin F. Nolting, *How Municipal Fire Defenses Affect Insurance Rates* (Chicago: International City Managers' Association, 1939), 7–21.

32. Dominge and Lincoln, *Insurance Inspection,* 807–11; Nolting, *Fire Defenses,* 7–21.

33. Riegel, *Underwriters' Associations,* 5

34. Robert Riegel, *Fire Underwriters Associations in the United States* (New York: Chronicle Company, 1916), 35–36; Schneiberg, "Political and Institutional Conditions," 81–82.

35. Schneiberg, "Political and Institutional Conditions," 81–82.

36. *Proceedings of the NBFU, 1901,* 22–23.

37. William Wandel, "The Control of Competition in Fire Insurance" (Ph.D. diss., Columbia University, 1935), 59–64; *Proceedings of the NBFU, 1915;* E. G. Richards, *The Experience Grading and Rating Schedule: A System of Fire Insurance Rate-Making Based upon Average Fire Costs* (New York: D. Van Nostrand, 1924), v–vii, 60, 87; Riegel, *Underwriters' Associations,* 22; Dominge and Lincoln, *Insurance Inspection,* 809; *Proceedings of the NBFU, 1925,* 59 ff.

38. Richards, *Experience Grading,* 120–21; E. G. Richards, *Fire Underwriting Profits as Related to Experience Rate-Making* (New York: Van Nostrand, 1924), 1.

39. Richards, *Experience Grading,* 110–11; Nolting, *Fire Defenses,* 46 ff.

40. *Proceedings of the National Board of Fire Underwriters, 1901,* 22–23; Schneiberg, "Political and Institutional Conditions," 97–98, n. 39; H. Roger Grant, *Insurance Reform: Consumer Action in the Progressive Era* (Ames: Iowa State University Press, 1979), 100–133.

41. Schneiberg, "Political and Institutional Conditions," 87; Nolting, *Fire Defenses,* 50 ff.

42. Schneiberg, "Political and Institutional Conditions," 87.

43. *Constitution and By-Laws of the Fire Insurance Society of Philadelphia* (1905).

44. *FISOP Bulletin* 6, no. 8 (January 1911), 2–3; *Proceedings of FISOP, 1904–1905; FISOP Bulletin* 9, no. 11–12 (November–December 1914); *FISOP Bulletin* 6, no. 7 (December 1910), 2–3.

45. Robert Coyle, "Office Practice," *FISOP Bulletin, Special Supplement* (May 1912), 7.

46. See, for instance, *FISOP Bulletin* 6, no. 6 (November 1910), 1; *FISOP Bulletin, Supplement* (April 14, 1913); for a definition of insurance engineering, see, for example, Frederick Cruger Moore, "Fire Insurance Engineering," in Lester Zartman, ed., *Yale Readings in Insurance: Property Insurance, Marine and Fire* (1909; reprint, New Haven, Conn.: Yale University Press, 1926), 348–60.

47. *FISOP Bulletin* 6, no. 7 (December 1910), 2–5.

48. *Annual Report of the President of FISOP, 1921;* Map of the Insurance Society of

Philadelphia and Insurance Row, circa 1901, FISOP Archives; *Annual Report of FISOP, 1990;* Robert Coyle, "Office Practice," *FISOP Bulletin, Special Supplement* (May 1912).

49. *FISOP Bulletin* 6, no. 5 (October 1910), 2–3. Moral hazard took on many dimensions, but became increasingly quantified; see, for example, Richard Bissell, "Rates and Hazards," in Zartman, *Yale Readings*, 148–80; Roy Foulke, *Relativity of the Moral Hazard* (New York: Dun and Bradstreet, 1940); Roy Foulke, *The Sinews of American Commerce* (New York: Dun and Bradstreet, 1941), 370–77.

50. *FISOP Bulletin* 6, no. 5 (October 1910), 2–3.

51. *FISOP Bulletin* 6, no. 8 (January 1911), 8.

52. On life insurance, see Angel Kwolek-Folland, *Engendering Business: Men and Women in the Corporate Office, 1870–1930* (Baltimore, Md.: Johns Hopkins University Press, 1994), 45–55, 70–93; Viviana Zelizer, *Morals and Markets: The Development of Life Insurance in the United States* (New York: Columbia University Press, 1979), 125–47; on manliness in this period, see Gail Bederman, *Manliness and Civilization: A Cultural History of Gender and Race in the United States, 1880–1917* (Chicago: University of Chicago Press, 1995), 5–20, 170–216; Michael S. Kimmel, "The Contemporary 'Crisis' of Masculinity in Historical Perspective," in Harry Brod, ed., *The Making of Masculinities: The New Men's Studies* (New York: Routledge, 1992), 121–69.

53. *Proceedings of the NBFU, 1917*, 158 ff.; *Proceedings of the NBFU, 1919*, 167 ff.; on the belief that building codes would not become fully effective until "the distant future," see *Proceedings of the NBFU, 1914*, 33–34.

54. *Proceedings of the NBFU, 1917*, 158, 159; *Proceedings of the NBFU, 1919*, 168; *SAAF* 6, no. 5 (May 1923).

55. See, for instance, *SAAF* 5, no. 10 (October 1922), 1–3.

56. NBFU, *Safeguarding the Home against Fire: A Fire Prevention Manual for the School Children of America* (New York: National Board of Fire Underwriters, 1920), 5–12; *Proceedings of the NBFU, 1919*, 82–83; *SAAF* 6, no. 3 (March 1923), 1 ff.; *SAAF* 6, no. 5 (May 1923), 1 ff.

57. See, for instance, *SAAF* 3, no. 4 (April 1920), 1 ff.

58. *SAAF* 6, no. 3 (March 1923), 1, 4; *SAAF* 6, no. 1 (January 1923), 1–6; *SAAF* 6, no. 5 (May 1923), 1 ff.

59. NBFU, *Dwelling Houses: A Code of Suggestions for Construction and Fire Protection . . . To Safeguard Homes and Lives against the Ravages of Fire* (New York: National Board of Fire Underwriters, 1916), 8.

60. *Proceedings of the NBFU, 1919*, 168; *SAAF* 6, no. 9 (September 1923), 2, 3 ff.

61. NBFU, *Safeguarding*, 8; *Proceedings of the NBFU, 1919*, 170, 167–70.

62. *SAAF* 6, no. 5 (May 1923), 2.

63. NBFU, *Dwelling Houses*, 8. On the development of the consciousness of home safety, see Joel A. Tarr and Mark Tebeau, "Housewives as Home Safety Managers: The Changing Perception of the Home As a Place of Hazard and Risk, 1870–1940," in Roger Cooter and Bill Luckin, eds., *Accidents, Fatalities, and Social Relations: Historical Orientations* (Amsterdam: Rodopi Press, Wellcome Institute Series in the History of Medicine, 1997), 196–234; on metropolitan life, see Zunz, *Making America Corporate*, 90–101.

64. On the history of the Underwriters' Laboratories, see Brearley, *A Symbol of Safety*, esp. 109–34, 203–20, 236; *Proceedings of the NBFU, 1915*, 7 ff.; also see *SAAF* 9, no. 11 (November 1926), 6–8.

EIGHT: Eating Smoke: Rational Heroes in the Twentieth Century

1. *St. Louis Republic,* July 16, 1909, Volume 2, CSS, MOHS.

2. *Safeguarding America Against Fire* 6, no. 8 (August 1923), 3.

3. *Fire and Water* 27 (1900), 192; on the IAFE, see Donald M. O'Brien, *"A Century of Progress through Service": The Centennial History of the International Association of Fire Chiefs* (International Association of Fire Chiefs, 1973).

4. On Progressivism, see for instance, Daniel T. Rodgers, "In Search of Progressivism," *Reviews in American History* 10 (1982), 113–32; Samuel Haber, *Efficiency and Uplift: Scientific Management in the Progressive Ear, 1890–1920* (Chicago: University of Chicago Press, 1964); Gabriel Kolko, *The Triumph of Conservatism: A Re-Interpretation of American History, 1900–1916* (Chicago: Quadrangle Books, 1963).

5. August 1898, Volume 1, CSS; James Neal Primm, *Lion of the Valley: St. Louis,* 2d ed. (Boulder, Colo.: Pruett Publishing, 1990), 345 ff.

6. August 1898, Volume 1, CSS.

7. August–December 1898, Volume 3, CSS.

8. *St. Louis Globe-Democrat,* January 18, 1902, *St. Louis Post Dispatch,* January 22, 1902, Volume 2, CSS.

9. *St. Louis Star,* January 19, 1902, Volume 2, CSS.

10. January–February 1902, Volume 2, CSS; Benjamin Fath, John Barry, Personnel Roster, STLFD; biographical information on John Barry, 1920, Folder 19, Box 1, STLVVFC, MOHS.

11. April–May 1903, Volume 2, CSS; Primm, *Lion,* 371 ff.

12. *St. Louis Globe,* November 25, 1903, also pp. 118, 121–25, 140, Volume 2, CSS.

13. *St. Louis Post Dispatch,* December 22, 1895, April–May 1898, October 1898, June 25, 1899, Volume 1, CSS; June 16, 1901, "Why St. Louis Has So Few" (n.d., p. 107), December 11, 1902, October 1898 (pp. 45–49), April–May 1903, August 8, 1903, January–February 1904, December 2, 1906, Volume 2, CSS; St. Louis Firemen's Fund, *History of the St. Louis Fire Department . . .* (1914), 192–95 (hereafter, *History STLFD*); for an overview of the IAFE agenda, see O'Brien, *Century of Progress,* 40–59.

14. *St. Louis Post Dispatch,* December 22, 1895, Volume 1, CSS; "Why St. Louis Has So Few" (n.d., p. 107), December 2, 1906, Volume 2, CSS; on the records systems, see card data from Personnel Files, STLFD; more generally, on fire department training nationwide, see *F & W* 34 (1903), 35; *F & W* 6 (1889), 62, 87; "The Chicago Training School," *Fireman's Herald* 24 (November 24, 1892); *Proceedings of the NAFE, 1887,* 65; *Proceedings of the NAFE, 1888,* 87.

15. *History STLFD,* 190–91; *St. Louis Post Dispatch,* December 22, 1895, Volume 1, CSS; on the record system, see Personnel Files, STLFD.

16. *Firemen's Journal,* December 27, 1879, 541–42; *Firemen's Journal,* April 23, 1881, 326; *St. Louis Post-Dispatch,* February 26, 1893; February 27, 1893; February, 1902, February, 1904, Volume 2, CSS; on salaries more generally, see United States Government, Department of Commerce, *Statistics of Fire Departments of Cities Having a Population of Over 30,000, 1917* (Washington, D.C.: Government Printing Office, 1918), tables 4 and 5, 47–60 (hereafter, *Statistics of Fire Departments*); on firefighter salaries in Philadelphia see Bureau of Municipal Research, *Standardization of Salaries and the Demands of Firemen* (1914); Bureau of Municipal Research, *How Would You Like to Be a Fireman?* (1914).

17. *Proceedings of the NAFE, 1880*, 15–17, 50; *History STLFD*, 188–92.

18. *History STLFD*, 188–200, 245–56.

19. *History STLFD*, 188 ff. Faith in technological solutions is deeply embedded in firefighting culture and mirrors values broadly held in American society, and this idiom of progress suffuses most, if not all, histories written and/or commissioned by firefighters. On the history of the idea of technological progress, see Merritt Roe Smith and Leo Marx, eds., *Does Technology Drive History?: The Dilemma of Technological Determinism* (Cambridge, Mass.: MIT Press, 1994).

20. "Kindly Caricatures," ca. 1906–1909, Volume 2, CSS; August 1901, Volume 1, CSS, on manliness in this era, see for instance Michael Kimmel, *Manhood in America* (New York: Free Press, 1996), 81–156, 191–222; Gail Bederman, *Manliness and Civilization: A Cultural History of Gender and Race in the United States, 1880–1917* (Chicago: University of Chicago Press, 1995), 170–239.

21. Quotation from January 19, 1902, Volume 2, CSS; on Swingley's heroics, see for instance December 31, 1901, February 5, 1902, October 23, 1903, July 16, 1909, *St. Louis Republic*, Volume 2, CSS.

22. December 22–23, 1910, Clippings Scrapbook, 1901–1911, FH; Injury Ledger, 1914–1925, 1941–1955, FH. In 1914, there were 357 reported injuries; 353 in 1919, 184 in 1924, and 669 in 1941.

23. *Telegraph*, March, 4, 1911, various from December 1910 to January 1911, Clippings Scrapbook, 1901–1911, FH.

24. Various, December 1910–March 1911, Clippings Scrapbook, 1901–1911.

25. For one view on the "deskilling" of American workplaces, see for example, Harry Braverman, *Labor and Monopoly Capital: The Degradation of Work in the Twentieth Century* (New York: Monthly Review Press, 1974); on manliness, see Bederman, *Manliness and Civilization.*

26. Walter Licht, *Getting Work: Philadelphia, 1840–1950* (Cambridge, Mass.: Harvard University Press, 1992), 174–220; more broadly, on scientific management see Daniel Nelson, *Frederick Taylor and the Rise of Scientific Management* (Madison: University of Wisconsin Press, 1980); Haber, *Efficiency*.

27. *Annual Report of the Civil Service Commission to His Honor the Mayor . . . December 31, 1906* (Philadelphia. 1907), 8–11, *Seventh Annual Report of the Civil Service Commission . . . December 31, 1912* (Philadelphia: 1913), 14–16; see, too, William Beyer, *Personnel Administration in Philadelphia* (Philadelphia, 1937).

28. Civil Service Commission of Philadelphia, *Some General Information for Applicants and the Public* (Philadelphia, 1906); Civil Service Commission of Philadelphia, *Manual of Civil Service Information* (Philadelphia, 1924); Civil Service Commission of Philadelphia, *General Information and Answers to Questions Most Frequently Asked* (Philadelphia, 1920).

29. James Bryson Gilbert, "An Effort to Portray . . . the Manner [of] the People of the Class to Which I Belonged Lived Their Lives," typescript, 148–54, HSP; Civil Service Commission of Philadelphia, *Some General Information and Answers to Questions.*

30. *Second Annual Message of Rudolph Blankenburg, Mayor of the City of Philadelphia with the Annual Reports of the Director of the Department of Public Safety . . . For the Year Ending December 31, 1912* (Philadelphia, 1913), 347; Licht, *Getting Work*, 182–90.

31. Quoted in *Fourth Annual Message of Rudolph Blankenberg, Mayor of Philadelphia. Volume 1. Containing the Mayor's Message and the Reports of the Departments of Public Safety, Civil*

Service Commission, City Transit for the Year Ending, December 31, 1914 (Philadelphia, 1915), 36–38; "Firemen Shortage Grave City Issue," *Philadelphia Public Ledger*, January 4, 1918; "Order 71," October 1, 1918, PFD Orders, PFDRG, PCA; "2-platoon system," May 1, 1918, PFD Orders; for historical staffing, PFD, Personnel Ledger, PFD, FH.

32. *Philadelphia Mayor's Message, 1912*, 234 ff.; *Statistics of Fire Departments*, 16–21; *Proceedings of the IAFE, 1910*, 23–45; firefighters adored their horses, which is expressed well in Natlee Kenoyer, *The Firehorses of San Francisco* (Los Angeles: Westernlore Press, 1970).

33. *F & W* 33 (1903), 64; "Press Release," January 25, 1954, Departmental Records, FH; on firefighting unions and municipal unions, see George J. Richardson, *Symbol of Action: A History of the International Association of Firefighters, AFL-CIO-CLC* (New York: International Association of Fire Firefighters, 1974); David Ziskind, *One Thousand Strikes of Government Employees* (New York: Columbia University Press, 1940), 53 ff.; for an overview of firefighters' unions in Philadelphia in the 1930s, see Mark Wilkens, "On Good and Welfare," seminar paper shared with author, 1995.

34. *Statistics of Fire Departments*, 20; "Firemen Ask Court to Enforce Law," *Philadelphia Public Ledger*, January 9, 1918; "Firemen Get Hot Shot in City Hall," *Philadelphia Public Ledger*, January 23, 1918; pp. 1–4, Working Hours, 2-Platoon to 3-Platoon, PEBFC, TUUA; "Order 103," October 24, 1916, December 18, 1916, General Orders, FH; Philadelphia Fire Department Historical Corporation, *Hike Out: The History of the Philadelphia Fire Department* (1999), 82.

35. *Philadelphia Mayor's Message, 1912*, 345; Philadelphia Department of Public Safety, *Information for Firefighters: A Story about the Way the Philadelphia Bureau of Fire Is Organized and Operated—Telling Something of Its History, Customs, and Regulation—with Hints as to How to Gain Advancement Therein* (Philadelphia: Department of Public Safety, 1915), 47–55 (hereafter, *Information for Firefighters*); also, January 1–4, 1916, March 1, 1916, May 4, 1916, June 21, 1916, General Orders, 1916–19, FH; March 23, 1927, June 7, 1927, November 22, 1927, January 21, 1929, "Instructions for Using Forms," Bureau of Fire—Headquarters Orders, 1926–1935, FH.

36. See various, General Orders, 1916–1919, FH; various, Bureau of Fire—Headquarters Orders, 1926–1935, FH; "Useful Information," "A Guide to Govern All Forms and Reports," Reports, n.d., FH; December 31, 1948, Reports, 1948–1951, FH; Lectures for Officers, ca. 1920–1930, typescript, p. 3 and "official reports," PFDRG, PCA.

37. Quoted February 23, 1932, April 15, 1926, Fire School—1952 and Prior, PEBFC; *Statistics of Fire Departments*, 20; Andrew Meserve, *The Fireman's Hand-book and Drill Manual* (Chicago: Stromberg, Allen, 1889); John Kenlon, *Fires and Fire-Fighters* (New York: George H. Doran, 1913); *Proceedings of the IAFE, 1920*, esp. 155–56; Fire and Water Engineering, *The New York Fire College Course* (New York: Fire and Water Engineering, 1920).

38. *Philadelphia Mayor's Message, 1914*, 36.

39. *Philadelphia Mayor's Message, 1912*, 239; February 5, 1913, May 6, 1913, May 2, 1913, January 17, 1915, May 23, 1923, Fire School—1952 and Prior, PEBFC; Paul Ditzel, *A Century of Service, 1886–1986: The Centennial History of the Los Angeles Fire Department* (Los Angeles: Los Angeles Firemen's Relief Association, 1986), 97 ff.; *Proceedings of the IAFE, 1920*, 89–477.

40. City of Philadelphia, *Classification of Positions in the Classified Service of the City of Philadelphia with Schedule of Compensation* (Philadelphia, 1920), 186–87; *Information for Firefighters*, 77–130; May 6, 1913, April 17, 1926, Fire School—1952 and Prior, PEBFC.

41. "Class Notes," Philadelphia Training School for Fire Service, n.d. (ca. 1918–1920), PFDRG, PCA; *The New York Fire College Course*, 4–20; *Proceedings of the IAFE, 1920*, 117–25, 155–308; miscellaneous, 1911–1958, Fire Regulation and Inspection, PEBFC.

42. "Class Notes," Philadelphia Training School for Fire Service, n.d. (ca. 1918–1920), PFDRG; "1922 Dates of Instruction and 1923 Company Attendance," "Dates of Instruction and Company Attendance For 1924," PFDRG; January 30, 1940, April 15, 1926, May 6, 1913, Fire School—1952 and Prior, PEBFC; for the connection to scientific management, see Frank Gilbreth, *Motion Study: A Method for Increasing the Efficiency of Workmen* (New York: D. Van Nostrand, 1911), xvii; Mike Mandel, *Making Good Time: Scientific Management, the Gilbreth's Photography, and Motion Futurism* (Riverside: California Museum of Photography, 1989); *The New York Fire College Course*, 15–20.

43. "Class Notes," Philadelphia Training School for Fire Service, n.d. (ca. 1918–1920), PFDRG; *Information for Firefighters*, 130 ff.

44. Compare the histories of firefighting in St. Louis and Philadelphia, for instance, *History STLFD; Information for Firefighters;* J. Albert Cassedy, *The Firemen's Record* (Philadelphia, n.d.); Edward Edwards, *History of the Volunteer Fire Department of St. Louis* (1906).

45. "Various Exercises for the Daily Roll Call Report," n.d., Reports, FH.

46. See for instance February 15, 1922, November 1, 1922, October 6, 1926, January 16, 1927, January 31, 1929, January 1, 1938, June 14, 1939, December 19, 1939, November 20, 1940, Reports, FH.

47. January 25, 1926, November 15, 1927, March 1, 1931, February 29, 1934, July 2, 1939, October 4, 1939, Reports, FH; August 11, 1928, November 15, 1934, November 13, 1941, Fire Regulation and Inspection, PEBFC.

48. April 27, 1927; December 4, 1938; December 25, 1938; April 5, 1939; October 2, 1939; "Fire Reports—Instructions," Reports, FH.

49. March 23, 1927, June 7, 1927, September 27, 1927, November 22, 1927, Bureau of Fire—Headquarters Orders, 1926 35, FH; April 27, 1927, April 5, 1939, Reports, FH.

50. January 17, 1926, Reports, FH; this report probably does not reflect an actual rescue, but is a concatenation of several incidents to provide an example of how to fill out the form.

51. The data for the discussion that follows is drawn from the following sources: Personnel Files, STLFD, PFD Personnel Ledger, FH, for more discussion of the data, including its limitations, see Appendix 1; Mark Tebeau, "'Eating Smoke': Masculinity, Technology, and the Politics of Urbanization, 1850–1950" (Ph.D. diss., Carnegie Mellon University, 1997), 431.

52. See, for example, tables in Appendix 2 and review in Appendix 1; note that the data on the PFD almost certainly overstates the proportion of firemen who died and, especially, those who were discharged. Extensive study of the data—on both the STLFD and the PFD—shows that dismissed firefighters had much shorter careers than those of *all* other firefighters, usually about one-third the length. Additionally, discharged firefighters were disproportionately represented among those firemen who had careers of fewer than five years. Therefore, because the extant data of the cohorts of firefighters who entered the PFD in the 1930s, 1940s, and 1950s is comprised, predominantly, of men with short careers, we would expect the proportion of firefighters discharged to be higher—much higher, in fact, than the data here indicate.

53. Alexander Keyssar, *Out of Work: The First Century of Unemployment in Massachusetts* (New York: Cambridge University Press, 1986), 260–61, tables A.4–A.6, pp. 314–26.

54. To create these figures, I aggregated data from Licht, *Getting Work*, table 7.1, pp. 34, 229; the author amassed the experience of *every* job that workers held and also collapsed four categories—laid off, fired, illness, and strike—into a single category of involuntary separation. Military service was not included. This produced 8,689 total job exits, of which 3,339 were made involuntarily. On Amoskeag, see Tamara K. Hareven, *Family Time and Industrial Time* (New York: Cambridge University Press, 1982), 237–40.

55. October 3, 1899, Volume 1, CSS, MOHS; D. R. Rowe, Personnel Files, STLFD.

56. William Connors, Personnel Files, STLFD; also on reduction, see biographical information on John Barry, 1920, Folder 19, Box 1, STLVVFC, MOHS; on pensions in fire departments, see *Statistics of Fire Departments*, table 9, and 18–23; on pensions for municipal workers more broadly, see Robert Fogelson, *Pensions: The Hidden Costs of Public Safety* (New York: Columbia University Press, 1984), 43–67; Lewis Meriam, *Principles Governing the Retirement of Public Employees* (New York, 1918), 3 ff.; Spencer Baldwin, "Retirement Systems for Municipal Employees," *Risks in Modern Industry: Annals of the American Academy of Political and Social Sciences* 38 (July 1911).

57. See Mark Tebeau, "Eating Smoke: Masculinity, Technology, and the Politics of Urbanization, 1850–1950" (Ph.D. diss., Carnegie Mellon University, 1997), fig. 8.2, p. 431.

58. *The New York Fire College Course*, 41–69; Fire Training School, Lectures for Officers, ca. 1930, PFDRG.

59. *Proceedings of the IAFE, 1920*, 120–21, 368–72; *Proceedings of the IAFC, 1932*, 139 ff.; *Proceedings of the IAFC, 1933*, 140; March 11, 1927, Rescue Squads, 1969 and Prior, PEBFC; entire, Rescue Squad, 1932, PEBFC; August 3, 1926, September 18, 1926, Clippings File–1926, FH; March 30, 1934, February 23, 1940, May 11, 1940, Reports, FH; "NYC subway fire," *New York Times*, January 8–10, 1915.

60. January 10, 1929, February 3, 1948, October 2, 1949, August 3, 1926, March 11, 1927, Rescue Squads, 1969 and Prior, PEBFC; March 18, 1926, August 3, 1926, September 18, 1926, CF–1926, FH.

61. Joseph Marshall, *Leather Lungs* (Philadelphia: Dorrance and Company, 1974), 5.

62. Marshall, *Leather Lungs*, 28–29; pp. 1–9, Firemen, Promotions, and Demotions, 1951 and Prior, PEBFC.

63. *Statistics of Fire Departments*, 20–22, and table 9; see also *Justifiably Proud*, 99; PFD, personnel ledger, FH; November 19, 1945, November 21, 1945, March 25, 1946, April 13–14, June 13–14, 1947, Working Hours, 2/3 Platoon System, PEBFC.

64. Marshall, *Leather Lungs*, 13–14; October 23, 1912, Awards, Honors, Memorials, PEBFC; "Various Exercises for the Daily Roll Call Report," n.d., Reports, FH; January 20, 1926, March 18, 1926, August 3, 1926, Clippings File–1926, FH.

Conclusion: Fighting Fire in Postwar America

1. *The White Fireman and His Work*, January 1928, INA, ACE.

2. "Will Feature 'The White Fireman,'" *Underwriters' Review*, December 22, 1927.

3. *The White Fireman and His Work*, January 1928, INA; "'The White Fireman' to Symbolize Safety," *United States Review*, December 17, 1927.

4. *The White Fireman and His Work*, January 1928, INA.

5. On firefighters' efforts to promote prevention, see NFPA, *Fire Prevention Week Handbook* (1926); Fire Prevention Week, 1951 and Prior, PFD, *Bulletin* Clippings, TUUA.

6. Evaluating the extent of fire danger—and whether it has actually diminished or simply changed is difficult; see Steven J. Pyne, *Fire in America: A Cultural History of Wildland and Rural Fire* (Princeton: Princeton University Press, 1982), 404–23; for several views of the threat, consider the data in the following: *Statistical Abstract of the United States, 1879–1960;* National Fire Protection Association, *Conflagrations in America Since 1900* (1951); L. E. Frost and E. L. Jones, "The Fire Gap and the Greater Durability of Nineteenth-Century Cities," *Planning Perspectives* 4 (1989), 333–47.

7. *CIGNA Historical Background Report* (March 1992), 33, ACE; William H. A. Carr, *Perils Named and Unnamed: The Story of the Insurance Company of North America* (New York: McGraw-Hill, 1967), 263.

8. Mike Davis, *The Ecology of Fear: Los Angeles and the Imagination of Disaster* (New York: Henry Holt, 1998), 147, 96–147; Malcolm Getz, *The Economics of the Urban Fire Department* (Baltimore, Md.: Johns Hopkins University Press, 1979).

9. Davis, *Ecology of Fear,* 143; Steven J. Pyne, *Fire in America*, xii, 404–23; Rutherford H. Platt, *To Burn or Not to Burn: Summary of the Forum on Urban/Wildland Fire*, January 26, 2001 (Washington, D.C.: National Academies Press, 2001); Steven J. Pyne, "Creating an Ecological Omelet: How We Got out of an Agrarian Frying Pan into an Exurban Fire," Presentation at National Disasters Roundtable Forum on Urban and Wildland Fire, Washington D.C., January 26, 2001; available on the web at http://books.nap.edu/books/N1000357/html/index.html; Patricia Nelson Limerick, "Lessons from Wildfires," *Colorado Central Magazine* 101 (July 2002), 37; or http://spot.colorado.edu/~daily/fire/Lessons.htm.

10. Steven Ruggles and Matthew Sobek et al., *Integrated Public Use Microdata Series: Version 2.0*, Minneapolis: Historical Census Projects, University of Minnesota, 1997 (http://www.ipums.umn.edu).

11. Ruggles et al., *Integrated Public Use Microdata Series.*

12. Robert McCarl, *The District of Columbia Firefighters' Project* (Washington, D.C: Smithsonian Institution Press, 1985), 100–102; Joseph Marshall, *Leather Lungs* (Philadelphia: Dorrance and Company, 1974).

13. Legal Department Files, Discrimination New York City Fire Department, 1942–1946, Segregation and Discrimination, Complaints and Responses, 1940–1955, Series A, Papers of the NAACP, Part 15.

14. Frank P. Sherwood and Beatrice Markey, *The Mayor and the Fire Chief: The Fight over Integrating the Los Angeles Fire Department* (Birmingham: University of Alabama Press, 1961).

15. McCarl, *Firefighters' Project*, 102, 101–4.

16. McCarl, *Firefighters' Project*, 108–10; many in the fire service have observed this to me informally, but "off the record"; Robert McCarl, "Exploring the Boundaries of Occupational Knowledge," in John Calagione, Doris Francis, and Daniel Nugent, eds., *Worker's Expressions: Beyond Accommodation and Resistance* (Albany: State University of New York Press, 1992); Carol Chetkovich, *Real Heat: Gender and Race in the Urban Fire Service* (New Brunswick, N.J.: Rutgers University Press, 1997), 83–154.

17. McCarl, *Firefighters' Project*, 58–60, 100–112; Chetkovich, *Real Heat*, 155–77; *NAACP vs. St. Louis Fire Department, 8th Circuit Court of Appeals, 1980.*

Essay on Sources

The relative lack of scholarly research on firefighting, fire insurance, and even fire protection more generally presented an early problem as I conceived this study, forcing me to cross the boundaries that usually separate fields of study within history, and the disciplines more broadly. By contrast, as I conceptualized the project and cast about for historical records, I was surprised to find a wide, almost overwhelming, range of source materials in historical societies, public libraries, corporate headquarters, and other archives in the nearly dozen states I visited during the life of this project. This essay focuses on those primary and secondary sources that contributed most directly to this history of how firefighters and underwriters confronted the problem of fire.

PRIMARY MATERIALS

Primary source materials on firefighting can be found in several places, some off the routes normally traveled by historians. Many local historical societies possess materials on firefighting, especially volunteer fire companies in the nineteenth century. Additionally, fire departments, city archives, and specialized fire museums also collected materials on fire protection, and W. Fred Conway, *Discovering America's Fire Museums* (1995) provides a comprehensive, though by no means complete, list of firefighting museums. Finally, professional firefighters and buffs often have collected firefighting materials and, even if they have not, they tend to know where records may be hidden. For *Eating Smoke*, the Thomas Collins Collection, St. Louis Veteran Volunteer Firemen's Collection, Tiffany Collection, and Newspaper Clippings Files at the Missouri Historical Society contained records on volunteer companies and the transition to a professional department. The scrapbooks of a St. Louis fire chief, the Charles Swingley Papers, shed light on the firefighting at the turn of the twentieth century. At the Historical Society of Pennsylvania (HSP) the Samuel Hazard Papers, the Campbell Collection, and the Philadelphia Fire Companies, Record Books, 1758–1904 held information on volunteer fire companies; the Philadelphia Fire Department, Record Books, 1871–1935, and the autobiography of a firefighter and reformer, James Bryson Gilbert, "An Effort to Portray . . . the Manner [of] the People of the Class to which I belonged Lived their Lives," proved invaluable to understanding professional firefighting. The Bucks County Historical Society has an excellent Firefighting Collection, which includes ledgers from volunteer fire companies and the department's management associations as well as an extensive collection of material culture, printed materials, and photographs.

Municipal publications, reports, and pamphlets also offered significant insight into the development of fire protection. Especially valuable at the Missouri Historical Society were the *Mayor's Messages and Accompanying Documents, City of St. Louis* (especially 1848–1865, 1872); *Annual Report of the St. Louis Fire Department; Report of the Fire Departments of Cincinnati and St. Louis and the Use of Steam Fire Engines* (1858). HSP, the Library Company of Philadelphia, and the Free Library of Philadelphia also have numerous pamphlets, which offer valuable insights; especially important were: *Annual Report of the Fire Marshal of Philadelphia* (especially 1859–1875), *Annual Report of the Committee of Legacies and Trusts of the Fire and Hose Establishment of the City of Philadelphia* (1838); *Statistics of Philadelphia Comprehending a Concise View of All the Public Institutions and the Fire Engine and Hose Companies of the City and County of Philadelphia, on the First of January, 1842* (1842); *Report of a Committee, Appointed at a Meeting Held on Friday Evening, December 3rd, 1852, to Consider the Propriety of Organising a Paid Fire Department* (1853); *Report of the Committee Appointed to Devise a Plan for the Better Organization of the Fire Department* (1853), *Letters of the Judges of the Court of Quarter Sessions, and the Marshal of Police, and Report of the Board of Trade* (1853); Philadelphia Hose Company, *Historical Sketches of the Formation and Founders of the Philadelphia Hose Company . . .* (1854); H. C. Watson, *Jerry Pratt's Progress; or Adventures in the Hose House* (1855); *Report of the Special Committee of the Select and Common Council in Relation to the Fire Alarm and Police Telegraph, Presented 12 October 1854* (1854); *Report of the Joint Special Committee of the Select and Common Councils on Steam Fire Engines* (1854); *Report of the Visiting Committee of the Board of Delegates of the Fire Association of Philadelphia, December 3, 1860* (1860); *The Philadelphia Fire Marshal Almanac and Underwriters' Advertiser, for the Year 1860* (1860). The St. Louis City Archives holds a wide range of materials related to municipal governance and the fire department, including departmental ledgers, expense books, and other official reports. In Philadelphia, HSP, the Free Library, and especially the Philadelphia City Archives (PCA) have extensive collections of materials pertaining to the city's political institutions, including annual reports of various mayors and of the fire department. In particular, the Philadelphia Fire Department Collection at the PCA has a full run of *Annual Reports of the Philadelphia Fire Department,* from 1871 forward. Also, it has an extensive collection of departmental orders, records of training school, and other administrative records. The Temple University Urban Archives holds records from the International Association of Fire Fighters, Local No. 22, for the period after 1940. In Philadelphia, the fire department operates the Fireman's Hall Museum, which houses a comprehensive collection of departmental orders, annual reports, clippings files, and images from the PFD, especially for the twentieth century.

Local newspapers routinely reported on fires and provide an excellent window into firefighting. The morgue of the *Philadelphia Evening Bulletin,* available at the Temple University Urban Archives, and the clippings files available at the Fireman's Hall Museum are especially rich collections. In addition, I systematically read short runs of several newspapers, including: *Niles Register,* the *Philadelphia Evening Bulletin,* the *Public Ledger,* the *Philadelphia Inquirer,* the *Missouri Republican,* the *St. Louis Post-Dispatch,* the *St. Louis Globe-Democrat.* I obtained these newspapers via interlibrary loan and from the following repositories: the American Philosophical Society, the Missouri Historical Society, the Historical Society of Pennsylvania, the City Archives of Philadelphia, the Free Library of Philadelphia, and the St. Louis Public Library. As this project neared conclusion, I also examined images in the morgue files of the *St. Louis Post-Dispatch* and the *Philadelphia Inquirer.*

Examining the official records of firefighting professional associations, periodicals, and training volumes proved exceptionally valuable. I read closely the *Proceedings of the Interna-*

tional Association of Fire Chiefs (and its predecessors), and especially the periodicals, *The Fireman's Herald, The Fireman's Journal, Fire and Water,* and *Fire and Water Engineering,* for the period from 1873 through 1950. Donald O'Brien, *"A Century of Progress through Service": A Centennial History of the International Association of Fire Chiefs, 1873–1973* (1973) offered an overview of the IAFC's history. As firefighting emerged as a formal occupation, firefighters increasingly had the opportunity to learn their craft in a number of detailed instructional manuals, including George Little, *The Fireman's Own Book* (1860), William C. Lewis, *A Manual for Volunteer and Paid Fire Organizations* (1877), Henry Champlin, *The American Fireman* (1880), Stephen Roper, *Handbook of Modern Steam Fire Engines* (1876), Andrew Isaac Meserve, *The Fireman's Hand-Book and Drill Manual* (1889), Charles Hill, *Fighting a Fire* (1898), John Kenlon, *Fires and Fire-Fighters: A History of Modern Firefighting . . .* (1913), and *The New York Fire College Course* (1920). Additionally, several other volumes contain useful information, though they are not instructional manuals per se; these include Clifford Thomson, *The Fire Departments of the United States* (New York: Fireman's Journal, 1879); Cairns and Brother, *The Fireman's Companion and Officers' Hand-Book* (1882); Joseph John Edward O'Reilly, *How to Become a Fireman* (New York: Chief Publishing Co., 1903); Horatio Bond and Warren Y. Kimball, *A Fire Department Manual: Hose and Ladder Work* (Boston: National Fire Protection Association, 1939). For a comparative overview of all of the urban fire departments in the United States, various census publications and the Statistical Abstracts contain much data, especially Bureau of Commerce, *Statistics of Fire Departments of Cities Having a Population of over 30,000, 1917* (1918).

Firefighters also have written, literally, the history of firefighting as part of the development of their occupational identity. Thomas Lynch, *The Volunteer Fire Department of St. Louis, 1819–1850* (1880), Edward Edwards, *History of the Volunteer Fire Department of St. Louis* (1906), St. Louis Firemen's Fund, *History of the St. Louis Fire Department . . .* (1914), and the St. Louis Fire Department, *History of the St. Louis Fire Department* (1977) each consider firefighting in St. Louis. J. Albert Cassedy, *The Firemen's Record* (Philadelphia, n.d.), Philadelphia Department of Public Safety, *Information for Firefighters . . .* (1915), and the Philadelphia Fire Department Historical Corporation, *Hike Out!: The History of the Philadelphia Fire Department* (1999) examine firefighting in that city. Additionally, there are a number of other useful fire department histories; see, for instance, A. W. Brayley, *A Complete History of the Boston Fire Department . . .* (Boston, 1889), Costello, *Our Firemen: A History of the New York Fire Department* (New York, 1887), and Thomas O'Connor, *History of the Fire Department New Orleans, from the Earliest Days to the Present Time . . .* (New Orleans, 1895). There are dozens, if not hundreds, of such histories written about cities throughout the nation, which are referenced in the usual places; however, an especially good reference is Thomas Scott's bibliography on the subject, *The Fire Buff's Bibliography Revised* (1993). Finally, this project made use of selected archival materials at the Hall of Flame (Phoenix), the Chicago Historical Society, the National Museum of American History, the Museum of the New York Fire Department, the Maryland Historical Society, the Brooklyn Historical Society, the Cincinnati Historical Society, and the American Imprints Series.

As part of this project, I studied the physical artifacts of firefighting—thousands of tools, images, and other items—and discussed my evaluations with curators and other experts in material culture. Although the insights from these explorations and discussions are rarely flagged in the text or notes, they significantly shaped this study. For example, my argument about the relation between the landscape, technology, and the increasing emphasis that firefighters placed on saving lives grew out of my contact with the tools of firefighting. I developed this hands-on experience in my work helping to catalog the Firefighting Collection and serving as curator for the exhibit "We Hazard Ourselves," both at the Bucks County Historical Society. A fellowship gave me the opportunity to view the material culture at the Missouri Historical Society, where I also curated an exhibit. Additionally, CIGNA and Fireman's Hall allowed me to examine their extensive art and artifact collections, and I studied photographs and items of material culture at the Historical Society of Pennsylvania, the National Museum of American History, and the headquarters of the STLFD. Anthony N. B. Garvan and Carol A. Wojtowicz, *The Catalogue of the Greentree Collection* (1977), and The Insurance Company of North America, *The Historical Collection of Insurance Company of North America* (1945) provide a view, albeit limited, of some of the material artifacts produced by firefighters.

The history of fire insurance is written in the ledgers and account books of insurance companies, in the proceedings of professional and business organizations, and in periodicals. When I conducted all my research for this study, CIGNA held the archival records of the Insurance Company of North America and the Aetna Fire Insurance Company, two of the firm's corporate predecessors. Those materials are now owned by the corporate archives of ACE America. The INA Collection and the Aetna Fire Insurance Collection provided remarkable documentation of two principal firms in the industry from the 1800s through the twentieth century. Several histories are helpful in understanding these two innovative firms, and serve as useful guides to the collections: Henry R. Gall and William George Jordan, *One Hundred Years of Fire Insurance: Being a History of the Aetna Insurance Company, Hartford, Connecticut, 1819–1919* (1919); Thomas Montgomery, *History of the Insurance Company of North America* (1885); Marquis James, *Biography of a Business, 1792–1942* (1942); William H. A. Carr, *Perils Named and Unnamed: The Dramatic History of the Insurance Company of North America* (1967). Horace Binney, *"Mr. Binney's Address," Centennial Meeting of the Philadelphia Contributionship for the Insurance of Houses from Loss by Fire, Second Monday of April, 1852* (1852; reprinted 1876, 1885), and J. A. Fowler, *History of Insurance in Philadelphia for Two Centuries, 1683–1882* (1888) provided critical insights in the industry's early history, glimpses into absent primary source materials, and commented on method and practice prior to and immediately following the Civil War. On the industry later in the nineteenth century, the *Proceedings of the National Board of Fire Underwriters*, formed in 1866, *Proceedings of the Underwriters Association of the Northwest*, and *Proceedings of the Underwriters Association of the Pacific* provided information about the industry's changing priorities. Harry Chase Brearley, *Fifty Years a Civilizing Force: The History of the National Board of Fire Underwriters* (1916), the National Board of Fire Underwriters, *Pioneers of Progress: A History of the National Board of Fire Underwriters* (1941), and A. L. Todd, *A Spark Lighted in Portland: The Record of the National Board of Fire Underwriters* (1966) provide an overview of the NBFU. As fire prevention came to the fore in the 1890s, the *Proceedings of the National Fire Protection Association*, formed in 1896, and documents in the appendices of the Proceedings pertaining to the Underwriters' Laboratories, formed in 1893, helped me to understanding the changing industry. A better source on the UL is Harry Chase Brearley, *A Symbol of Safety: An Interpretive Study of a Notable Institution Organized for Service—Not Profit* (1923). The fire insurance industry also produced a voluminous number of pamphlets, available in public libraries and specialized insurance libraries that still exist. For instance, the American Insurance Association (the modern heir of the NBFU), the Fire Insurance Society of Philadelphia, and the Insurance Library of Boston have extensive pamphlet collections, including those issued by the NBFU's influential Committee of Twenty, and FISOP also has a record of its meetings and organizational minutes.

An active insurance press developed, and reported on the changing industry, including *The Annual Cyclopedia of Insurance in the United States, Best's Insurance Reports, The Western Underwriter, The Weekly Underwriter, The Insurance Chronicle, The Insurance Monitor, United States Review and Insurance World.* Likewise, the insurance press published numerous books dealing with industry practice: J. B. Bennett, *Aetna Guide to Fire Insurance: For the Representatives of the Aetna Insurance Company of Hartford, Conn.; Cincinnati Branch* (1867); C. C. Hine, *Fire Insurance: A Book of Forms, Containing Forms of Policies, Endorsements, Certificates, and Other Valuable Matter, for the Use of Agents and Others* (1865); C. C. Hine, *Fire Insurance: A Book of Instructions for the Use of the Agents of the United States* (1870); C. C. Hine, *Fire Insurance, Policy Forms, and Policy Writing* (1882); Francis Cruger Moore, *Fires: Their Causes, Prevention, and Extinction; Combining Also a Guide to Agents Respecting Insurance against Loss by Fire* . . . (1877; reprint, 1881); Francis Cruger Moore, *Standard Universal Schedule for Rating Mercantile Risks* (1892); Francis Cruger Moore, *Fire Insurance and How to Build: Combining Also a Guide to Insurance Agents Respecting Fire Prevention and Extinction, Special Features of Manufacturing Risks, Writing of Policies, Adjustment of Losses, Etc., Etc.* (1903); Fontaine Talbott Fox Jr., *A Treatise on Warranty in Fire Insurance Contracts* (1883); the Insurance Society of New York, *The Fire Insurance Contract: Its History and Interpretation* (1922); Charles C. Dominge and Walter O. Lincoln, *Fire Insurance Inspection and Underwriting: An Encyclopedic Handbook* . . . (1919; reprint,1929). F. C. Oviatt, "Historical Study of Fire Insurance in the United States," *Annals of the American Academy of Political and Social Science* 26 (1905); Lester Zartman, *Property Insurance* (1909) and *The Yale Readings in Insurance* (1915); Frank Knight, *Risk, Uncertainty, and* Profit (1921); Robert Riegel, *Fire Insurance Associations in the United States* (1916); E. R. Hardy, *The Making of the Fire Insurance Rate* (1926); and William Wandel, *The Control of Competition in Fire Insurance* (1935) stood at the forefront of a wave of research by economists early in the twentieth century that illuminated important industry issues and helped to refine daily practice.

As I studied the tools of firefighting, I began to conceive of fire insurance maps as tools used by underwriters to conceptualize and to manage risk. Fire insurance maps are widely available, and I examined maps created by several firms: Western and Bascome, William Perris, Charles Hexamer, Hexamer and Locher, the Sanborn Fire Insurance Mapping Company, and the Whipple Insurance Mapping Agency. I used the atlases at the Missouri Historical Society, the Free Library of Philadelphia, the Historical Society of Pennsylvania, the Library of Congress, and the University of Pennsylvania Libraries. Walter W. Ristrow, in his introduction to *Fire Insurance Maps in the Library of Congress: Plans of North American Cities and Towns Produced by the Sanborn Map Company* (1981) and "United States Fire Insurance and Underwriters Maps, 1852–1968," *Quarterly Journal of the Library of Congress* (1968), provides an overview of insurance mapping. Helena Wright, in "Insurance Mapping and Industrial Archeology," *IA: The Journal of the Society for Industrial Archeology* 9, no. 1 (1993), studies these maps as tools for scholarship; she also graciously shared her research notes. Sanborn Maps are the most common and are explained in detail in the company's *Description and Utilization of the Sanborn Map* (1926). The *Whipple Daily Fire Reporter*, published by the Whipple Fire Insurance Mapping Agency, offered insight into the connections between mapmakers and underwriters, and into the manner in which the fire insurance industry's conceptualization of danger and safety mattered. The *Daily Fire Reporter* is available for much of the period from 1875 to 1905 at the St. Louis City Archives.

SECONDARY SOURCES

Stephen Pyne, *Fire in America: A Cultural History of Wildland and Rural Fire* (1982), stands alone in its critical examination of fire as a physical, environmental, and social phenomenon. Walter Hough, *Fire As an Agent in Human Culture* (1926), Gaston Bachelard, *The Psychoanalysis of Fire* (1964), and Johan Goudsblom, *Fire and Civilization* (1992), also examine the material and cultural dimensions of how fire has shaped human experience. Margaret Hindle Hazen and Robert M. Hazen, in *Keepers of the Flame: The Role of Fire in American Culture, 1775–1925* (1992), document the significant role that fire has played in everyday experience, as a good servant and bad master. Malcolm Getz, *The Economics of the Urban Fire Department* (1979), argues that nineteenth-century patterns of fire department organization have persisted until the late twentieth century. Mike Davis, in the chapter "The Case for Letting Malibu Burn," *The Ecology of Fear: Los Angeles and the Imagination of Disaster* (1998), 96–147, contends that the organization of contemporary fire protection privileges wealthy suburban dwellers over inner-city urbanites. Christine Meisner Rosen, *The Limits of Power: Great Fires and the Process of City Growth in America* (1986), examines the broad issues surrounding urban redevelopment after conflagrations, and Karen Sawislak, *Smoldering City: Chicagoans and the Great Fire, 1871–1874* (1994), follows Rosen's model, with a more social-historical focus. Carl Smith, *Urban Disorder and the Shape of Belief: The Great Chicago Fire, the Haymarket Bomb, and the Model Town of Pullman* (1995), offers a new perspective on disastrous fires and late-nineteenth-century urban life—one that emphasizes disorder as a normative condition.

Excellent overviews of the history of firefighting can be found in Rebecca Zurier, *The Firehouse: An Architectural and Social History* (1982), and Paul Ditzel, *Fire Engines, Fire Fighters* (1976). Also see Dennis Smith, *History of Firefighting in America* (1978), and Robert S. Holzman, *The Romance of Firefighting* (1956). More in-depth insights into firefighting—especially how gender and race matter—can be found in Robert McCarl, *The District of Columbia's Firefighting Project* (1990), Miriam Kaprow, "Magical Work: Firefighters in New York," *Human Organization* 50, no. 1 (1991), and Carol Chetkovich, *Real Heat: Gender and Race in the Urban Fire Service* (1997). Norman McLean's elegiac account of a disastrous 1948 wildfire, *Young Men and Fire* (1992), inspired me to think about the dichotomy between technological skill and physicality in firefighting work. Amy Greenberg, *Cause for Alarm: The Volunteer Fire Department in the Nineteenth-Century City* (1998), has written the first book-length historical treatment of volunteer firefighting, reinterpreting it through the lens of gender. Her argument offers a provocative reassessment of previous scholarship on this subject, including Bruce Laurie's "Fire Companies and Gangs in Southwark: The 1840s," in Allen F. Davis and Mark Haller, eds., *The Peoples of Philadelphia: A History of Ethnic Groups and Lower Class Life, 1790–1840* (1973), which nonetheless remains, to my mind, the point of departure for research in this area. A number of scholars have also considered volunteer firefighting in the context of larger works; among the best is Richard Stott, *Workers in the Metropolis: Class, Ethnicity, and Youth in Antebellum New York City* (1990). For an engaging cross-cultural study of firemen as cultural icons late in the nineteenth century, which links that symbolism to a broader crisis of manhood, see Robyn Cooper, "The Fireman: Immaculate Manhood," *Journal of Popular Culture* 28, no. 4 (Spring 1995), 139–70. An astounding number of essays in local historical magazines and dissertations outline the broader history of firefighting. (*American History and Life* should be consulted for a thorough listing.) On Philadelphia, see Andrew Neilly, *The Violent Volunteers: A History of the Volunteer Fire Department of Philadelphia, 1736–1871* (1959). On St. Louis, see Arlen Ross Dykstra, *A History of St. Louis Firefighting . . . 1850–1880* (1970), and Anthony Lampe, *The St. Louis Fire Department, 1820–1950* (1966). I also found several other theses to be especially helpful, including Richard Boyd Calhoun, "From Community to Metropolis: Fire Protection in New York City, 1790–1875" (Ph.D. diss., Columbia University, 1973), and Kathleen J. Kiefer, "A History of the Cincinnati Fire Department in the Nineteenth-Century City" (M.A. thesis, University of Cincinnati, 1967).

Fire insurance has garnered less attention than firefighting, but there is some literature in this area; Clive Trebilcock, *Phoenix Assurance and the Development of British Insurance: Volume 1, 1782–1870* (1985), provides an excellent and detailed analysis of the early British Fire Insurance Industry. Robert Bernard Considine, *Man against Fire: Fire Insurance—Protection against Disaster* (1955), and Andrew Tobias, *The Invisible Bankers: Everything the Insurance Industry Never Wanted You to Know* (1982) provide overviews of the industry, and H. Roger Grant, *Insurance Reform: Consumer Action in the Progressive Era* (1979), explores reform in the early twentieth century. By contrast, the life insurance industry has received more attention, and this research offered much guidance. Viviana Zelizer, *Morals and Markets: The Development of Life Insurance in the United States* (1979), provides significant insights into the connections between insurance, everyday experience, and historical change. Morton Keller, *The Life Insurance Enterprise: 1885–1910* (1963), affords an excellent overview of the industry from a business history perspective. More recently, Angel Kwollek-Folland, *Engendering Business: Men and Women in the Corporate Office, 1870–1930* (1994), argues persuasively that gender shaped the experiences of workers in the life insurance industry, as well as in American business more broadly; her insights provided critical guidance to studying how gender mattered among fire underwriters. On the topic of moral hazard, I learned much from Robin Pearson, "Moral Hazard and the Assessment of the Insurance Risk in Eighteenth- and Early-Nineteenth-Century Britain," *Business History Review* 76 (Spring 2002), 1–35. Several recent dissertations were helpful as well. Most important was Marc Schneiberg, "Private Order and Inter-Firm Governance in the American Fire Insurance Industry, 1820–1950" (1994), which outlined the industry's history. His recently published "Political and Institutional Conditions for Governance by Association: Private Order and Price Controls in American Fire Insurance," *Politics and Society* 27, no. 1 (1999), provided more in-depth and exceptionally clear discussions of complex industry issues. Betsey Bahr, in her doctoral dissertation, "New England Mill Engineering: Rationalization and Reform in Textile Mill Design, 1790–1920" (1988), offered insight into the development of fire prevention engineering; Sara E. Wermiel, *The Fireproof Building Technology and Public Safety in the Nineteenth-Century American City* (2000), follows in this path, exploring how safety engineers and others worked to develop innovations in building design that were crucial to the broader efforts at preventing conflagration.

In recent years, scholars have begun to study risk, danger, and safety, underscoring the importance of the risk of fire as the central analytic tension to this narrative. Mary Douglas, *Risk Acceptability According to the Social Sciences* (1985), and *Risk and Blame: Essays in Cultural Theory* (1992), and especially Ulrich Beck, *Risk Society: Towards a New Modernity* (1992), helped to establish what has been dubbed the sociology of risk. Historians, too, have taken up the topic of risk, especially studying accidents and safety. Recently, Bill Luckin and Roger Cooter, *Accidents in History* (1997), collected a series of essays that explored how danger and safety have been constructed in very concrete contexts. Historical research remains strongest in the area of workplace accidents and safety; especially notable are Arthur McEvoy, "Working Environments: An Ecological Approach to Industrial Health and Safety," *Technology and Culture* Supplement, 36 (1995), and Mark Aldrich, *Safety First: Technology, Labor, and Business in the Building of American Work Safety, 1870–1939* (1997). And, I too have considered risk in another spatial context, considering the home as a dangerous environment: Mark Tebeau and Joel Tarr, "Managing Danger in the Home Environment, 1900–1940," *Journal of Social History* 29 (1996). Finally, it is important to note that risk, as a subject of study, remains both undertheorized and understudied in a historical context, although Peter Bernstein, *Against the Gods* (1998), has explored the history of the mathematical underpinnings of contemporary risk theory in what is a highly readable and at turns fascinating study.

Contemporary risk theory is based on statistical and quantitative reasoning, which gained ascendance in the nineteenth century. See, for example, Theodore Porter, *Trust in Numbers: The Pursuit of Objectivity in Science and Public Life* (1995) and Lorenz Kruger, Lorraine J. Daston, and Michael Heidelberger, eds., *The Probabilistic Revolution* (1987). The broad literature of urban history and urban studies provided important guidance in thinking about fire protection, especially the literature on the urban infrastructure. Joel Tarr's work at the borders of urban history, the history of technology, and environmental history, in both *In Search of the Ultimate Sink: Urban Pollution in Historical Perspective* (1996) and *Technology and the Rise of the Networked City in Europe and America* (1988), offers guidance in crossing boundaries and raises many of the broader issues in which this work is situated. Likewise, Martin Melosi, *The Sanitary City* (2000), and Stanley Schultz, *Constructing Urban Culture: American Cities and City Planning, 1800–1920* (1989), provide models for studying the urban technological infrastructure through their studies of the waste infrastructure and professional engineering in cities. More broadly, the comparative research of Paul M. Hohenberg and Lynn Hollen Lees, *The Making of Urban Europe, 1000–1950* (1985), and the broad narrative of urbanization in the United States by Eric Monkkonen, *America Becomes Urban, 1780–1980* (1988), helped me to locate this work in the broad story of American urbanization. Several of Jon Teaford's books chronicling the history of U.S. cities offered guidance, but especially valuable was *The Unheralded Triumph: City Government in America, 1870–1900* (1984). If David Harvey, *The Urbanization of Capital: Studies in the History and Theory of Capitalist Accumulation* (1985), has underscored the significance of capitalism in structuring cities, William Cronon, *Nature's Metropolis: Chicago and the Great West* (1991), provides the trenchant historical example of how capitalism organized the American landscape into city and country, ordering the "natural" environment as surely as it structured urban life. Likewise, historic studies of order and disorder helped me to think about the chaotic city posited by Georg Simmel, "The Metropolis and Mental Life," in Kurt Wolff, trans. and ed., *The Sociology of Georg Simmel* (1950). Among others, Robert H. Wiebe, *The Search for Order, 1877–1920* (1967), Paul Boyer, *Urban Masses and Moral Order in America, 1820–1920* (1978), and Richard Sennett, *The Uses of Disorder: Personal Identity and City Life* (1970), have pointed to the ways that bringing order to the unruly cultural, social, political, and economic vibrancy of the city has been a long historical process. Indeed, when Carl Smith, in *Urban Disorder and the Shape of Belief*, suggested that disorder may well have been the normative condition of cities, I think he got it right. Of course, order and disorder are undercurrents in a host of works on public parks and public health, as well as the provision of other municipal services, such as policing. Indeed, policing can be understood as central to the political and social processes of creating order, as suggested by Roger Lane, *Policing the City: Boston, 1822–1855* (1967). Of course, the police, like firefighters, do more than provide social control, and the literature in this area is vast. For starters, Samuel Walker, *Popular Justice: A History of American Criminal Justice* (1980), provided an overview of the extensive historical literature and explores the many facets of policing, including reform, work, and issues of municipal governance. However, this project benefited most from the historical studies that examined the work of police—labor that has taken them throughout city streets and into urban neighborhoods. In particular, Eric Monkkonen, *Police in Urban America, 1860–1920* (1981) and Robert Fogelson, *Big City Police* (1977), as well the ethnographic work of Jonathan Rubinstein, *City Police* (1973), provided insights into the day-to-day labor of police, offering a model for studying firefighters.

To the degree that I explored the development of the urban infrastructure and the experience of working in cities, I sought to capture different experiences of the city—to show how several sets of men imagined, labored, and produced the spaces of cities—emulating the creative approach of Italo Calvino's *Invisible Cities* (1986). More concretely, I drew insight from scholarship that examines how communities and societies "produce space," to borrow from Henri Lefebvre, *Production of Space* (1991). Other scholars, including Gaston Bachelard, *The Poetics of Space* (1964), Michel de Certeau, *The Practice of Everyday Life* (1984), or Allan Pred, *Making Histories and Constructing Human Geographies* (1990), also offered insights into the connections between everyday experience and mental and physical spaces. The role that economic forces have had in organizing cities and space long has been a critical vein in urban studies. It has been represented well in the work of, among others, David Harvey, *The Urbanization of Capital*, and has recently been recast in terms of "postmodern" experience in Edward Soja, *Postmodern Geographies: The Reassertion of Space in Critical Social Theory* (1989). The consequences of the so-called "linguistic turn" on studying cities is discussed adroitly in Timothy Gilfoyle, "White Cities, Linguistic Turns, and Disneylands: The New Paradigms of Urban History," *Reviews in American History* 26, no. 1 (1998), 175–204. I would certainly agree with Gilfoyle's conclusion that urban history remains vibrant albeit without a center—precisely because of the heterodoxy of method and theme.

Conceiving the physical and cultural infrastructure as a technological system helped me to make sense of several disparate traditions within history, and was especially critical in helping me to connect the seemingly unrelated work and tools of firefighters and insurers into a single narrative. Over a decade ago, historians of technology, represented well by Wiebe E. Bijker, Thomas P. Hughes, and Trevor J. Pinch, eds., *The Social Construction of Technological Systems: New Directions in the Sociology and History of Technology* (1987), reimagined the relation between technology and everyday life, arguing that technology is "socially constructed." To borrow Bruno Latour's phrase, "technology is society made durable"; see the similarly titled essay in John Law, ed., *A Sociology of Monsters: Essays on Power, Technology, and Domination* (1991). Several recent studies of electricity as a technology constructed within social contexts provide the best model of how the infrastructure can be studied: how it existed within particular community contexts, as told by Harold Platt, *The Electric City: Energy and the Growth of the Chicago Area, 1880–1930* (1991); how it has been understood and represented within the broader social and cultural attitudes as told by David Nye, *Electrifying America: Social Meanings of a New Technology* (1990); and how the electrical infrastructure developed in international comparative context as a result of the intersection of technological factors, political interests, and business dynamics. One of the many excellent recent studies by historians of technology, Donald Reid, *Paris Sewers and Sewermen: Realities and Representations* (1991), provides an example of how to bring disparate methods and historical actors into a single narrative, showing how an urban technological system developed as a result of complex negotiations between politics, workers (sewermen), sanitation practices, and popular beliefs (and fears) about waste and sewers.

However, the idea of including fire insurers in this story, as well as guidance in how to study them, came from business historians, who persistently argue that understanding business development is integral to understanding broader historical processes. When Louis Galambos asked the question, "What Makes Us Think We Can Put Business Back into American History?" (*Business and Economic History* 20 [1990], 1–11), he recalled Alfred Chandler's remarkable history of managerial capitalism, *The Visible Hand* (1976), which offered a convincing portrait of how business activity structured the American past. More recently business historians have argued that the history of business can and should be analyzed using the same methods that are used to study society more broadly, as pointed out succinctly by Kenneth Lipartito, "Culture and the Practice of Business History," *Business and Economic History* 24 (Winter 1995), 1–41. Of particular relevance to the history of fire insurance is a rapidly growing literature on how businesses have managed information—gathering, processing, organizing, and analyzing it. For instance, Naomi Lamoreaux, *Insider Lending: Banks, Personal Connections, and Economic Development in Industrial New England* (1994), shows how personal connections mediated the exchange and management of information by banks in preindustrial and industrial New England, before the advent of impersonal bureaucratic systems and new communication technologies, such as telegraphs and railroads. James Beniger, *The Control Revolution: Technological and Economic Origins of the Information Society* (1986), and, especially, JoAnne Yates, *Control through Communication: The Rise of System in American Management* (1989), explored the transformation of administrative practices and bureaucratic routines in American business, providing a view into how the daily work activities of corporate managers changed from the nineteenth into the twentieth centuries as impersonal and bureaucratic structures developed to manage and disseminate information. Of course, as Michel Foucault, *The Order of Things* (1970), aptly demonstrates, how a given society processes, organizes, and classifies information is not isolated to business communities. Such day-to-day practices are discussed in much greater detail in the work of Geoffrey C. Bowker and Susan Leigh Star, *Sorting Things Out: Classification and Its Consequences* (1999), who examine how the meanings, strategies, and particular practices of classification operate in multiple contexts—educational, governmental, and business.

The study of work and workers, first outlined by E. P. Thompson, *The Making of the English Working Class* (1963), and later, in the American context, by Herbert Gutman, *Work, Culture, and Society in Industrializing America: Essays in American Working-Class and Social History* (1976), has added much to the historical record. Building on this tradition, Stuart Blumin, *The Emergence of the Middle-Class: Social Experience in the American City, 1760–1900* (1989) explores the formation of the middle class; David Roediger, *Wages of Whiteness: Race and the Making of the American Working Class* (1991), examines the importance of race in creating "white" identity among workers; Ava Baron, *Work Engendered: Toward a New History of American Labor* (1991), studies how gender has mattered in various workplaces. For the most part, these studies—and much of the labor history that followed Thompson's incisive work—emphasizes class formation and focuses less on the particulars of ordinary workplaces. On the whole, sociologists and ethnographers who study work, skill, and occupation, have done a better job than historians of understanding daily work practices and procedures, as illustrated in Douglas Harper, *Working Knowledge: Skill and Community in a Small Shop* (1987), Robert McCarl, *The District of Columbia's Firefighting Project* (1990), and George Sturt, *The Wheelwright's Shop* (reprint, 1993). The sociological literature on the dynamic interplay of work and skill, and the nature of occupation and profession is vast; Andrew Abbott, *The System of Professions: An Essay on the Division of Expert Labor* (1988), offers a critical overview of the development of this tradition. Kai Erikson and Steven Peter Vallas, eds., *The Nature of Work: Sociological Perspectives* (1990), and a special issue of *Work and Occupation* 17, no. 4 (November 1990), provide insight into diverse approaches to examining skill, occupation, and technology; Patrick Joyce, ed., *The Historical Meanings of Work* (1987), offers a sweeping overview of different work environments. David Montgomery, *Workers' Control in America: Studies in the History of Work, Technology, and Labor Struggles* (1979), and Harry Bravermen, *Labor and Monopoly Capital* (1975), provide two historical perspectives on changes in work, technology, and skill. Theodore Hershberg, ed., *Philadelphia: Work, Space, Family, and Group Experience in the Nineteenth Century* (1981), has examined the lives of nineteenth-century workers in Philadelphia. Walter Licht, *Getting Work: Philadelphia, 1840–1950* (1992), has studied how workers gained employment in the early twentieth century, and Tamara Hareven, *Family Time and Industrial Time: The Relationship between the Family and Work in a New England Industrial Town* (1982), explores work in the textile industry in extraordinary detail. Alexander Keyssar, *Out of Work: The First Century of Unemployment in Massachusetts* (1986), has examined the history of unemployment and reveals much about the work patterns and routines of a wide spectrum of American laborers. Each of these latter four studies served as examples of using quantitative evidence to make historical arguments, not to mention providing comparative data that helped me to understand firefighting careers.

Any research on firefighters or fire underwriters would be remiss without considering the argument that gender matters, made abundantly clear by three decades of work in women's history. A short account of the field's development can be found in Nancy Hewitt's introductory essay to Ellen Carol DuBois and Vicki L. Ruiz, *Unequal Sisters: A Multi-Cultural Reader in U.S. Women's History* (1990.) For a review of the importance of gender, as a category of analysis, see Joan Scott, *Gender and the Politics of History* (1988), and to understand how gender has mattered in various workplaces, see Ava Baron, *Work Engendered: Toward a New History of American Labor*. Research on manhood has become an important area of study, and among the first to examine this were Peter Filene, *Him/Her/Self: Sex Roles in Modern America* (1975), and Peter Stearns, *Be a Man!: Males in Modern Society* (1979). Elliot Gorn, *The Manly Art*, is perhaps the best historical case study of changing definitions of manhood in the nineteenth century; also, Mark Carnes and Clyde Griffen, eds., *Meanings for Manhood: Constructions of Masculinity in Victorian America* (1990), has become standard reading for scholars exploring the nature of manhood in that period. Gail Bederman's provocative cultural history, *Manliness and Civilization: A Cultural History of Gender and Race in the United States, 1880–1917* (1995), has offered an excellent view of manliness in the United States from the late nineteenth through the early twentieth century. E. Anthony Rotundo, *American Manhood* (1993), and Michael Kimmel, *American Manhood: Transformations in Masculinity from the Revolution to the Modern Era* (1996), recently have provided overviews of the American past in light of changing notions of manhood, taking account of its variations according to class, race, and ethnicity.

In pulling together so many disparate fields of research, this project employed a diverse methodology—using the methods of social historians, quantitative sociology, and cultural studies. The approaches of quantitative historians were especially helpful in evaluating the personnel files of the St. Louis and Philadelphia fire departments. John Modell, *Into One's Own, from Youth to Adulthood in the United States* (1990), Theodore Hershberg, ed., *Philadelphia*, Olivier Zunz, *The Changing Face of Inequality: Urbanization, Industrial Development, and Immigrants in Detroit, 1880–1920* (1982), and Alexander Keyssar, *Out of Work*, provide exemplary models of how to use quantitative evidence in historical research. However, the practicalities of developing and analyzing a quantitative database developed from Modell's more personal guidance, and is discussed separately in my doctoral dissertation, "'Eating Smoke': Masculinity, Technology, and the Politics of Urbanization, 1850–1950" (1997), and also in Appendix 1: Firefighting by the Numbers.

The methods of anthropologists, museum professionals, and cultural theorists had a distinct influence on this project. Broadly speaking, anthropologists' concern with methodological sophistication shaped the interdisciplinary approach to culture and society used here. Nicholas Dirks, Geoff Eley, and Sherry Ortner, eds., *Culture / Power / History: A Reader in Contemporary Social Theory* (1994), provides one overview of such methodological issues. Another approach, specifically emphasizing power relations, exists in the work of poststructural theorists like Pierre Bourdieu, *Outline of a Theory of Practice* (1977), and Michael Foucault, *Discipline and Punish* (1976). In particular, Foucault's description of Bentham's panopticon inspired me to think about the mapping regime and managerial discipline developed by insurers as a form of surveillance.

I began this project planning to examine primarily, as do most historians, documentary sources, but I soon discovered that material culture held numerous insights. As I wrote *Eating Smoke*, I benefited enormously from the research of specialists in the study of material culture—insights that complement the traditions of anthropology and new work in cultural theory. John Modell, ed., *Theory, Method, and Practice in Social and Cultural History* (1992), and Lynn Hunt, *The New Cultural History* (1989), provided thoughtful discussions of method, as well as practical guidance. Barbara Melosh, *Engendering Culture: Manhood and Womanhood in New Deal Public Art and Theater* (1991), offered insights into studying gender iconography in art. The long and excellent tradition of research in material culture, including studies of images and architecture such as the classics—Henry Glassie, *Folk Housing in Middle Virginia: A Structural Analysis of Historic Artifacts* (1975), and Dell Upton and John Vlach, eds., *Common Places: Readings in American Vernacular Architecture* (1986)—provided exceptional assistance in understanding material culture through their study of the built environment. Steve Lubar offered professional guidance in thinking about and studying historical artifacts. Steve Lubar, "Representations and Power," *Technology and Culture* Supplement, 36 (1995), S54–S81, Steven Lubar and W. David Kingery, eds., *History from Things: Essays on Material Culture* (1993), and Thomas Schlereth, ed., *Material Culture Studies in America* (1982), also provide examples of the sort of direction that I received directly from Lubar and other museum folks over the years.

Finally, although far from least in importance, this project benefited enormously from many studies on Philadelphia and St. Louis, including especially those listed below. For an overview of Philadelphia's history see J. Thomas Scharf and Thompson Westcott, *History of Philadelphia, 1609–1884* (1884), and Russell F. Weigley, *Philadelphia: A Three-Hundred-Year History* (1982). Allen Steinberg, *The Transformation of Criminal Justice: Philadelphia, 1800–1880* (1989), offers guidance to understanding the city during the mid-nineteenth century, especially comprehending some of the complexities behind municipal consolidation. Sam Bass Warner, *The Private City: Philadelphia in Three Periods of Its Growth* (1968), provided insights into the changing landscape during the nineteenth century, as did Davis, *Parades and Power* (1986), Hershberg, ed., *Philadelphia*, and Blumin, *The Emergence of the Middle Class* (1990). For an overview of St. Louis's economic and political history see James Neal Primm, *Lion of the Valley* (1981), J. Thomas Scharf, *History of Saint Louis City and County . . .* (1883), William Hyde and Howard Conard, *Encyclopedia of the History of Saint Louis* (1899), Walter B. Stevens, *St. Louis, the Fourth City, 1764–1909* (1909). Jeffrey Scott Adler, *Yankee Merchants and the Making of the Urban West: The Rise and Fall of Antebellum St. Louis* (1991) offers an incisive portrait of nineteenth-century St. Louis; for portraits of St. Louis in the twentieth century, see George Lipsitz, *Sidewalks of St. Louis: Places, People, and Politics in an American City* (1991). Both Andrew Hurley and Eric Sandweiss taught me much about St. Louis's history. Their insights are published in Andrew Hurley, *Common Fields: An Environmental History of St. Louis* (1997) and Eric Sandweiss, *St. Louis: The Evolution of an American Urban Landscape* (2001). In addition, the Missouri Historical Society has published a long-running historical journal, now titled *Gateway Heritage*, which contains numerous articles on the city's history.

Index

Page numbers in *italics* refer to illustrations or tables.